Reliability Characterisation of Electrical and Electronic Systems

Related titles

Electromigration in thin films and electronic devices
(ISBN 978-1-84569-937-6)

Circuit analysis
(ISBN 978-1-78242-044-6)

Woodhead Publishing Series in Electronic and Optical Materials: Number 74

Reliability Characterisation of Electrical and Electronic Systems

Edited by

Jonathan Swingler

ELSEVIER

AMSTERDAM • BOSTON • CAMBRIDGE • HEIDELBERG
LONDON • NEW YORK • OXFORD • PARIS • SAN DIEGO
SAN FRANCISCO • SINGAPORE • SYDNEY • TOKYO
Woodhead Publishing is an imprint of Elsevier

WP WOODHEAD PUBLISHING

Woodhead Publishing is an imprint of Elsevier
80 High Street, Sawston, Cambridge, CB22 3HJ, UK
225 Wyman Street, Waltham, MA 02451, USA
Langford Lane, Kidlington, OX5 1GB, UK

British Library Cataloguing-in-Publication Data
A catalogue record for this book is available from the British Library.

Library of Congress Control Number: 2014954005

ISBN 978-1-78242-221-1 (print)
ISBN 978-1-78242-225-9 (online)

For information on all Woodhead Publishing publications visit our website at http://store.elsevier.com/

Typeset by Spi Global
www.spi-global.com
Printed and bound in the United Kingdom

Contents

List of contributors **ix**
Woodhead Publishing Series in Electronic and Optical Materials **xi**
Foreword **xv**

1 Introduction **1**
J. Swingler
1.1 Introduction **1**
1.2 The focus of the book **1**
1.3 Reliability science and engineering fundamentals (Chapters 2–4) **3**
1.4 Reliability methods in component and system development (Chapters 5–9) **6**
1.5 Reliability modelling and testing in specific applications (Chapters 10 and 11) **8**
1.6 Conclusion **9**
References **10**

2 Reliability and stupidity: mistakes in reliability engineering and how to avoid them **11**
R.W.A. Barnard
2.1 Introduction **11**
2.2 Common mistakes in reliability engineering **12**
2.3 Conclusion **24**
References **24**

3 Physics-of-failure (PoF) methodology for electronic reliability **27**
C. Hendricks, E. George, M. Osterman, M. Pecht
3.1 Introduction **27**
3.2 Reliability **27**
3.3 PoF models **29**
3.4 PoF reliability assessment **32**
3.5 Applications of PoF to ensure reliability **34**
3.6 Summary and areas of future interest **37**
References **38**

4 Modern instruments for characterizing degradation in electrical and electronic equipment **43**
P.D. Goodman, R. Skipper, N. Aitken
4.1 Introduction **43**

4.2 Destructive techniques **43**
4.3 Nondestructive techniques **52**
4.4 *In situ* measurement techniques **57**
4.5 Conclusions **61**
References **62**

5 Reliability building of discrete electronic components **63**
T.-M.I. Băjenescu, M.I. Bâzu
5.1 Introduction **63**
5.2 Reliability building **63**
5.3 Failure risks and possible corrective actions **67**
5.4 Effect of electrostatic discharge on discrete electronic components **78**
5.5 Conclusions **79**
References **79**

6 Reliability of optoelectronics **83**
J.-S. Huang
6.1 Introduction **83**
6.2 Overview of optoelectronics reliability **84**
6.3 Approaches and recent developments **85**
6.4 Case study: reliability of buried heterostructure (BH) InP semiconductor lasers **90**
6.5 Reliability extrapolation and modeling **98**
6.6 Electrostatic discharge (ESD) and electrical overstress (EOS) **101**
6.7 Conclusions **109**
References **110**

7 Reliability of silicon integrated circuits **115**
A.S. Oates
7.1 Introduction **115**
7.2 Reliability characterization approaches **116**
7.3 Integrated circuit (IC) wear-out failure mechanisms **118**
7.4 Summary and conclusions **133**
Acknowledgments **135**
References **135**

8 Reliability of emerging nanodevices **143**
N. Raghavan, K.L. Pey
8.1 Introduction to emerging nanodevices **143**
8.2 Material and architectural evolution of nanodevices **146**
8.3 Failure mechanisms in nanodevices **148**
8.4 Reliability challenges: opportunities and issues **160**
8.5 Summary and conclusions **163**
References **163**

9 Design considerations for reliable embedded systems **169**
R.A. Shafik, A. Das, S. Yang, G. Merrett, B.M. Al-Hashimi
9.1 Introduction **169**
9.2 Hardware faults **170**
9.3 Reliable design principles **173**
9.4 Low-cost reliable design **180**
9.5 Future research directions **187**
9.6 Conclusions **190**
References **190**

10 Reliability approaches for automotive electronic systems **195**
D. Medhat
10.1 Introduction **195**
10.2 Circuit reliability challenges for the automotive industry **195**
10.3 Circuit reliability checking for the automotive industry **196**
10.4 Using advanced electronic design automation (EDA) tools **200**
10.5 Case studies and examples **208**
10.6 Conclusion **212**
Acknowledgment **212**
References **212**

11 Reliability modeling and accelerated life testing for solar power generation systems **215**
F. Schenkelberg
11.1 Introduction **215**
11.2 Overview **215**
11.3 Challenges **218**
11.4 Modeling **222**
11.5 Accelerated life testing (ALT) **226**
11.6 ALT example: how to craft a thermal cycling ALT plan for SnAgCu (SAC) solder failure mechanism **238**
11.7 How to craft a temperature, humidity, and bias ALT plan for CMOS metallization corrosion **243**
11.8 Developments and opportunities **247**
11.9 Conclusions **248**
11.10 Sources of further information **248**
References **248**

Index **251**

List of contributors

N. Aitken ERA Technology Ltd, Leatherhead, United Kingdom

B.M. Al-Hashimi University of Southampton, Southampton, United Kingdom

R.W.A. Barnard Lambda Consulting, Pretoria, South Africa

T.-M.I. Băjenescu C.F.C., La Conversion, Switzerland

M.I. Bâzu National Institute for R&D in Microtechnologies, Bucharest, Romania

A. Das University of Southampton, Southampton, United Kingdom

E. George Center for Advanced Life Cycle Engineering (CALCE), University of Maryland, College Park, MD, USA

P.D. Goodman ERA Technology Ltd, Leatherhead, United Kingdom

C. Hendricks Center for Advanced Life Cycle Engineering (CALCE), University of Maryland, College Park, MD, USA

J.-S. Huang Emcore Corporation, Alhambra, CA, USA

D. Medhat Mentor Graphics, Cairo, Egypt

G. Merrett University of Southampton, Southampton, United Kingdom

A.S. Oates Taiwan Semiconductor Manufacturing Company Ltd, Hsinchu, Taiwan

M. Osterman Center for Advanced Life Cycle Engineering (CALCE), University of Maryland, College Park, MD, USA

M. Pecht Center for Advanced Life Cycle Engineering (CALCE), University of Maryland, College Park, MD, USA

K.L. Pey Singapore University of Technology and Design (SUTD), Singapore

N. Raghavan Singapore University of Technology and Design (SUTD), Singapore

F. Schenkelberg FMS Reliability, Los Gatos, CA, USA

R.A. Shafik University of Southampton, Southampton, United Kingdom

R. Skipper ERA Technology Ltd, Leatherhead, United Kingdom

J. Swingler Heriot-Watt University, Edinburgh, United Kingdom

S. Yang University of Southampton, Southampton, United Kingdom

Woodhead Publishing Series in Electronic and Optical Materials

1 **Circuit analysis**
J. E. Whitehouse

2 **Signal processing in electronic communications: For engineers and mathematicians**
M. J. Chapman, D. P. Goodall and N. C. Steele

3 **Pattern recognition and image processing**
D. Luo

4 **Digital filters and signal processing in electronic engineering: Theory, applications, architecture, code**
S. M. Bozic and R. J. Chance

5 **Cable engineering for local area networks**
B. J. Elliott

6 **Designing a structured cabling system to ISO 11801: Cross-referenced to European CENELEC and American Standards Second edition**
B. J. Elliott

7 **Microscopy techniques for materials science**
A. Clarke and C. Eberhardt

8 **Materials for energy conversion devices**
Edited by C. C. Sorrell, J. Nowotny and S. Sugihara

9 **Digital image processing: Mathematical and computational methods Second edition**
J. M. Blackledge

10 **Nanolithography and patterning techniques in microelectronics**
Edited by D. Bucknall

11 **Digital signal processing: Mathematical and computational methods, software development and applications Second edition**
J. M. Blackledge

12 **Handbook of advanced dielectric, piezoelectric and ferroelectric materials: Synthesis, properties and applications**
Edited by Z.-G. Ye

13 **Materials for fuel cells**
Edited by M. Gasik

14 **Solid-state hydrogen storage: Materials and chemistry**
Edited by G. Walker

15 **Laser cooling of solids**
S. V. Petrushkin and V. V. Samartsev

16 **Polymer electrolytes: Fundamentals and applications**
Edited by C. A. C. Sequeira and D. A. F. Santos

17 **Advanced piezoelectric materials: Science and technology**
Edited by K. Uchino

18 **Optical switches: Materials and design**
Edited by S. J. Chua and B. Li

19 **Advanced adhesives in electronics: Materials, properties and applications**
Edited by M. O. Alam and C. Bailey

20 **Thin film growth: Physics, materials science and applications**
Edited by Z. Cao

21 **Electromigration in thin films and electronic devices: Materials and reliability**
Edited by C.-U. Kim
22 ***In situ* characterization of thin film growth**
Edited by G. Koster and G. Rijnders
23 **Silicon-germanium (SiGe) nanostructures: Production, properties and applications in electronics**
Edited by Y. Shiraki and N. Usami
24 **High-temperature superconductors**
Edited by X. G. Qiu
25 **Introduction to the physics of nanoelectronics**
S. G. Tan and M. B. A. Jalil
26 **Printed films: Materials science and applications in sensors, electronics and photonics**
Edited by M. Prudenziati and J. Hormadaly
27 **Laser growth and processing of photonic devices**
Edited by N. A. Vainos
28 **Quantum optics with semiconductor nanostructures**
Edited by F. Jahnke
29 **Ultrasonic transducers: Materials and design for sensors, actuators and medical applications**
Edited by K. Nakamura
30 **Waste electrical and electronic equipment (WEEE) handbook**
Edited by V. Goodship and A. Stevels
31 **Applications of ATILA FEM software to smart materials: Case studies in designing devices**
Edited by K. Uchino and J.-C. Debus
32 **MEMS for automotive and aerospace applications**
Edited by M. Kraft and N. M. White
33 **Semiconductor lasers: Fundamentals and applications**
Edited by A. Baranov and E. Tournie
34 **Handbook of terahertz technology for imaging, sensing and communications**
Edited by D. Saeedkia
35 **Handbook of solid-state lasers: Materials, systems and applications**
Edited by B. Denker and E. Shklovsky
36 **Organic light-emitting diodes (OLEDs): Materials, devices and applications**
Edited by A. Buckley
37 **Lasers for medical applications: Diagnostics, therapy and surgery**
Edited by H. Jelínková
38 **Semiconductor gas sensors**
Edited by R. Jaaniso and O. K. Tan
39 **Handbook of organic materials for optical and (opto)electronic devices: Properties and applications**
Edited by O. Ostroverkhova
40 **Metallic films for electronic, optical and magnetic applications: Structure, processing and properties**
Edited by K. Barmak and K. Coffey
41 **Handbook of laser welding technologies**
Edited by S. Katayama
42 **Nanolithography: The art of fabricating nanoelectronic and nanophotonic devices and systems**
Edited by M. Feldman
43 **Laser spectroscopy for sensing: Fundamentals, techniques and applications**
Edited by M. Baudelet

44 **Chalcogenide glasses: Preparation, properties and applications**
Edited by J.-L. Adam and X. Zhang
45 **Handbook of MEMS for wireless and mobile applications**
Edited by D. Uttamchandani
46 **Subsea optics and imaging**
Edited by J. Watson and O. Zielinski
47 **Carbon nanotubes and graphene for photonic applications**
Edited by S. Yamashita, Y. Saito and J. H. Choi
48 **Optical biomimetics: Materials and applications**
Edited by M. Large
49 **Optical thin films and coatings**
Edited by A. Piegari and F. Flory
50 **Computer design of diffractive optics**
Edited by V. A. Soifer
51 **Smart sensors and MEMS: Intelligent devices and microsystems for industrial applications**
Edited by S. Nihtianov and A. Luque
52 **Fundamentals of femtosecond optics**
S. A. Kozlov and V. V. Samartsev
53 **Nanostructured semiconductor oxides for the next generation of electronics and functional devices: Properties and applications**
S. Zhuiykov
54 **Nitride semiconductor light-emitting diodes (LEDs): Materials, technologies and applications**
Edited by J. J. Huang, H. C. Kuo and S. C. Shen
55 **Sensor technologies for civil infrastructures**
Volume 1: Sensing hardware and data collection methods for performance assessment
Edited by M. Wang, J. Lynch and H. Sohn
56 **Sensor technologies for civil infrastructures**
Volume 2: Applications in structural health monitoring
Edited by M. Wang, J. Lynch and H. Sohn
57 **Graphene: Properties, preparation, characterisation and devices**
Edited by V. Skákalová and A. B. Kaiser
58 **Silicon-on-insulator (SOI) technology**
Edited by O. Kononchuk and B.-Y. Nguyen
59 **59 Biological identification: DNA amplification and sequencing, optical sensing, lab-on-chip and portable systems**
Edited by R. P. Schaudies
60 **High performance silicon imaging: Fundamentals and applications of CMOS and CCD sensors**
Edited by D. Durini
61 **Nanosensors for chemical and biological applications: Sensing with nanotubes, nanowires and nanoparticles**
Edited by K. C. Honeychurch
62 **Composite magnetoelectrics: Materials, structures, and applications**
G. Srinivasan, S. Priya and N. Sun
63 **Quantum information processing with diamond: Principles and applications**
Edited by S. Prawer and I. Aharonovich
64 **Advances in non-volatile memory and storage technology**
Edited by Y. Nishi
65 **Laser surface engineering: Processes and applications**
Edited by J. Lawrence, C. Dowding, D. Waugh and J. Griffiths

66 **Power ultrasonics: Applications of high-intensity ultrasound**
Edited by J. A. Gallego-Juárez and K. F. Graff

67 **Advances in delay-tolerant networks (DTNs): Architectures, routing and challenges**
Edited by J. J. P. C. Rodrigues

68 **Handbook of flexible organic electronics: Materials, manufacturing and applications**
Edited by S. Logothetidis

69 **Machine-to-machine (M2M) communications: Architecture, performance and applications**
Edited by C. Anton-Haro and M. Dohler

70 **Ecological design of smart home networks: Technologies, social impact and sustainability**
Edited by N. Saito and D. Menga

71 **Industrial tomography: Systems and applications**
Edited by M. Wang

72 **Vehicular communications and networks: Architectures, protocols, operation and deployment**
Edited by W. Chen

73 **Modeling, characterization, and production of nanomaterials: Electronics, photonics and energy applications**
Edited by V. Tewary and Y. Zhang

74 **Reliability characterisation of electrical and electronic systems**
Edited by J. Swingler

75 **Handbook of industrial wireless sensor networks: Monitoring, control and automation**
Edited by R. Budampati S. Kolavennu

76 **Epitaxial growth of complex metal oxides: Techniques, properties and applications**
Edited by G. Koster, G. Rijnders and M. Huijben

77 **Semiconductor nanowires: Materials, synthesis, characterization and applications**
Edited by J. Arbiol and Q. Xiong

Foreword

Reliability of systems and their sub-elements – be they optical, optoelectronic, electrical, or electronic – is a crucially important topic in industry for enabling the classification of the faults and failure modes of devices and systems during their operational lifetime and the quantification of their residual useful life. Unfortunately it is a topic absent in the majority of syllabi in universities worldwide. As electrical and electronic devices become more complex, students ought to understand, let alone apply, design-for-reliability and the various testing methods that they might employ in industry.

In that respect, this book fills an important gap in the market. Its unashamedly empirical approach to the topic and the case studies presented in some of its chapters offer the reader a unique opportunity to relate to his/her own experience in the field.

The contributors to the book rightly took a practical view to reliability while presenting physics of failure and accelerated lifetime testing. More importantly, they dissected how knowledge of reliability could be applied not only across various classes of devices—microcircuits, diode, transistors, and embedded systems—but also across many applications ranging from oil and gas, automotive, and photovoltaics. This pedestrian approach is certainly welcome and will help those engineers who require specific information related to their main concerns.

In summary, I am delighted to give this book my personal recommendation and congratulate Dr Swingler and his contributors on providing a book that will be a very useful addition to the body of knowledge already present in the field.

Marc Desmulliez
Heriot-Watt University, Edinburgh, Scotland, United Kingdom

Introduction

1

J. Swingler
Heriot-Watt University, Edinburgh, United Kingdom

1.1 Introduction

This book has been put together to equip the new graduate and the practicing engineer in the importance of understanding electrical and electronic system reliability from the perspective of those working in the area on particular components and systems. It explores the subtle significance and consequence of building and assessing reliability of a system whether starting with defining it (or not) in the specification stage of a system concept through making particular guaranteed levels at the final implementation stage.

The basics of reliability engineering are not necessarily addressed, even though some chapters may touch on these, and the reader is therefore directed to such texts as *Practical Reliability Engineering* by O'Connor and Kleyner [1] or *Accelerated Quality and Reliability Solutions* by Klyatis and Klyatis [2].

The book repeatedly highlights how reliability engineering has moved from the use of, sometimes arbitrary, standards approach to an empirical scientific approach of understanding operating conditions, failure mechanisms, and the need for testing for a more realistic characterisation.

The book brings together a number of experts and key players in the discipline to concisely present some of the fundamental background of reliability theory, elaborate on the current thinking and developments behind reliability characterisation from components to systems, and give a detailed account of issues across a range of example applications.

1.2 The focus of the book

Firstly, to aid in explaining the focus of the book, a working definition of reliability is presented here. However, a number of more detailed definitions and descriptions, with their particular subtleties, have been presented throughout the book. But here, the working definition of reliability is given as:

> *An expectation of a system's level of functionality at a particular time.*

Functionality is the key. Is the system functioning or partly functioning at a level that matches the specification? How does that functionality change with time?

Reliability Characterisation of Electrical and Electronic Systems. http://dx.doi.org/10.1016/B978-1-78242-221-1.00001-0

1.2.1 Reliability characterisation

The wording of the title of the book combines "reliability" with the noun "characterisation" and that is deliberately done. The word "characterisation," can not only convey the idea of an act of describing essential attributes but also convey the idea of an act of creating or building a character or attribute (e.g., in a literary work). Both ideas are emphasized in this book. The engineer and scientist are aiming to build a specific reliability and assess it to be able to provide evidence that verifies that reliability. These should be two inseparable activities when delivering a system with a reliability character, which is reasonable.

Reliability characterisation could imply that reliability is a feature of a system, which cannot be controlled, and all the developer can do is to characterise its attributes. However, the reliability characterisation of a system should not be seen as something that the engineer conducts passively once there is a product to analyse and describe. As many of the chapters in the book show, reliability is not a given attribute of a final system but should be defined in the specification, continuously built (and enhanced), and characterised throughout all levels of the technology development. That is, from the initial design to the final mass production and beyond, the reliability as with the functionality should be compliant to the specification. Figure 1.1 shows the relationship between the technological readiness levels and reliability building and assessing leading to a "Model of understanding of the system functionality over time," a model of its reliability character.

The earliest step is to define the reliability and life expectancy in the "reliability specification." Designs should be informed from past experience or any prior relevant data, but as Figure 1.1 illustrates, empirical studies of failure mechanisms need to be conducted to get proper understanding of the system's reliability. This is emphasized in this book. In Figure 1.1, this is separated into the influences of application stresses (stresses associated with its function) and operational stresses (stresses associated with its environment). In the development cycle, operational stresses can often be investigated later, but it must be considered that synergistic effects can occur by combining stress types. Accelerated life testing (ALT) and highly accelerated life testing are used to check and make sure that the product will behave as expected. These are repeatedly

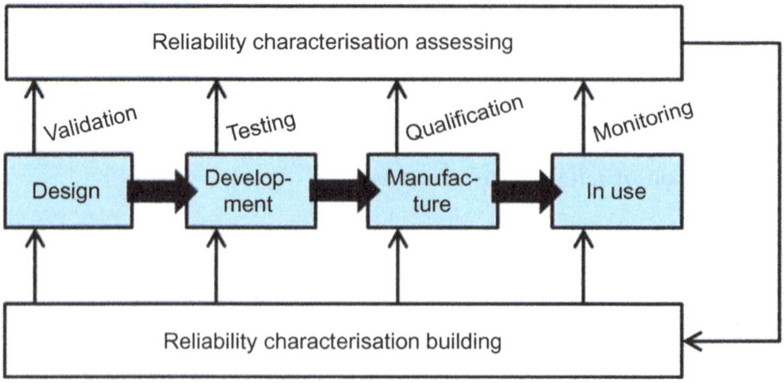

Figure 1.1 Reliability characterisation and technology readiness levels.

addressed throughout the book. What is becoming increasingly of interest is condition monitoring of the system in operation to identify their health. This adds vital real-life data to the understanding of its behaviour. All these enables a "Model of understanding of system functionality over time," which in turn can give a prediction of the reliability, preferably the minimum, mean, and maximum expected lifetime.

1.2.2 Electrical and electronic systems

Both electrical and electronic systems are named in the title to identify and broaden the scope of systems addressed. How the difference between an electrical system and an electronic system is defined may depend on the engineer's background, but a useful definition incorporates at least two concepts: the capability and the scale. Electrical systems tend to have no decision processing capability and tend to be of "higher" voltage scale (say 5 V and up), whereas an electronic system tends to have decision processing capability (handles input signals and delivers an output signal) and tends to be of "lower" voltage scale.

The focus of the book starts from these electrical/electronic systems and works out to their reliability character and how to build and assess it for those specific components/systems. This is the distinct difference with other texts. A simple search on published books in the area of reliability of electrical and electronic systems produces a valuable list of publication. Just to name a few, these are the following:

(1) *Design for Reliability* (Electronics Handbook Series) by Crowe and Feinberg [3].
(2) *Reliability Technology: Principles and Practice of Failure Prevention in Electronic Systems* by Pascoe [4].
(3) *Reliability of Safety-Critical Systems: Theory and Applications* by Rausand [5].

These texts give a wealth of information concerning reliability of electrical and electronic systems. Where the current book is concerned, it is distinguished by giving a survey of components/systems and special application areas of experts working in these particular areas.

1.2.3 The readers and the contributing authors

The book is aimed at the new graduate or the practicing engineer who are perhaps for the first time thinking about the reliability characterisation of a product. The reader is assumed to have some knowledge of reliability issues but wishes to dip into the expertise of others. The contributing authors are practicing scientists and engineers working in industry and academia. They are truly international coming from four continents.

1.3 Reliability science and engineering fundamentals (Chapters 2–4)

Three chapters present current thinking in the context of moving the science forward. The first of these chapters is eye-catching because we are all prone, if we are not careful, to fall into flawed thinking and practice particularly due to time pressures. The following chapter focuses on developing the science explicitly with physics-of-failure (PoF) thinking. In this methodology, real data are acquired on how the system or subsystem behaves under known conditions. The third chapter introduces some of the instruments, which can be used for investigating this behaviour.

The purpose is to acquire a fuller understanding of the system, its operation, and its environment to characterise how its functionality behaves with time. This is essentially an empirical scientific approach. A hypothesis may be presented, models may elegantly describe the detail of the hypothesis, computer simulations may beautifully depict the effect on the system, but if the hypothesis is not grounded in the reality of conducted physical experiments to dig out the cause of the phenomena observed, a fuller understanding cannot be gained. As Figure 1.2 suggests, continual physical experimental investigation is necessary to mitigate against any surprises.

Figure 1.2 is based on a figure by Dummer and Winton [6] where they present several cartoons that illustrate the importance of reliability and how to handle this discipline with care.

1.3.1 Reliability and stupidity

Mistakes in reliability engineering and how to avoid them are addressed in the first chapter contributed by the invited authors—this is up front for good reason. It is placed right at the beginning, and given a gripping title of *Reliability and Stupidity*, to draw attention to the critical aspects in reliability engineering. Chapter 2 ably crystallizes and presents these aspects.

Here, we define what stupidity is in reliability thinking, namely, "being afflicted by nonsense," "contravening deductive logic," and "unreasonable and irrational activities." Barnard identifies that "stupidities" within reliability engineering can be related to three activities:

(1) The correct selection of reliability engineering activities.
(2) The correct execution of those activities.
(3) The correct timing for execution of selected activities.

Barnard discusses various aspects of reliability engineering that should be considered during product development, since many industry-standard practices may be misleading or simply a waste of time and money. A list of nine common mistakes that should

Figure 1.2 One test is worth 1001 opinions.

be avoided in reliability engineering is addressed in detail by Barnard. A quick test is given here for you to judge how you are faring on this:

(1) What level of reliability engineering are you applying at each stage of the design, development, and implementation cycle?
(Inadequate integration of reliability engineering with product development)
(2) Are you focusing on probabilities? Why?
(Unhealthy focus on "probability" in conventional definition of reliability engineering)
(3) Are you only interested in quantifying reliability? Why?
(Distraction of quantification of reliability)
(4) Do you investigate the cause of failure?
(Ignoring cause-and-effect relationships in reliability engineering)
(5) What does MTBF mean in reality to your product? Are you sure?
(Incorrect understanding of the meaning of MTBF)
(6) What level of failure testing do you conduct during product development?
(Inadequate failure testing during product development)
(7) When should you conduct reliability activities?
(Incorrect timing of reliability engineering activities)
(8) Who conducts reliability activities?
(Inappropriate personnel performing reliability engineering activities)
(9) Why are you conducting reliability activities when you do?
(Non-value-adding reliability engineering activities)

What do you think? It is hoped that this book will help answer questions the reader might have and at least point in the right direction on reliability science and engineering.

1.3.2 Physics-of-failure thinking

The reliability science and engineering of a product require an understanding of how that product is designed to function and the duration of that function. Reliability predictions on its useful lifetime can be found from the theory of probability resulting from possible variations in manufacturing processes and operating conditions. Reliability data from life testing are used to obtain distributions of failure times, which are designed to capture information about these variations. However, if reliability predictions are based on life testing that does not mimic conditions or failures seen in the real application, the reliability predictions may be inaccurate with a low level of certainty. As Chapter 3 makes clear, accurate reliability predictions require in-depth knowledge that can be gained through the PoF approach.

It must be stated that there is a cause as to when a system will fail. It does not just happen "randomly"; otherwise, why try to deliver the right reliability for the right cost? Randomisation in the mathematics of reliability should be used to model the unknown in the variations in manufacturing processes and operating conditions. The more thorough the understanding of the failure mechanisms coming from these variations in manufacturing processes and operating conditions, the smaller the uncertainty of the reliability distribution. There is a cause-and-effect. Chapter 3 focuses on these causes and effects using the PoF-based methodology.

The "Center for Advanced Life Cycle Engineering" has greatly benefited reliability science and engineering in establishing and consolidating PoF importance.

1.3.3 Acquiring observational evidence

Modern analytic instrumentation for characterising degradation and failure in electrical and electronic systems is given its own chapter. Chapter 4 reviews several types of instruments and highlights how developments in the electronics industries may impact on the types of instrument that can be used. Goodman et al. classify the instruments into three types, destructive, non-destructive, and *in situ* measurement techniques.

1.4 Reliability methods in component and system development (Chapters 5–9)

A series of chapters (Chapters 5–9) are dedicated to exploring reliability methods used in the development of the individual discrete component through sophisticated complex systems. A general theme can be identified where at some or all stages of the development process, there is a concentration on building reliability, "Reliability Characterisation Building," and then assessing reliability of the product, "Reliability Characterisation Assessing." These two elements of the reliability character of the product in the development process are illustrated in Figure 1.3. There is the important feedback from the output of the assessing element to the building element. Figure 1.3 divides the development process into four major stages, but obviously, this can be elucidated into much finer detail depending on the system and processes involved.

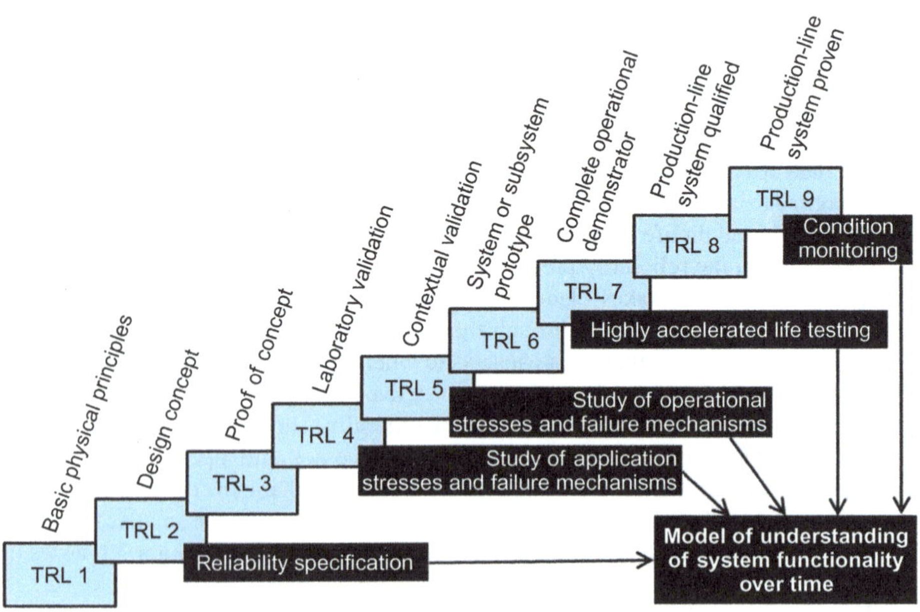

Figure 1.3 Reliability characterisation building and assessing.

1.4.1 Components and devices

The components discussed in Chapter 5 are some of the most representative used in electrical and electronic systems. These include aluminium and tantalum families of capacitors, silicon and non-silicon families of diodes, and bipolar (silicon/non-silicon) and MOS (silicon/non-silicon) families of transistors. Chapter 5 gives a comprehensive analysis of these components, their typical failure mechanisms, and corresponding remedies to build reliability. The design for reliability (DfR) method is presented in detail as an important technique for building reliability along with process reliability, concurrent engineering, screening, and burn-in methods.

Optoelectronics devices compared with electronics devices significantly have fewer reliability research publications. Chapter 6 explains that this is probably because of a shorter history and smaller number of applications for optoelectronics devices. Huang presents details on the approaches and recent developments to these types of devices and focuses on methods in reliability modelling. The reliability of laser diodes is used as a typical example, and case study materials are presented.

Interestingly, both Chapters 5 and 6 draw our attention to electrostatic discharge (ESD) and electrical overstress (EOS) as two causes of catastrophic random failures.

1.4.2 Micro- and nanointegrated circuits

A review of the current understanding of the reliability issues that generally impact silicon circuits of all types is presented in Chapter 7. The focus of the chapter is on degradation mechanisms occurring on the silicon device, that is, the degradation of the transistor and gate dielectrics, the metal conductors, and insulating dielectrics. The incidence of radiation and its influence on induced "soft errors" is also highlighted.

Chapter 7 introduces the two distinct approaches or methods that have been developed to assess the reliability of silicon-integrated circuits. The probabilistic method determines failure rates while not attempting to specify the detailed mechanisms involved. The goal with this method is to describe complete electronic system failure rates and to achieve reliability improvement by accommodating failure rates of individual components. The second approach or method is the deterministic method with the goal to understand the mechanisms responsible for failure and to model their physical and statistical properties. Reliability developments are achieved by root-cause elimination of failure mechanisms during circuit design and manufacture. The focus of Chapter 7 is on this second method.

The developments in semiconductor technology and emerging nanodevices in terms of new materials and device architectures are presented in Chapter 8 along with the challenges for establishing a robust reliability test and assessment methodology. Raghavan highlights that the philosophy of reliability has undergone significant change in recent times with an increased emphasis on DfR at an early stage of product design and development. Reliability is proactively in-built in the design stage rather than being confined to passive postproduction investigations with feedback from accelerated failure tests on the final product. There are an improved choice of materials and device architectures made available by modelling and simulation methods.

1.4.3 More complex systems

The reliability of embedded systems is introduced in Chapter 9. With technology scaling, this has enabled the fabrication of efficient and low-power devices to give massive system integration capability. Shafik et al. discuss the major challenges with such integration as there are increased numbers of hardware faults with logic upsets or timing violations. These are exacerbated by manufacturing imperfections and harsh operating environments. These hardware faults can be overcome with the appropriate methods. Shafik et al. present the reliable system design principles and practices, which usually involved hardware redundancy or error-hardened hardware design.

Shafik et al. highlights that the traditional methods employed involve hardware redundancy designs to achieve the desired reliability of a system. These include triple modular redundancy, higher-order techniques with N-modular redundancy, and standby-sparing designs. The issues with these methods are significantly high silicon footprints and power consumption. Shafik et al. also highlight the error-hardening methods for transient fault, which can be implemented at either physical (shielded against radiation) or device level (using novel substrate materials). The trade-off for each of these methods is discussed.

The chapter closes with the challenges of current and future generations of embedded systems. Shafik et al. discuss possible motivations for future directions of research.

1.5 Reliability modelling and testing in specific applications (Chapters 10 and 11)

The previous series of chapters are the heart of the book focusing on the current thinking on reliability for specific component/system type. The final two chapters conclude the discussion on a few of the application areas.

1.5.1 Application examples

Chapter 10 addresses the reliability of electrical and electronic systems in the automotive vehicle application. The challenge is that the vehicle owner expects everything in their car to work first time every time. This is a challenge because there are an ever-increasing number of ever more complex systems that the owner wants. This coupled with harsh operating conditions sets up a conflict between complexity and dependability, which forces the automotive electronics industry to constantly look for ways to enhance the design and verification of electronic components and systems to improve their dependability, operating efficiency, and functional lifespan.

Chapter 11 addresses solar power facilities and identifies the crucial need to understand the reliability character and lifetime of such a system. What is interesting about this application area is that these facilities are expected to operate for the long term, 25 years plus, and thus, a clear understanding of the reliability characteristics is required. Initial costs tend to be high compared to alternatives, and as Schenkelberg reports, with the average cost per unit of energy over expected operating lifetime of photovoltaic electricity generation expected to be $0.130/kWh, whereas for a coal-fired plant, the cost is $0.096/kWh, stakeholders need to consider the approach carefully. The "bankability" of such a facility is explored in Chapter 11 as stakeholders will want to understand reliability and robustness before they invest in setting up such a system.

1.5.2 Verification techniques

Medhat reports that the automotive electronics industry is constantly looking for ways to enhance the design and verification of components and systems to improve reliability. Electronic design automation techniques are developed to overcome these challenges including circuit reliability verification. The circuit is verified for robustness of schematic design and layout design by checking electrical and physical characteristics against specified design rules. Characteristics checking presented in Chapter 10 includes voltage dependence, negative voltages and reverse current, ESD and latch-up, and EOS susceptibility.

The Calibre PERC reliability verification platform is discussed in detail as a verification technique example with case studies.

1.5.3 Block modelling with ALT techniques

Chapter 11 looks at reliability block modelling to provide the means to identify system weaknesses and to estimate and describe system reliability. This block modelling is supported and developed with carefully executed accelerated life tests. The objective is for the solar power system to have a long lifetime with minimised initial and maintenance costs. ALT permits the exploration and understanding of the design decisions relevant for a specific system and location to address these objectives. It should be noted that each accelerated life test is a tool for the estimation of reliability for a specific failure mechanism. The results from these tests are then included in an overall system reliability model to improve reliability estimates for the system. The block modelling is described in detail in Chapter 11 with the focus on ALT.

ALT for two failure mechanisms is presented in Chapter 11 as an example of how to craft such ALT plans: "A Temperature, Humidity, and Bias ALT Plan for CMOS Metallization Corrosion" and "A Thermal Cycling ALT Plan for SAC Solder Failure Mechanism."

1.6 Conclusion

This introductory chapter describes the focus, rationale, and scope of the book. It introduces the material contributed by the invited authors but starts with some initial ideas to set the scene.

The relationship between the technological readiness levels of a system under development and how to achieve an understanding of the system functionality over time, a model of its reliability character, is shown. Critical to this is acquiring empirical evidence from experimental work, which will always have a place to verify the understanding of the developer. Therefore, understanding the reliability character of the product under development requires the activities, in some form, of reliability characterisation building and reliability characterisation assessing. These concepts are introduced in this chapter.

Also, a brief introduction to the subject areas that the invited authors address has been given. These subject areas have been divided into three themes: reliability science and engineering fundamentals, reliability methods in component and system development, and reliability characterisation for specific applications.

References

[1] P. O'Connor, A. Kleyner, Practical Reliability Engineering, fifth ed., Wiley-Blackwell, USA, 2012.
[2] L.M. Klyatis, E. Klyatis, Accelerated Quality and Reliability Solutions, Elsevier Science, USA, 2006.
[3] D. Crowe, A. Feinberg, Design for Reliability, CRC Press, USA, 2001.
[4] N. Pascoe, Reliability Technology: Principles and Practice of Failure Prevention in Electronic Systems, Wiley, USA, 2011.
[5] M. Rausand, Reliability of Safety-Critical Systems: Theory and Applications, Wiley-Blackwell, USA, 2014.
[6] G. Dummer, R. Winton, An Elementary Guide to Reliability, Pergamon Press, Oxford, UK, 1969.

Reliability and stupidity: mistakes in reliability engineering and how to avoid them

2

R.W.A. Barnard
Lambda Consulting, Pretoria, South Africa

2.1 Introduction

> *Unfortunately, the development of quality and reliability engineering has been afflicted with more nonsense than any other branch of engineering. This has been the result of the development of methods and systems for analysis and control that contravene the deductive logic that quality and reliability are achieved by knowledge, attention to detail, and continuous improvement on the part of the people involved.*
>
> *Patrick O'Connor*

These statements, written by a well-known reliability engineering author, provide for an excellent introduction on stupidity in reliability engineering. At the same time, they explain why executing industry-standard reliability engineering activities is no guarantee that high product reliability will be achieved in operations.[1] In order to understand this paradox in reliability engineering, let us consider a few fundamental concepts. We need to understand that all product failures are caused and that all product failures can be prevented. Crosby [1] stated that "All nonconformances are caused. Anything that is caused can be prevented." Based on this statement and applying common sense to real-life experience, reliability may simply be defined as the absence of failures, and reliability engineering as the management and engineering discipline that prevents the creation of failures.

These simple definitions imply that a product is reliable if it does not fail and that this failure-free state can only be achieved if failure is prevented from occurring. What is required to prevent failure? Firstly, engineering knowledge to understand the applicable failure mechanisms and, secondly, management commitment to mitigate or eliminate them. Proactive prevention of failure should be the primary focus of reliability engineering, and never reactive failure management or failure correction. Figure 2.1 shows part of a typical product development process, with an emphasis on design and production verification. The iterative nature of product development is evident. It can also be seen that reliability engineering changes from "proactive" during design (i.e., failure prevention) to "reactive" during production and especially

[1]Although the word "product" is frequently used in this chapter, information equally applies to systems.

Reliability Characterisation of Electrical and Electronic Systems. http://dx.doi.org/10.1016/B978-1-78242-221-1.00002-2

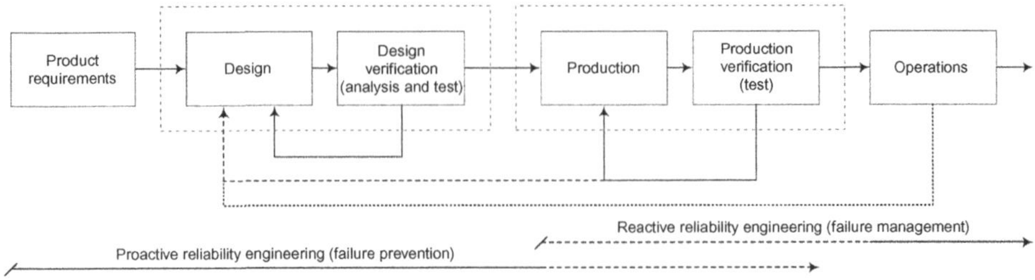

Figure 2.1 Design and production verification.

during operations (i.e., failure management). Reactive reliability engineering should be avoided due to the very high cost of corrective actions, which may be required (e.g., redesign and product recalls).

It is important to understand that reliability is a nonfunctional requirement during design and that it becomes a characteristic of a product during operations. "Analysis" and "test" are two primary verification methods used in engineering. Reliability engineering activities to perform reliability analyses and tests are well documented in various textbooks on reliability engineering [2]. Product reliability is the result of many management and technical decisions taken during all product development stages (i.e., concept, definition, design, and production).

The worst mistake in reliability engineering that a company can make is to ignore reliability during product development. For any product of reasonable complexity, the outcome almost certainly will be unacceptable reliability, resulting in an inferior product. In theory, a formal reliability engineering program is not a prerequisite for successful product development, but in practice, it is highly recommended.

Stupidity in reliability engineering implies that some activities can be considered as nonsense, ineffective, inefficient, incorrect, wasteful, etc. To decide which reliability engineering activities fall into this category is not trivial, since it depends on many factors such as product complexity, product costs, technology maturity, and failure consequences.

This chapter discusses various aspects of reliability engineering that should be considered during product development, since many industry-standard practices may be misleading or simply a waste of time and money. An attempt is made to relate these aspects to the ultimate goal of reliability engineering, namely, the prevention or avoidance of product failure. It will become evident that "stupidities" experienced with reliability engineering are related to (1) the correct selection of reliability engineering activities, (2) the correct execution of those activities, or (3) the correct timing for execution of selected activities.

2.2 Common mistakes in reliability engineering

The following aspects, discussed in no particular order, are common mistakes that should be avoided in reliability engineering.

2.2.1 Inadequate integration of reliability engineering with product development

What is the goal of reliability engineering? We need to distinguish between tasks that are often useful way stations and the ultimate goal. The ultimate goal is to have the product meet the reliability needs of the user insofar as technical and economic constraints allow. The ultimate goal surely is not: To generate an accurate reliability number for the item.

Ralph Evans

Reliability engineering activities are often neglected during product development, resulting in a substantial increase in the risk of project failure or customer dissatisfaction. In recent years, the concept of design for reliability has been gaining popularity and is expected to continue for years to come [3]. Reliability engineering activities should be formally integrated with other product development processes. A practical way to achieve integration is to develop a reliability program plan at the beginning of the project.

Appropriate reliability engineering activities should be *selected* and *tailored* according to the objectives of the specific project and should be documented in the reliability program plan. The plan should indicate which activities will be performed, the timing of these activities, the level of detail required for the activities, and the persons responsible for executing the activities. Raheja [4] wrote, "Reliability is a process. If the right process is followed, results are likely to be right. The opposite is also true in the absence of the right process. There is a saying: 'If we don't know where we are going, that's where we will go.'"

ANSI/GEIA-STD-0009-2008 [5], *Reliability Program Standard for Systems Design, Development, and Manufacturing*, can be referenced to develop a reliability program plan. This standard addresses not only hardware and software failures but also other failure causes during manufacturing, operations, maintenance, and training.

ANSI/GEIA-STD-0009-2008 supports a life cycle approach to reliability engineering, with activities logically grouped under the following objectives:

- Understand customer/user requirements and constraints
- Design and redesign for reliability
- Produce reliable systems/products
- Monitor and assess user reliability

The first category (i.e., understanding of requirements and constraints) is of significant importance. Yet, design teams rarely pay enough attention to formal requirements analysis. Inadequate understanding of requirements frequently results in design modifications and production rework, or, even worse, premature product failure. Complete and correct requirements are therefore necessary at the beginning of product development. These requirements should include all intended environmental and operational conditions. It is not possible to design and produce a reliable product if you do not know where and how the product will be used. It is also a bad practice to "copy and paste" requirements from one development specification to another. Rather, spend more time upfront, and avoid costly redesign and rework later.

Due to a multitude of reliability engineering activities available, inexperienced engineers may find it difficult to develop an efficient and effective reliability program plan. For example, should failure mode and effects analysis (FMEA) be considered, or should fault tree analysis (FTA) rather be considered, or perhaps both analyses? Figure 2.2 indicates a few relevant questions, which may be used to guide the development of a reliability program plan for a specific project [6].

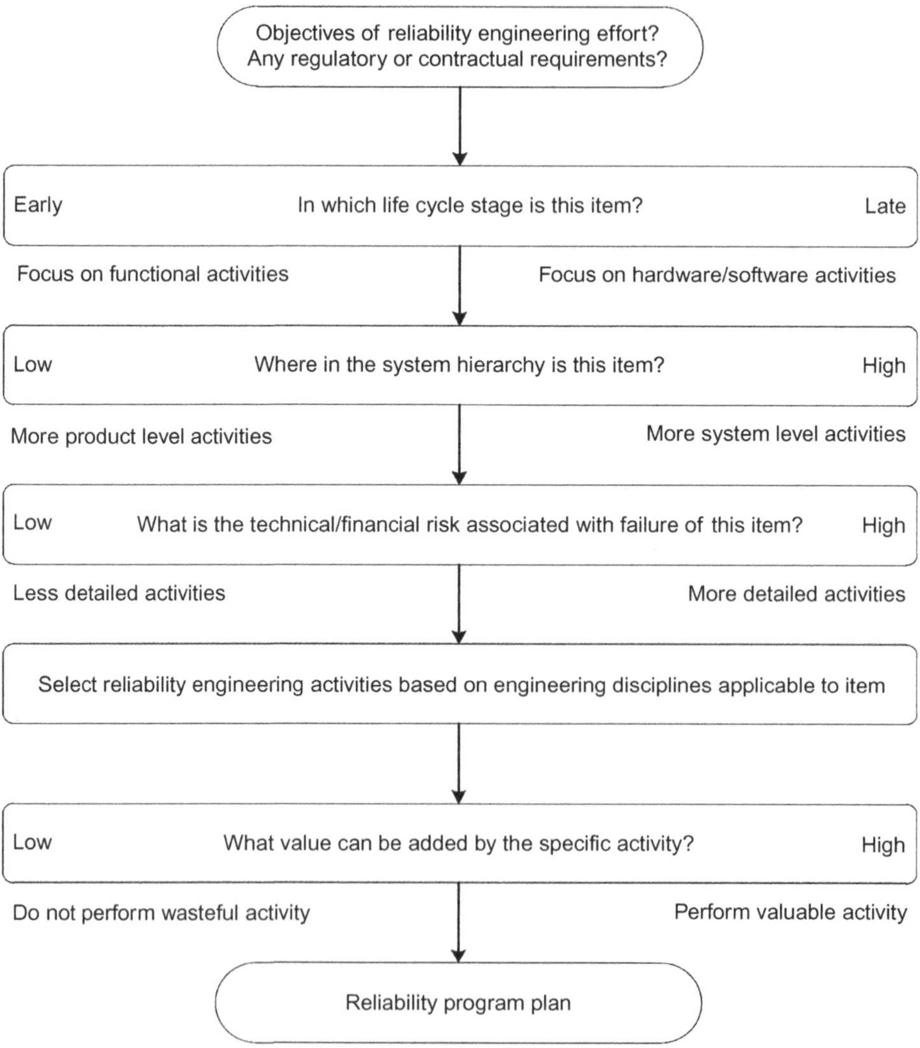

Figure 2.2 Reliability program plan development.

2.2.2 Focus on "probability" in conventional definition of reliability engineering

Reliability engineering is everything you do today to prevent product failure tomorrow.

Albertyn Barnard

According to the conventional definition [2], reliability is defined as "the probability that an item will perform a required function without failure under stated conditions for a stated period of time." While this definition is of course totally correct, the focus on probability has over the years resulted in major emphasis on various aspects of mathematics and statistics in reliability engineering. In many cases, the other three parts of the definition (i.e., required function, stated conditions, and period of time) have been severely neglected. Ironically, these other aspects relate more to the "engineering" in "reliability engineering," since they describe what the product should do, where it will be used and what the expected life should be. Proper answers to these questions provide essential design input and form part of requirements analysis performed at the start of product development.

However, since the beginning of this century, reliability engineering has evolved more into the engineering discipline necessary to deliver high-quality products to customers. Without reliability, no company can compete successfully in today's marketplace. Although not the same, quality and reliability are often used interchangeably. In the simplest terms [7], these can be defined as the following:

- Quality is conformance to specifications.
- Reliability is conformance to specifications *over time.*

Reliability is thus the continuation of quality over time. Note that probability is not part of this definition.

As stated previously, the common sense definitions of reliability and reliability engineering are [8] the following:

- Reliability is the absence of failures in products.
- Reliability engineering is the management and engineering discipline that prevents the creation of failures in products.

The focus on probability is clearly also absent from the common sense definition of reliability. It can be argued that the average customer is far more interested in buying and using hassle-free products than knowing the theoretical probability of failure of those products.

In an attempt to shift the focus away from quantitative reliability, some authors seem to prefer the word "ability" instead of "probability." For example, ANSI/GEIA-STD-0009-2008 defines reliability as "the *ability/probability* of failure-free performance of the system/product, over the expected service use profile and environmental conditions over a given period of time." Other authors define reliability as "the *ability* of a product or system to perform as intended (i.e., without failure and within specified performance limits) for a specified time, in its life cycle conditions" [9].

It may not be immediately evident, but a company's approach to reliability engineering will depend to a large extent on which definition of reliability they support. The conventional definition requires much more emphasis on, for example, failure data analysis (using various statistical methods), while the other definitions place more emphasis on in-depth understanding and knowledge of possible failure mechanisms. Companies supporting the common sense definition of zero failures will invest in technical and management processes to design and produce failure-free products.

The following "reliability equation" clearly illustrates the difference between these approaches to reliability engineering:

$$\text{Reliability} = f\begin{bmatrix}\text{material characteristics, part tolerances,}\\ \text{temperature coefficients, design parameters,}\\ \text{operating stresses, environmental conditions,}\\ \text{customer usage, duty cycles, system integration,}\\ \text{interfaces between parts and subsystems,}\\ \text{quality of parts and production processes, etc.}\end{bmatrix}$$

While it may be useful to calculate, measure, or monitor reliability (i.e., the left side of the equation), it does little or nothing to achieve or improve reliability. Rather, focus on defining and controlling the factors that directly influence reliability (i.e., the right side of the equation) (derived from Carlson [10]).

2.2.3 Quantification of reliability

Quantification of reliability is in effect a distraction to the goals of reliability.

Ted Kalal

As mentioned, the conventional definition of reliability requires a major focus on probability of failure. Although useful information can be obtained from proper statistical analyses (e.g., analysis of part failures), in many cases, the reliability effort simply degrades to "playing the numbers game." This is almost always the case where a quantitative requirement is set by the customer, and the developer has to show compliance with this requirement. Reliability prediction and reliability demonstration are prime examples of activities where quantification of reliability should be avoided during product development.

It is not possible to determine product reliability by merely adding part failure rates obtained from any document (such as Mil-Hdbk-217). If you claim to be able to predict reliability, you imply that you know what will fail and when it will fail. If this were possible, why don't you rather use that knowledge to prevent the failure from occurring? The only valid reliability prediction is for wear-out failure modes, where the specific failure mechanisms are well understood.

Many reliability engineering activities therefore become part of a historical documentation process, which cannot possibly contribute to higher product reliability. Assumptions are often "adjusted" by the reliability engineer to achieve the desired "result." The results of these activities are only useful to satisfy a project manager (who wants to "tick the box"). Remember that product reliability can only be improved by changing a design or production process, where required. In other words, not only the project manager but also the product should "know" that an activity has been performed.

Experience has shown that using a qualitative approach (compared to a quantitative approach) is usually sufficient and accurate enough to guide reliability engineering efforts. For example, while no design engineer can predict product reliability during development, all engineers can perform relative assessments of technical risks. A qualitative analysis (e.g., based on a simple 1-10 scale) can normally be used to rank probability of failure. The usefulness of such approaches can be even further improved when ratings from different knowledgeable persons are combined.

However, quantification of the reliability of a product during operations is not only valid but also necessary. Without a proper failure reporting system, it is impossible to know what the actual field reliability of a product is. Monitoring or measuring of reliability during operations is thus totally different to predicting and demonstrating reliability during product development.

2.2.4 Ignoring cause and effect relationship in reliability engineering

All failures in electronic equipment can be attributed to a traceable and preventable cause, and may not be satisfactorily explained as the manifestation of some statistical inevitability.

Norman Pascoe

Most engineers believe that all parts eventually wear-out and that all parts therefore have a property such as failure rate. While this statement is theoretically correct, it is absolutely incorrect to assume that product failure is simply caused by individual part failures. When applied and used correctly, modern electronic parts can function for extremely long periods of time, even under severe environmental conditions. Many modern electronic products are replaced by customers due to technological obsolescence and not due to wear-out of parts.

When product failure occurs and a root cause analysis is performed, it becomes evident that failures are created, primarily due to errors made by people such as design and production personnel. However, many engineers do not understand that root cause analysis should proceed to the identification of the real cause of failure. For example, when a part fails due to a very high junction temperature, high dissipated power may easily be seen as the cause of failure, while the real cause may be inferior thermal design.

Products seldom fail due to part failure, but often fail due to incorrect application and integration of those parts. Many so-called part failures can be traced to inadequate design margins (to provide robustness against manufacturing and operating variations). It should also be noted that many electronic part failures are caused by mechanical failure mechanisms. For example, a designer cannot expect reliable operation of any electronic part if the specific printed circuit board is allowed to resonate at some frequency during operations. Failures observed during development tests therefore provide valuable design input, and design teams should be careful not to discard these simply as "random failures."

2.2.5 Incorrect understanding of the meaning of MTBF

MTBF is the worst four letter acronym in our profession.

Fred Schenkelberg

Mean time between failures (or MTBF) is probably the most widely used reliability parameter. It is frequently specified as requirement in product development specifications, it is published in both marketing and technical product (and even part) brochures, it is discussed by engineers in technical meetings, etc. Yet, when engineers are asked to explain the meaning of MTBF, only a small minority of them actually understands it correctly. Most people, including many engineers, think that MTBF is the same as "average life" or "useful life." This is of course totally incorrect and frequently leads to inferior design decisions.

Theoretically, MTBF is the mean value of the exponential failure distribution. For time t as an independent variable, the probability density function $f(t)$ is written as

$$f(t) = \lambda e^{-\lambda t} \tag{2.1}$$

The probability of no failures occurring before time t (i.e., reliability) is obtained by:

$$R(t) = 1 - \int_0^t f(t)dt \tag{2.2}$$

$$R(t) = e^{-\lambda t} \tag{2.3}$$

For items that are repaired, λ is called the failure rate, and $1/\lambda$ is called the MTBF. Using Equation (2.3), reliability at time t=MTBF can be calculated as only 36.8%. In other words, 63.2% of items will have failed by time t=MTBF. This characteristic of the exponential failure distribution suggests that MTBF may not be a very useful reliability parameter.

Reliability calculations using a constant failure rate (as associated with the exponential distribution) are relatively easy to perform. Many engineers therefore assume an exponential failure distribution, even when it is not applicable (e.g., wear-out failures). Using a mean value provides no engineering knowledge necessary to prevent failures, and therefore, it should not be used. Some companies are using other metrics instead (such as probability of success (i.e., reliability) or Bx-life or similar) and have terminated the use of MTBF entirely. Engineers should never use an average or a mean life and should rather promise failure-free (or minimum) life to customers [4].

2.2.6 Inadequate failure testing during product development

To understand reliability, one must study 'unreliability' and the root cause of field failures.

Kirk Gray

Product development processes should include design and production verification prior to full-scale production, as shown in Figure 2.1. Many companies therefore perform formal qualification tests to demonstrate that the product complies with all specifications. However, how do you verify compliance with a reliability specification? Testing of a prototype (or a sample of prototypes) is simply not practical due to the extremely long test time involved. The only feasible alternative is to employ some kind of accelerated life testing, such as HALT (highly accelerated life testing).

However, to obtain life acceleration, one has to test prototypes under severe electrical and environmental conditions, which inevitably may lead to product failure. Some engineers are against accelerated testing and argue that test conditions in excess of product specifications are "unrepresentative" of normal conditions and therefore will produce invalid results. Other engineers argue that one has to test to failure (or at least beyond specification levels), since without malfunction or physical failure, no new knowledge on potential failure mechanisms is generated. For example, any failure mode observed during HALT becomes an opportunity for improvement, through root cause analysis and corrective action [11]. If you do not know and understand the failure mechanisms applicable to your product, how can you claim to know anything about product reliability?

It is interesting to note that while the goal of design for reliability is to design reliability into the product, it is often in practice achieved by actually removing unreliability from the product. This necessitates a closed-loop corrective action system to implement (and verify) design or process improvements. Remember that no reliability analysis or test can improve product reliability; only people can implement the necessary design and process changes.

The concept of failure testing (vs. success testing) is a well-known part of the Toyota product development process [12]. If performance specifications are set in advance and prototypes are tested for compliance to those specifications, learning is minimal due to the pass-fail approach. "*Ijiwaru* testing on the other hand is the practice of testing subsystems to failure. By testing these subsystems under both normal and abnormal conditions and pushing designs to the point of failure Toyota engineers gain a great deal of insight into both current and future designs and materials by understanding the absolute physical limitations of their subsystems."

2.2.7 Reliability engineering activities performed at incorrect time during development

Reliability engineering often becomes a "numbers game" after the real game is over. Reliability cannot be economically added after the system has been conceived, designed, manufactured and placed in operation.

Roy Billinton

Execution of reliability engineering activities at the incorrect time during product development is a common mistake made by many companies. Incorrect time means that the activity is performed too early or too late. Product development is an iterative process, where design is followed by verification, which may result in redesign. For obvious reasons, a specific reliability analysis can only be performed after a specific design cycle has been completed, and a reliability test can only be performed after a prototype has been manufactured. If either of these activities is performed too early, the analysis or test may provide incomplete or even invalid information and may have to be repeated at a later stage.

A more serious (and frequent) problem seen in industry is that reliability activities are performed too late during development. Due to the ever-increasing pressure on timescales, there is usually a short "window of opportunity" to perform reliability engineering activities between design and production. Only experienced project managers will make adequate provision for executing these activities. The author has identified numerous design errors (e.g., overstressed parts) while performing reliability analyses on new designs, but corrective action was often not possible since "the printed circuit boards have already been ordered"!

Finding and correcting design or production problems during later life cycle stages can be very expensive. Consider a product whose product cost is Pc. The costs due to failure at the various stages of the product's life cycle have been investigated [13], and in terms of Pc, they have been found to be the following:

0.1 Pc: internal failure cost due to rework at end of production line
Pc: external failure cost for return from customer inspection
10 Pc: external failure cost for warranty return due to failure with customer in use

This relationship is commonly known as the "10x rule" and demonstrates how a design or production error, if not discovered, will give rise to ten times the original elimination costs in a later stage of the product life cycle. Other surveys have found that these costs could be much higher. It therefore makes a lot of sense to practice proactive reliability engineering and to invest more resources early during product development.

2.2.8 Reliability engineering activities performed by incorrect personnel

By the work one knows the workman.

Jean de La Fontaine

Reliability engineering is related to other disciplines such as logistics engineering, maintenance engineering, safety engineering, and quality engineering, since these disciplines are all "concerned about failures." Some companies therefore delegate reliability engineering to these other engineering disciplines. However, the differences in viewpoint between these disciplines may lead to ineffective and inefficient execution of reliability engineering activities. For example, the objective of a logistics engineer is to provide cost-effective logistic support in the event of failure, while the objective of a reliability engineer is to prevent that failure from occurring.

A more subtle problem is that these engineering disciplines may all perform apparently similar activities. For example, a logistics FMEA, which is an essential part of the logistics engineering process, is very different to a design FMEA. The format and results look similar due to the FMEA process, but the focus and objectives are totally different.

It is essential that reliability engineering activities be performed by or under the guidance of an experienced and competent reliability engineer. This person needs a sufficiently high level of technical know-how to be respected by the design team. He or she must in fact sometimes be more competent than individual designers to understand how the design works and to understand how the design does not work (i.e., how it can fail). You cannot expect positive results when you give inexperienced or incompetent personnel the responsibility for reliability. The following important questions show how technical detail (which can definitely cause product failure) is far outside the competency of the average logistics, maintenance, safety, or quality engineer:

- Which part will have the highest electrical stress in the intended operating environment?
- Which part will have the highest thermal stress in the intended operating environment?
- What will the natural frequency of the printed circuit board assembly be?

Although some design analyses are closely related to reliability engineering (e.g., finite element analysis and thermal analysis), it may not be a good idea to have these activities performed by design engineers. Many design errors may go unidentified if an analysis is performed by the same person who has made the error in the first instance. This fact may necessitate third-party analysis and test or at least involvement of personnel other than the design engineer (e.g., peer review).

2.2.9 Non-value adding reliability engineering activities

Reliability in a product is not what you put into it. It is what the customer gets out of it.[2]

Albertyn Barnard

The diagram to guide the development of a reliability program plan (shown in Figure 2.2) refers to the "value" of a specific reliability activity. If the value of a specific activity is high, it makes sense to perform the activity. If not, it should not be performed due to potential waste of resources. But what is value? "Value is what the customer says it is, considers important, and is willing to pay for" [14]. Value can thus only be determined by the customer and not by the producer. In order to make a decision on the value of a reliability activity, we need to define value in this context.

Value in reliability engineering is created by:

- performing an activity that, through any design or process improvement, results in the elimination (or reduction in the probability) of product failure or
- performing an activity that, through verification of the absence (or low probability) of potential failure modes, results in a reduction in uncertainty.

[2]Adaptation of quotation about quality by Peter Drucker.

Based on this definition, it can be argued that many companies often perform non-value adding activities. If a specific activity does not contribute to the ultimate goal of designing and producing failure-free products, it should simply not be performed.

Activities that may be considered as non-value adding include

- reliability prediction (using failure rate models published in various databases (such as Mil-Hdbk-217)),
- reliability demonstration (using Mil-Hdbk-781),
- reliability block diagram analysis (using MTBF values published in part manufacturer's data sheets),
- reliability block diagram analysis (performed on series systems),
- derating analysis (using part maximum ratings listed for 25 °C ambient temperature),
- FMEA (without in-depth understanding of cause of failure modes),
- FTA (performed on series systems (i.e., without using AND gates)),
- reliability analysis based on hardware only (i.e., ignoring software),
- Weibull analysis (ignoring requirements for valid data analysis),
- thermal analysis (performed on low-power electronic designs),
- HALT (without design or process improvement when required).

Incorrect assumptions that can easily invalidate otherwise valuable activities include

- constant failure rate (i.e., exponential distribution),
- electrical stress in a stress analysis of an electronic design,
- thermal stress in a stress analysis of an electronic design,
- structural stress in a stress analysis of a mechanical design,
- redundancy,
- environmental and operating conditions,
- mission profiles (e.g., duty cycles),
- single-point failures.

It is important to observe that even reliability engineering activities that are normally classified as valuable (e.g., FMEA and HALT) are considered to be non-value adding when they are not correctly executed or when incorrect assumptions are made during their execution.

2.2.10 Incorrect viewpoint on cost of reliability

> *When you focus on cost, quality goes down; when you focus on quality, cost goes down.*
>
> *Unknown (translated from Dutch)*

Execution of the activities described in a reliability program plan can be expensive, depending on the specific product and its reliability requirements. This is due to costs involved with design improvements, reliability personnel, reliability analyses, reliability tests, test equipment, test units, etc. The conventional viewpoint on reliability costs is shown in Figure 2.3. It suggests that for higher reliability, reliability costs will increase and failure costs will obviously decrease. This results in an "optimum"

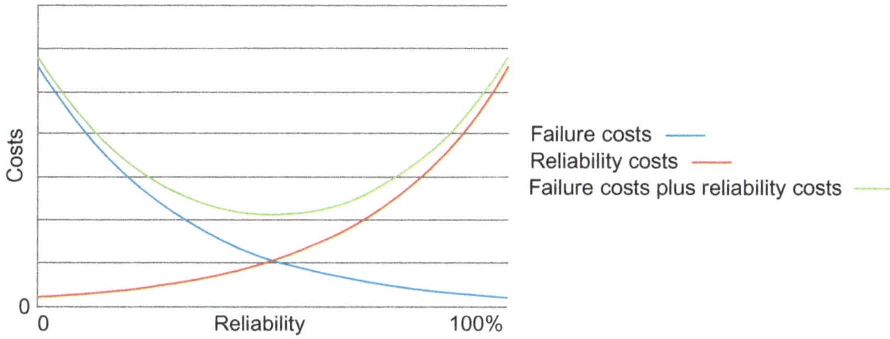

Figure 2.3 Conventional viewpoint on reliability costs.

reliability value where total costs are minimized. Reliability costs include all costs associated with achieving reliability, while failure costs refer to all costs related to unreliability (e.g., rework, repair, and warranty claims).

However, since less than 100% reliability is the result of failures, all of which have preventable causes, this viewpoint may be misleading. All efforts to improve reliability by identifying and removing potential causes of failures should result in cost savings later in the product life cycle [2]. The conventional viewpoint ignores other important "bigger picture" aspects such as enhanced brand reputation, increased market share, more focus of engineers on new product development (and not repairing failed units), and improved personnel morale. At higher levels of reliability, the return on investment becomes progressively larger than the cost of achieving that higher reliability (i.e., incremental profit) [15]. This viewpoint on reliability costs is in agreement with many well-known quality philosophies (e.g., corrective action and "kaizen" or continuous improvement). The correct viewpoint on reliability costs is shown in Figure 2.4.

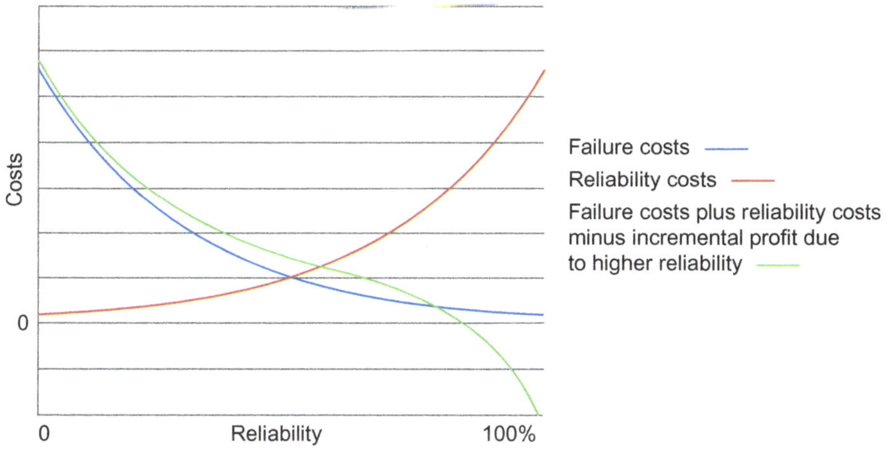

Figure 2.4 Correct viewpoint on reliability costs.

It may be tempting for management to argue in favor of an optimum reliability value. However, the correct viewpoint provides a sound argument that expenses to improve reliability are indeed investments and not costs. It is important to communicate these concepts clearly to project management. If not, they may be reluctant to support proactive reliability activities and therefore only "pay lip service" to reliability efforts.

2.3 Conclusion

To efficiently meet any reliability objective requires comprehensive knowledge of the relationships between failure modes, failure mechanisms and mission profile.
German Electrical and Electronic Manufacturers' Association [16]

A successful product has to be reliable. A product is not reliable simply because reliability engineering activities have been performed. It is reliable if it was designed and manufactured to be reliable. This requires that the correct reliability engineering activities are correctly executed at the correct time during product development. It is reliable when management provides the necessary resources to enable engineering to achieve the goal of zero failures. Reliability is simply a consequence of good engineering and good management.

References

[1] P.B. Crosby, Quality Without Tears: The Art of Hassle-Free Management, McGraw-Hill, New York, 1995.
[2] P.D.T. O'Connor, A. Kleyner, Practical Reliability Engineering, fifth ed., John Wiley, UK, 2012.
[3] A. Kleyner, Foreword, in: D. Raheja, L.J. Gullo (Eds.), Design for Reliability, John Wiley, Hoboken, NJ, 2012, p. xv.
[4] D. Raheja, Design for reliability paradigms, in: D. Raheja, L.J. Gullo (Eds.), Design for Reliability, John Wiley, Hoboken, NJ, 2012, pp. 4–5.
[5] ANSI/GEIA-STD-0009-2008, Reliability Program Standard for Systems Design, Development, and Manufacturing, Information Technology Association of America, 2008.
[6] R.W.A. Barnard, Lean reliability engineering, in: INCOSE EMEA Systems Engineering Conference, South Africa, 2014.
[7] M.A. Levin, T.T. Kalal, Improving Product Reliability: Strategies and Implementation, John Wiley, UK, 2003.
[8] R.W.A. Barnard, What is wrong with reliability engineering? in: 18th Annual International Symposium of INCOSE, The Netherlands, 2008.
[9] K.C. Kapur, M. Pecht, Reliability Engineering, John Wiley, Hoboken, NJ, 2014.
[10] C.S. Carlson, Effective FMEAs: Achieving Safe, Reliable, and Economical Products and Processes Using Failure Mode and Effects Analysis, John Wiley, Hoboken, NJ, 2012.
[11] G.K. Hobbs, Accelerated Reliability Engineering: HALT and HASS, John Wiley, UK, 2000.

[12] J.M. Morgan, J.K. Liker, The Toyota Product Development System: Integrating People, Process, and Technology, Productivity Press, New York, 2006.
[13] J.D. Booker, M. Raines, K.G. Swift, Designing Capable and Reliable Products, Butterworth-Heinemann, UK, 2001.
[14] B.W. Oppenheim, Lean for Systems Engineering with Lean Enablers for Systems Engineering, John Wiley, Hoboken, NJ, 2011.
[15] C. Haibel, Organizing for Success, Hobbs Engineering Webinar, Colorado, USA, 2012.
[16] German Electrical and Electronic Manufacturers' Association, Electronic Components and Systems Division, in: Handbook for Robustness Validation of Automotive Electrical/Electronic Modules, Frankfurt am Main, Germany, 2013.

Physics-of-failure (PoF) methodology for electronic reliability

3

C. Hendricks, E. George, M. Osterman, M. Pecht
Center for Advanced Life Cycle Engineering (CALCE), University of Maryland, College Park, MD, USA

3.1 Introduction

Ensuring product reliability often requires significant testing to identify the expected time to failure (TTF) for a population of devices. To assess reliability, several approaches are taken, such as accelerated product design and testing. Accelerated design and testing require insight into the various ways that external stresses affect the reliability of a device. A physics-of-failure (PoF) approach to reliability uses knowledge of the mechanisms and processes that degrade a product to estimate the life of that product in its usage conditions.

This chapter focuses on using a PoF-based methodology for accelerated reliability qualification and continuous reliability and health assessment [1,2]. In Section 3.2, reliability is defined, potential costs of unreliability are examined, and challenges with reliability estimation are presented. In Section 3.3, PoF models are described with case studies to demonstrate their characteristics and use. In Section 3.4, the implementation of failure modes, mechanisms, and effects analysis (FMMEA) for PoF reliability assessment is discussed. Applications of PoF-based methodology are provided in Section 3.5. In Section 3.6, a summary of the state of the art of PoF methodologies is provided and areas of future research are identified.

3.2 Reliability

Reliability is defined as the ability of a product or system to perform its intended function for a specified time under its expected operating conditions [3]. This definition of reliability necessitates an understanding of how a product will be used and for what length of time. The information on product use conditions and the duration of exposure enables manufacturers to make choices about warranties while considering user expectations. Due to variations in manufacturing processes and operating conditions, a product's reliability is founded on the theory of probability. Failure data from life testing are used to obtain a distribution (or multiple distributions) of failure times. The probability distribution that best fits the experimental data allows manufacturers to quantify their product's reliability by using a metric such as mean time to failure (MTTF) and assign

Reliability Characterisation of Electrical and Electronic Systems. http://dx.doi.org/10.1016/B978-1-78242-221-1.00003-4

a level of statistical confidence to the reliability estimates. If failure predictions are based on test results that do not mimic conditions or failures seen in the field, the reliability of the product could be either understated or overstated. Accurate failure predictions require in-depth knowledge that can be gained through the PoF approach.

Unreliable products may result in financial losses and tarnish the goodwill of a manufacturer. These losses can be either direct or indirect. If the reliability of a product is unknown or misjudged, warranty claims could result in added costs for a company through the costs of repair and replacement. For example, Toyota experienced problems with gas pedals in 2009 and 2010 that caused cars to accelerate out of control, resulting in several deaths [4]. Some of the early failures were linked to the gas pedal getting lodged in the downward position due to faulty floor mats. This finding resulted in a recall of 4.2 million floor mats in September 2009. However, questions remained as to whether the floor mats were actually the root cause or if the drive-by-wire design was to blame. In January 2010, after a crash where floor mats were ruled out as a cause, Toyota was forced to recall 2.3 million cars. This recall came after many public statements by Toyota claiming that incorrect installation of floor mats was the sole reason for the acceleration issues. Toyota also temporarily halted all sales of the affected models and shut down assembly lines. Not only did these reliability issues hurt Toyota's sales, but also they damaged its public image.

Another potential cost of unreliable products is compensation. A device manufacturer may be responsible for reimbursing their customers for lost revenue due to excessive downtime. For example, lithium-ion battery failures on two separate Boeing 787 Dreamliners in January 2013 resulted in the grounding of the entire Dreamliner fleet [5]. All Nippon Airways (ANA) has one of the largest fleets of Dreamliners and has relied on the fuel-efficient jet for long-distance flights. As a result, ANA had to change flight routes and reported losing over US $1 million per day due to the reliability issues in Dreamliners's batteries [6]. Boeing will likely need to compensate its customers for this loss of income [7].

Due to the pressures of a competitive, global economy, engineers often need to assess reliability in a short period of time. For a device with a projected lifetime of 10 years or more, a company cannot afford to spend 10 years testing the device to determine its reliability over all stages of use. Instead, the company needs to rely on accelerated tests to assess reliability quickly and accurately. Accelerated tests can be based on time compression or stress elevation. With time compression methods, the stress levels are maintained within operational limits, but the number of operations is increased to match the expected number of lifetime operations. With stress elevation, one or more load levels, such as temperature or voltage, are increased beyond those anticipated in operation. Operation under the elevated stress level is expected to cause failure in a shorter time frame. Therefore, a model for transforming the time under test to the time in the field (acceleration model) is needed. PoF provides a basis for creating acceleration models. Only when an acceleration model is known can a test be defined as an accelerated life testing (ALT). Unfortunately, the term ALT is often improperly applied to elevated stress tests (ESTs). Without a proper understanding of PoF, the tests performed cannot assess the field reliability of a product.

3.3 PoF models

PoF reliability estimation utilizes on physics-based models designed to estimate the TTF for the product or component under investigation. Models isolate specific mechanisms that produce device failure and quantify TTF through a fundamental understanding of the materials and stress conditions that a device will encounter [8].

The failure mode refers to the actual observation of failure. For example, the failure of a passive component such as a resistor can be identified through a change in electrical resistance. This change can be measured, and it is indicative of some underlying physical problem. It could also be observed through visual inspection of a crack that results in an electrical open. While a failure mode identifies a state, it does not necessarily identify the underlying cause of failure.

The underlying process that causes failure is known as the failure mechanism. A failure mechanism can include physical, chemical, thermodynamic, or other processes that result in failure. In the case of a resistor mounted on a printed circuit board (PCB), the failure mechanism could be electrochemical migration induced by residual ionic contamination from the assembly process in the presence of an electric field or fatigue fracture in solder interconnects as a result of thermomechanical stresses. Failure due to electrochemical migration may result in an electrical short, while failure due to fatigue fracture may result in an electrical open. Either one of these failure mechanisms could be observed through changes in resistance or optical inspection during failure analysis.

Failures can occur in different places on a device, so the location needs to be identified and defined as well. The failure site is the location of the failure. In the case of a resistor, electrochemical migration may occur between the solder pads on which the resistor is mounted, across the top surface of the resistor between the resistor terminals, or even within a resistor [9,10]. Solder fracture could occur at the intermetallic compounds on the solder-component or PCB-solder interfaces or in the bulk of the solder material.

PoF also makes a distinction between the types of failure mechanisms. Overstress failure mechanisms represent abrupt, catastrophic failures, while wear-out mechanisms occur after a period of use due to environmental, operational, and architectural factors. The two types of failure mechanisms do not need to be mutually exclusive. Wear-out can lead to an overstress failure under normal loads as a result of localized faults. Even though the stresses remain within the overstress limits, discontinuities could result in an effective stress that is amplified. Table 3.1 provides examples of different overstress and wear-out failure mechanisms and the types of loads that initiate each mechanism. A device's overstress limits describe the magnitude of stress that is required to induce sudden failure. Identifying the maximum or minimum operating conditions and operating within a margin of safety can help to mitigate overstress-related failures. ALT is applied at stresses beyond the manufacturer's recommended limits, but within the overstress limits. In ALT, the device undergoes repeated stresses to induce wear-out failure. This type of failure is associated with progressive degradation that occurs over time under normal operating conditions. While

Table 3.1 **Examples of overstress and wear-out failure mechanisms in electronic products [11]**

	Overstress failure mechanism	Wear-out failure mechanism
Mechanical	Delamination, fracture	Fatigue, creep
Thermal	Excessive heating of component past glass transition temperature	Diffusion voiding, creep
Electrical	Dielectric breakdown, electrostatic discharge	Electromigration
Radiation	Single-exposure failure	Embrittlement, charge trapping
Chemical	Etching	Dendrites, whiskers, corrosion

device overstress limits can typically be controlled during field use conditions, wear-out will always occur. However, the rate at which wear-out occurs can be accounted for. ALT is used to determine life expectancies, provided that PoF models are available to define the acceleration factor between the specific test condition and the anticipated field use condition. For example, fatigue can be modeled by a power law based on the cyclic strain range expected in the section of material susceptible to failure. Fatigue failure can be accelerated in experiments by using PoF models to determine a larger cyclic strain range that would produce failures in a shorter duration. Without the model, the application of the larger cyclic range would cause early failures during testing, but an analyst would not be able to relate the accelerated TTF to the product's anticipated use condition.

In addition to the use environment and operational loads, the architecture of a product, materials used, and manufacturing processes influence device failure. For example, the vibration-induced stresses experienced by a device located at the center of a PCB can be different from that in a device on the edge of the board. The rigidity of the board also influences how the device responds to applied stresses. The manufacturing processes could also induce various stresses. The temperatures in the solder reflow process may go up to 250 °C, resulting in high-temperature-induced mechanical loads not typically experienced during use [12,13]. Table 3.2 outlines various examples of wear-out failure mechanisms in PCBs and semiconductor components, sites of possible failures, loads causing failure, and common models relating failure mechanisms to failure.

One of the most widely used models to quantify fatigue in wire bonds, solder joints, traces, and plated through holes (PTHs) is the Coffin-Manson equation. The Coffin-Manson equation relates cycles to failure to a strain range metric using a power law relationship. To provide a quick estimate of the cycles to failure, researchers have expanded the Coffin-Manson equation to analytically obtain the strain range metric. In Ref. [15], torsional loads were applied to a PCB test vehicle with eight plastic ball grid array components mounted on the board to induce fatigue in the solder balls. A Coffin-Manson-Basquin PoF model was used to relate the number of cycles to

Table 3.2 **Failure mechanisms, sites, loads, and models [14]**

Failure mechanism	Failure site	Relevant loads	Sample model
Fatigue	Die attach, wire bond/ tape-automated bonding (TAB), solder leads, bond pads, traces, vias/PTHs, interfaces	ΔT, T_{mean}, dT/dt, t_{dwell}, ΔH, ΔV	Power law (Coffin-Manson)
Corrosion	Metallizations	M, T, ΔV	Eyring (Howard)
Electromigration	Metallizations	T, J	Eyring (Black)
Conductive filament formation	Between metallizations	M, ΛV	Power law (Rudra)
Stress-driven diffusion voiding	Metal traces	s, T	Eyring (Okabayashi)
Time-dependent dielectric breakdown	Dielectric layers	V, T	Arrhenius (Fowler-Nordheim)

Δ, cyclic range; Λ, gradient; V, voltage; M, moisture; T, temperature; s, strain; H, humidity; J, current density.

failure, N_f, to high-cycle elastic damage and low-cycle plastic damage through the following equation:

$$\frac{\Delta\gamma}{2} = \frac{\sigma_f}{E}(2N_f)^b + \epsilon_f(2N_f)^c$$

where $\Delta\gamma$ is the total cyclic strain range in the solder balls, E is the modulus of elasticity of the solder material, σ_f is the stress strength coefficient, b is the fatigue strength exponent, c is the fatigue ductility exponent, and ϵ_f is the fatigue ductility coefficient. Model constants specific to the solder type are obtained from the literature. If model constants are not available in the literature, model constants may be determined from physical testing.

A commonly used PoF model (see Table 3.2) for electromigration is Black's model [16], which is based on Eyring's equation. Black's model uses current density, J, and temperature, T, to determine the MTTF of a trace on a semiconductor die due to electromigration [17]. The MTTF is expressed as

$$\text{MTTF} = A(J^{-n})e^{\left(\frac{E_a}{KT}\right)}$$

where A is a factor that takes into account several physical properties of the metal, E_a is the activation energy, K is Boltzmann's constant, and n is an experimentally determined constant. In Black's formulation, n was set to 2, but other researchers have determined that this value depends on a variety of factors, such as residual stress [18]. The selection of n and other parameters requires experimental data to determine the form that best fits. Black demonstrated in Ref. [16] that current density and temperature were two parameters that influence the TTF due to electromigration in a metal trace. Black found that aluminum traces cracked and formed crystal whiskers due to applied current densities larger than 10^5 A/cm^2 and temperatures above 100 °C as a result of mass transport.

Time-dependent dielectric breakdown is a function of the applied voltage and temperature. Silicon devices such as MOSFETS can fail due to damage on the dielectric oxide layer by electron trapping in the oxide layer and weakened covalent bonds between silicon and oxygen [19,20]. A PoF model for time-dependent dielectric breakdown, based on an Arrhenius relationship, is given by

$$t_{\mathrm{bd}} = A\mathrm{e}^{-\frac{\gamma V}{X}}\mathrm{e}^{\frac{E_{\mathrm{a}}}{KT}}$$

where t_{bd} is the time to breakdown of the dielectric, A is the model breakdown coefficient, γ is the electric field acceleration factor, V is the voltage on the oxide layer, X is the thickness of the oxide layer, E_{a} is the activation energy, K is Boltzmann's constant, and T is the operating temperature. High temperatures and voltages decrease the time to dielectric breakdown, resulting in earlier failure. There has been some controversy as to whether the V relationship is valid or if it follows a $1/V$ trend [21]. The V-model is based on electric field and dipole interactions, while the $1/V$-model is based on Fowler-Nordheim conduction [21]. The $1/V$-model matches experimental data under high electric fields and high current testing; however, it lacks a physics-based explanation for phenomena at lower electric fields. Alternately, the V-model fits low electric field experimental data. In Ref. [21], the unification of these two alternate explanations was proposed by allowing both electric field and current-induced degradation to occur at the same time.

A variety of PoF models have been developed to describe various failure mechanisms [22–39]. A prerequisite for these models is an understanding of the failure mechanism(s), loading conditions that cause the failure mechanism(s), and the design choices that impact the susceptibility of a device to failure. To develop a PoF model, a rigorous methodology to identify the relevant failure modes, failure mechanisms, failure sites, and stresses that cause failure is necessary. In the next section, the role of FMMEA in the PoF reliability assessment process is described.

3.4 PoF reliability assessment

While failure mechanisms, failure modes, and failure sites can be determined through past experience and existing failure models, a structured methodology is necessary to identify the contributors to unreliability. FMMEA is an extension of the failure modes and effects analysis (FMEA) framework [14]. FMEA is used to evaluate potential failures that can occur in a product and to develop actions to reduce or eliminate the probability of failure. While FMEA identifies the failure modes of components in a system and the consequence of failure at different system levels, FMMEA additionally

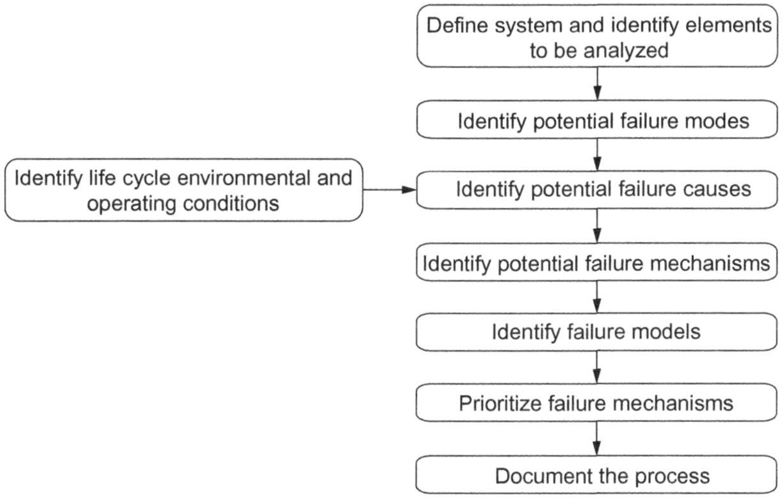

Figure 3.1 Failure modes, mechanisms, and effects analysis (FMMEA) methodology [14].

identifies failure mechanisms and models and prioritize failure mechanisms based on the consequence and frequency of occurrence. Without this additional information, it is very difficult to perform root cause analysis and develop accurate accelerated test programs. Steps in a generic FMMEA methodology are shown in Figure 3.1.

The first step in an FMMEA process is to define the system to be analyzed. To accomplish this, the system is broken down into its constituent parts. Each of these parts is divided until the lowest sublevel is reached. This fundamental unit is known as an element or component of the overall system. The system can be broken down based on function, location, or a combination of the two. Furthermore, it is necessary to identify the structure of the device in terms of materials and geometry.

For each element, potential failure modes are identified. Failure modes are the actual observations of failure, and they can be caused by one or multiple failure mechanisms. To understand how failure modes manifest themselves, it is necessary to identify all of the possible causes of failure. This process includes identifying the environmental and operating conditions to which the system is expected to be subjected over its life cycle. A careful analysis of the loading conditions present during the lifetime of the device and the resulting stresses is essential. The type, severity, and frequency of the loading conditions and stresses impact the reliability and the TTF of the device. An understanding of the loading conditions, system materials, and geometries will assist in identifying the physical processes that cause failure, also known as failure mechanisms.

After the identification of failure mechanisms, appropriate failure models are identified and failure mechanisms are prioritized. The entire FMMEA process is documented during the course of the process. Based on the identified failure mechanisms, failure sites, and load profiles, PoF models are used to quantify the TTF. This information is used to update the design, develop product qualification processes, and develop failure mitigation strategies. Examples of some widely used PoF models used in the electronic packaging industry were given in Section 3.3.

The first step in prioritizing the failure mechanisms involves quantifying the likelihood and severity of failure. Failure mechanisms that are more severe and more likely to occur are given a high-risk rating, while less severe and unlikely failure mechanisms are given a low-risk rating. Identifying mechanisms with the highest risk, along with an understanding of the expected life cycle loading conditions, allows for a more reliable design that focuses on mitigating the most common and detrimental failures. The documentation process allows the company to develop a historical record of the FMMEA process undertaken for each system. A well-documented FMMEA process saves the company time and resources by allowing them to revisit the FMMEA process for new products without having to start from the beginning. The list of failure mechanisms may be updated based on field data and experience. Improper documentation defeats the purpose of performing FMMEA in the first place.

3.5 Applications of PoF to ensure reliability

The focus of PoF is to apply knowledge gained through product design and analysis to improve or predict the reliability of a system. PoF is applied to assess and improve reliability at all stages of a device's life cycle history, from the design phase to field operation.

One application of PoF is the design of ALTs to assess the life expectancy and reliability of a new product. While applying ESTs can be used to ruggedize a product, it does not allow for assessment of product life expectancy or reliability. By themselves, ESTs do not relate to the actual operating conditions and cannot be used to estimate reliability [40]. To relate ESTs to field or use conditions, PoF must be applied. PoF allows for the definition of an acceleration transform that relates the test condition to a use condition. With an acceleration transform, an EST may be referred to as an accelerated test. Incorrect application of an acceleration factor could result in unexpected field failures, resulting in financial loss. However, if accelerated tests are performed correctly, the time to market can be reduced, and more confidence can be placed on the reliability of the product.

To implement PoF-based accelerated tests, system manufacturers should identify the device architecture and the effect of life cycle loads on the reliability of the product. For example, the PCB material properties need to be well known to see how the stiffness of the board influences other parameters. The locations of components, types of components, and types of solder used all affect the reliability of a product and are related to the loading conditions that the device experiences during its life cycle.

Virtual qualification is the first step of a PoF-based accelerated test [41,42]. It is applied early in the product development process and, thus, helps in the implementation of the reliability assessment in the design phase [43]. For example, George et al. [44] used virtual qualification and testing based on PoF principles to identify the top reliability concerns in communications hardware. Virtual qualification takes advantage of computer-aided engineering (CAE) software to estimate TTF, thus permitting components and systems to be qualified based on an analysis of the relevant failure mechanisms and associated failure modes. calcePWA software [45], which provides the ability to model printed wiring assemblies (PWAs), conducts thermal and dynamic analyses on the modeled PWAs, and estimates life expectancy based on PoF models and user-defined life cycle loading conditions, is an example of a CAE software for conducting virtual qualification. In the calcePWA software used in Refs. [44–50], a database of already validated PoF models is available to identify potential design weaknesses. The impact of specific design choices can then be assessed to determine the best course of action. Through the modeling analysis in Ref. [44], vibration and shock reliability were seen as less of a concern than thermal cycling. Based on these findings, recommendations were made to mitigate the risk of thermal cycling-induced failures.

The second step in PoF-based accelerated stress testing is to plan and develop accelerated tests. The failure mechanisms and failure sites identified in the virtual qualification process are used in planning the accelerated tests. Using past experience and information obtained through virtual qualification, a test matrix is developed that stresses the most dominant failure mechanisms. For example, surface mount solder failure is typically due to a creep-fatigue interaction. This failure mechanism can be aggravated by both cyclic thermomechanical and vibration-induced stresses [40]. To investigate this type of failure, combinations of thermal cycling and vibrational loads can be used, with the actual magnitudes of the stress test loads identified through PoF analysis and preliminary tests.

Once the test loads are identified, a test vehicle can be designed and characterized for virtual and ALT. The design of a test vehicle depends on the application, but in the case of a PCB, it can be a nonfunctional duplicate of the actual circuit board. If a nonfunctional duplicate is used to study solder interconnect failure, the interconnects can be daisy-chained together to allow for the monitoring of electrical parameters. Ideally, the test vehicle should be less expensive than the actual device because it does not incorporate the full functionality of the actual device, but it should still have the same characteristics, material set, and assembly process as the functional assembly. Otherwise, no useful information can be gained from the accelerated tests. As part of the test development process, the test vehicle response to loading is characterized by instrumenting sensors such a strain or thermocouples and subjecting the test vehicle to step stress to determine the test vehicles limits. These limits may be based on functionality or survivability. This process provides another opportunity to determine if failure modes and mechanisms identified in the virtual qualification occur at the higher loading conditions. With knowledge gained from the test vehicle characterization, the test levels for the accelerated test can be established.

Once the test vehicle is physically characterized and the test conditions are established, virtual testing is carried out to predict testing outcome. For detailed stress analysis, finite-element analysis programs may be used to identify the device's response to proposed test condition [51,52]. The outcomes of the virtual testing results in combination with the virtual qualification results establish the acceleration factor for the accelerated life test.

Finally, the ALT is carried out and documented. The test results provide verification of the virtual test results and a distribution of failure and a confidence level based on the selected sample size. If carried out correctly, the results can be extrapolated to estimate the TTF under ordinary operating conditions using the established acceleration factor. With consideration of failure distribution and sample size, the reliability of the product can be established. This approach to reliability assessment can save time and money through the use of computer-aided tools and fewer test vehicles than an assessment technique that blindly chooses test conditions without consideration of the dominant failure mechanisms and life cycle loads.

Another application of PoF utilizes prognostics and health management (PHM), which is defined as "an enabling discipline consisting of technologies and methods to assess the reliability of a product in its actual life cycle conditions to determine the advent of failure and mitigate system risk" [53]. For example, a PoF approach to reliability can be integrated with PHM through the use of sensors to monitor the *in situ* conditions that a device undergoes and allow for models to be updated as the environmental stresses change. This method of degradation tracking can provide a customer with accurate, real-time information that allows for better decision making. Through a PoF understanding of how systems and components degrade under various operating conditions, condition-based maintenance can be implemented. Rather than wait for a failure or perform regularly scheduled maintenance, maintenance can be performed only when it is necessary through the use of sensors and prognostic predictions based on PoF models [54]. The TTF can be predicted and even updated to provide a remaining useful performance (RUP) metric. With this information, unexpected in-use failures can be avoided and operating conditions can be altered to preserve critical functionality until replacement or repair is possible.

One method for avoiding in-use failures is the implementation of a PHM "canary" device that is designed to provide premature warning of failure through the knowledge of failure mechanisms, modes, and sites [54]. Actual canaries were once used in mine shafts to detect the presence of dangerous gases and serve as a warning of unsafe conditions. Because canaries are more sensitive to gases than humans, a sick or dead canary indicated that it was time to evacuate the mines. The same concept has been successfully applied in electronic systems [55–65]. A canary device is incorporated into a product to provide advance warning of failure. Canaries can be expendable, much like a fuse. They can also be nonfunctional units that provide advance warning of functional degradation in another part of the device. Devices can even incorporate multiple canaries that provide several levels of warning to more accurately determine the potential TTF.

To implement a canary device correctly, a detailed understanding of how a device fails is essential. With this knowledge, the geometry, materials, and loading conditions of a canary can be modified to cause the canary to fail earlier than the actual device. Without PoF, a canary might be designed that would not mimic the true failure mechanism. This mismatch could then provide an unrealistic TTF estimation, which would defeat the purpose of including a canary.

Chauhan et al. [55] demonstrated the use of PoF to develop a canary for Ball grid array (BGA) solder ball failure. The BGA components were identified as critical components on a sample PCB assembly. Additionally, Engelmaier's equation, a PoF model, is used to relate TTF to thermal stresses. Due to the mismatch in the coefficient of thermal expansion between the BGA chip and the PCB, stresses on the solder balls connecting the BGA package to the PCB will arise when the system undergoes or is subjected to a temperature excursion. Depending on the BGA construction, solder balls further from the center of the component will generally experience higher stresses than solder balls closer to the center of the component [55]. In BGA constructions where the outer solder balls are more likely to fail earlier, the outer solder balls can be used as a canary to predict the failure of the inner solder balls. In one test, experimental validation demonstrated that the outer solder balls failed first 99.5% of the time, indicating that this technique could successfully be applied to predict failure for the inner solder balls [55]. Canaries can be used in conjunction with PoF models to improve TTF estimates and account for unit-to-unit variations and changing loading conditions.

Alternatively, sensors can be integrated into a product to monitor the device's exposure to various stresses throughout its life cycle. Sensors can also be used with a PoF-based model to determine how much life has been consumed and how much life remains in a product. This approach can be used to provide an RUP prediction that can influence maintenance and usage decisions. For example, Mishra et al. [66,67] utilized sensors to perform life consumption monitoring on a PCB mounted under the hood of an automobile. These sensors enabled real-time collection of data from the ambient environment to enhance PoF model life predictions. A car accident that happened to occur during the data collection manifested itself in the shock recording and led to a decrease in the remaining life of the PCB assembly. Through the use of PHM and PoF, changes in damage accumulation were captured and incorporated into the model predictions. The role of PoF in PHM ultimately enables better decision making by providing a constant source of data for model updating and estimation improvement.

3.6 Summary and areas of future interest

Central to the PoF approach is the availability and utilization of models that determine the effect of stresses on component or device lifetime. These models are developed through knowledge of the device structure and materials, the failure causes, failure modes, failure mechanisms, and life cycle loading conditions. Using a physics-based model allows for improved reliability assessment through the design of accelerated life tests that can cause the same types of failure mechanisms in a shorter time frame. This type of testing alone can help bring a product to market more quickly without compromising reliability.

Work is still needed in capturing the variety of failure mechanisms seen in new technologies. As advances are made in the electronic industry, PoF approaches may need to be updated. Multichip modules and systems on a chip are used to consolidate many different functions and components onto a single chip. The interactions between the various components should be considered and incorporated into PoF models. Additionally, electronic devices will be exposed to a number of different stresses throughout their life cycle. Determining damage accumulation models under multiple loading conditions and accounting for degradation path dependency require engineering effort. The PoF-based methodology can be combined with PHM and failure prediction, leading to many more opportunities for future progress. PoF can dictate the design of canaries for advanced failure detection, or it can be used to monitor the *in situ* life cycle conditions for model updating and prediction improvement. Further research into canaries and on how to alter their physical properties to ensure earlier TTF is needed. Finally, there are new advances in fusion approaches to PHM. A fusion approach combines the benefits of PoF-based models with data-driven methods for anomaly detection and prognostics. This combination can help account for unit-to-unit manufacturing variations and deal with some of the uncertainty present in trying to model device failure.

PoF can be applied to a wide variety of electronic devices to predict failure and improve their design. Due to rapid advances in technology, time-to-market is critical. A PoF-based methodology not only helps in the production of reliable products in a timely manner but also allows a device to provide sufficient warning of impending failure for repair and maintainability actions.

References

[1] M. Pecht, A. Dasgupta, D. Barker, C.T. Leonard, The reliability physics approach to failure prediction modeling, Qual. Reliab. Eng. Int. 6 (4) (1990) 273–276.

[2] M. Pecht, A. Dasgupta, Physics-of-failure: an approach to reliable product development, in: Proceedings of the Integrated Reliability Workshop, 1995, pp. 111–117.

[3] Institute for Electrical and Electronics Engineers (IEEE), IEEE 1413: IEEE Standard Framework for Reliability Prediction, second ed., New York, NY, 2010.

[4] A. MacKenzie, S. Evans, The Toyota Recall Crisis, Motor Trend, Retrieved March 12, 2013 from: http://www.motortrend.com/features/auto_news/2010/112_1001_toyota_recall_crisis/vieview.html, January 2010.

[5] N. Williard, W. He, C. Hendricks, M. Pecht, Lessons learned from the 787 Dreamliner issue on lithium-ion battery reliability, Energies 6 (9) (2013) 4682–4695, http://dx.doi.org/10.3390/en6094682.

[6] J. Topham, ANA Counts Cost of Grounded 787, Not Yet Seeking Damages, Reuters, Retrieved March 12, 2013 from: http://www.reuters.com/article/2013/01/31/us-boeing-dreamliner-ana-january-idUSBRE90U06T20130131, January 31, 2013.

[7] J. Bachman, Disgruntled Airline Speaks Up About Boeing 787 Payout, Buisnessweek, Retrieved May 20, 2013 from: http://www.businessweek.com/articles/2013-07-30/disgruntled-airline-speaks-up-about-boeing-787-payout, July 30, 2013.

[8] J. Hu, D. Barker, A. Dasgupta, A. Arora, The role of failure mechanism identification in accelerated testing, J. Inst. Environ. Sci. 36 (4) (1993) 39–45.

[9] Association Connecting Electronics Industries (IPC), IPC-TR-476A: Electrochemical Migration: Electrically Induced Failures in Printed Wiring Assemblies, first ed., Bannockburn, IL, 1997

[10] S. Zhan, M.H. Azarian, M.G. Pecht, Surface insulation resistance of conformally coated printed circuit boards processed with no-clean flux, IEEE Trans. Electron. Packag. Manuf. 29 (3) (2006) 217–223, http://dx.doi.org/10.1109/TEPM.2006.882496.

[11] I. Snook, J. Marshall, R. Newman, Physics of failure as an integrated part of design for reliability, in: Proceedings of the IEEE Annual Reliability and Maintainability Symposium, 2003, pp. 46–54, http://dx.doi.org/10.1109/RAMS.2003.1181901.
[12] C. Yin, H. Lu, C. Bailey, Y.-C. Chan, Effects of solder reflow on the reliability of flip-chip on flex interconnections using anisotropic conductive adhesives, IEEE Trans. Electron. Packag. Manuf. 27 (4) (2004) 254–259, http://dx.doi.org/10.1109/TEPM.2004.843152.
[13] J. Pang, L. Xu, X. Shi, W. Zhou, S. Ngoh, Intermetallic growth studies on Sn-Ag-Cu lead free solder joints, J. Electron. Mater. 33 (10) (2004) 1219–1226, http://dx.doi.org/10.1007/s11664-004-0125-7.
[14] S. Ganesan, V. Eveloy, D. Das, M.G. Pecht, Identification and utilization of failure mechanisms to enhance FMEA and FMECA, in: Proceedings of the IEEE Workshop on Accelerated Stress Testing & Reliability (ASTR), Austin, TX, 2005.
[15] V. Srinivas, S. Menon, M. Osterman, M.G. Pecht, Modeling the rate-dependent durability of reduced-Ag SAC interconnects for area array packages under torsion loads, J. Electron. Mater. 42 (8) (2013) 2606–2614, http://dx.doi.org/10.1007/s11664-013-2575-2.
[16] J.R. Black, Mass transport of aluminum by momentum exchange with conducting electrons, in: Sixth Annual Reliability Physics Symposium, Los Angeles, CA, 1967, http://dx.doi.org/10.1109/IRPS.1967.362408.
[17] A. Thaduri, A.K. Verma, V. Gopika, R. Gopinath, U. Kumar, Reliability prediction of semiconductor devices using modified physics of failure approach, Int. J. Syst. Assur. Eng. Manag. 4 (1) (2013) 33–47, http://dx.doi.org/10.1007/s13198-013-0146-9.
[18] J.B. Bernstein, et al., Electronic circuit reliability modeling, Microelectron. Reliab. 46 (2006) 1957–1979, http://dx.doi.org/10.1016/j.microrel.2005.12.004.
[19] N. Raghavan, K. Prasad, Statistical outlook into the physics of failure for copper low-k intra-metal dielectric breakdown, in: IEEE International Reliability Physics Symposium, 2009, pp. 819–824, http://dx.doi.org/10.1109/IRPS.2009.5173358.
[20] F. Lu, J. Shao, X. Liu, X. Wang, Validation test method of TDDB physics-of-failure models, in: Prognostics & System Health Management Conference, 2012, pp. 1–4, http://dx.doi.org/10.1109/PHM.2012.6228800.
[21] J.W. McPherson, R.B. Khamankar, A. Shanware, Complementary model for intrinsic time-dependent dielectric breakdown in SiO_2 dielectrics, J. Appl. Phys. 88 (2000) 5351–5359, http://dx.doi.org/10.1063/1.1318369.
[22] H. Huang, P.A. Mawby, A lifetime estimation technique for voltage source inverters, IEEE Trans. Power Electron. 28 (8) (2013) 4113–4119, http://dx.doi.org/10.1109/TPEL.2012.2229472.
[23] M. Musallam, C. Yin, C. Bailey, C.M. Johnson, Application of coupled electro-thermal and physics-of-failure-based analysis to the design of accelerated life tests for power modules, Microelectron. Reliab. 54 (1) (2014) 172–181, http://dx.doi.org/10.1016/j.microrel.2013.08.017.
[24] M. Baybutt, C. Minnella, A.E. Ginart, P.W. Kalgren, Improving digital system diagnostics through prognostics and health management (PHM) technology, IEEE Trans. Instrum. Meas. 58 (2) (2009) 255–262, http://dx.doi.org/10.1109/AUTEST.2007.4374265.
[25] C. Kulkarni, G. Biswas, X. Koutsoukos, K. Goebel, J. Celaya, Physics of failure models for capacitor degradation in DC-DC converters, in: The Maintenance & Reliability Conference, MARCON 2010, Knoxville, TN, February 2010.
[26] C.S. Gray, S.J. Watson, Physics of failure approach to wind turbine condition based maintenance, Wind Energy 13 (5) (2010) 395–405, http://dx.doi.org/10.1002/we.360.
[27] E. Habtour, B. Werner, A. Hilburn, H.I. Saraidaridis, Physics of failure for portable electronic devices in military applications, in: Reliability and Maintainability Symposium (RAMS), Orlando, FL, January 2013, http://dx.doi.org/10.1109/RAMS.2013.6517620.

[28] L. Yang, P.A. Agyakwa, C.M. Johnson, Physics-of-failure lifetime prediction models for wire bond interconnects in power electronic modules, IEEE Trans. Device Mater. Reliab. 13 (1) (2013) 9–17, http://dx.doi.org/10.1109/TDMR.2012.2235836.

[29] P.O. Quintero, F.P. McCluskey, Temperature cycling reliability of high-temperature lead-free die-attach technologies, IEEE Trans. Device Mater. Reliab. 11 (4) (2011) 531–539, http://dx.doi.org/10.1109/TDMR.2011.2140114.

[30] J. Fan, K.C. Yung, M. Pecht, Physics-of-failure-based prognostics and health management for high power white light-emitting diode lightning, IEEE Trans. Device Mater. Reliab. 11 (3) (2011) 407–416, http://dx.doi.org/10.1109/TDMR.2011.2157695.

[31] W. Lee, L. Nguyen, G. Selvaduray, Solder joint fatigue models: review and applicability to chip scale packages, Microelectron. Reliab. 40 (2000) 231–244, http://dx.doi.org/10.1016/S0026-2714(99)00061-X.

[32] R. Tilgner, Physics of failure for interconnect structures: an essay, Microsyst. Technol. 15 (2009) 129–138, http://dx.doi.org/10.1007/s00542-008-0630-3.

[33] J. Clement, Electromigration modeling for integrated circuit interconnect reliability analysis, IEEE Trans. Device Mater. Reliab. 1 (1) (2001) 33–42.

[34] J. Lee, I. Chen, C. Hu, Modeling and characterization of gate oxide reliability, IEEE Trans. Electron Devices 35 (22) (1988) 2268–2278, http://dx.doi.org/10.1109/16.8802.

[35] B. Rudra, D. Jennings, Tutorial: failure-mechanism models for conductive filament formation, IEEE Trans. Reliab. 43 (3) (2004) 354–360, http://dx.doi.org/10.1109/24.326425.

[36] J. Li, A. Dasgupta, Failure mechanism models for material aging due to inter-diffusion, IEEE Trans. Reliab. 43 (1) (1994) 2–10.

[37] H. Huang, P.A. Mawby, A lifetime estimation technique for voltage source inverters, IEEE Trans. Power Electron. 28 (8) (2013) 4113–4119, http://dx.doi.org/10.1109/TPEL.2012.2229472.

[38] W. Yao, C. Basaran, Computational damage mechanics of electromigration and thermomigration, J. Appl. Phys. 114 (10) (2013) 1–12, http://dx.doi.org/10.1063/1.4821015.

[39] H. Wang, M. Liserre, F. Blaabjerg, P. Rimmen, J.B. Jacobsen, T. Kvisgaard, J. Landkildehus, Transitioning to physics-of-failure as a reliability driver in power electronics, IEEE J. Emerg. Sel. Topics Power Electron. 2 (1) (2014) 97–113, http://dx.doi.org/10.1109/JESTPE.2013.2290282.

[40] K. Upadhyayula, A. Dasgupta, PoF guidelines for accelerated qualification of electronic systems, Qual. Reliab. Eng. Int. 14 (1998) 433–447, http://dx.doi.org/10.1002/(SICI)1099-1638.

[41] S. Menon, E. George, M. Osterman, M. Pecht, Physics of failure based reliability assessment of electronic hardware, in: 6th Annual World Congress on Engineering Asset Management, Cincinnati, OH, October 2-5, 2011, http://dx.doi.org/10.1007/978-1-4471-4993-4_23.

[42] M.-L. Wu, E. George, M. Osterman, M. Pecht, Simulation assisted physics of failure based reliability assessment, in: ICEP-IAAC 2012, Joint Conference of International Conference on Electronics Packaging and IMAPS All Asia Conference, Tokyo Big Sight, Tokyo, Japan, April 17-20, 2012.

[43] J. Cunningham, R. Valentin, C. Hillman, A. Dasgupta, M. Osterman, A demonstration of virtual qualification for the design of electronic hardware, in: Proceedings of the Institute of Environmental Sciences and Technology Meeting, April 24, 2001.

[44] E. George, D. Das, M. Osterman, M. Pecht, C. Otte, Physics of failure based virtual testing of communications hardware, in: ASME International Mechanical Engineering Congress and Exposition, Lake Buena Vista, FL, November 13-19, 2009, http://dx.doi.org/10.1115/IMECE2009-12181.

[45] M. Osterman, T. Stadterman, Failure assessment software for circuit card assemblies, in: Proceedings of Reliability and Maintainability Symposium, January 18-21, 1999, pp. 269–276, http://dx.doi.org/10.1109/RAMS.1999.744130.
[46] M. Osterman, A. Dasgupta, B. Han, A strain range based model for life assessment of Pb-free SAC solder interconnects, in: Proceedings of 56th Electronic Components and Technology Conference, May 30-June 2, 2006, pp. 884–890, http://dx.doi.org/10.1109/ECTC.2006.1645760.
[47] M. Osterman, A. Dasgupta, Simulation improved testing of electronic hardware, in: Proceedings of the 48th ATM, April 1, 2002.
[48] M. Osterman, M. Pecht, Strain range fatigue life assessment of lead-free solder interconnects subject to temperature cycle loading, Sold. Surf. Mt. Tech. 19 (2) (2007) 12–17, http://dx.doi.org/10.1108/09540910710836494.
[49] H. Qi, W. Chris, M. Osterman, M. Pecht, Failure analysis and virtual qualification of PBGA under multiple environmental loadings, in: Proceedings of 54th Electronic Components and Technology Conference, vol. 1, June 2004, pp. 413–420.
[50] P. McCluskey, M. Pecht, S. Azarm, Reducing time-to-market using virtual qualification, in: Proceedings of the Institute of Environmental Sciences Conference, 1997, pp. 148–152.
[51] M.A.J. van Gils, W.D. van Driel, G.Q. Zhang, H.J.L. Bressers, R.B.R. van Silfhout, X. J. Fan, J.H.J. Janssen, Virtual qualification of moisture induced failures of advanced packages, Microelectron. Reliab. 47 (2007) 273–279, http://dx.doi.org/10.1016/j.microrel.2006.09.011.
[52] G.A.O. Davies, J. Ankersen, Virtual testing of realistic aerospace composite structures, J. Mater. Sci. 43 (2008) 6586–6592, http://dx.doi.org/10.1007/s10853-008-2695-x.
[53] S. Cheng, M.H. Azarian, M.G. Pecht, Sensor systems for prognostics and health management, Sensors 10 (2010) 5774–5797, http://dx.doi.org/10.3390/s100605774.
[54] A. Dasgupta, R. Doraiswami, M. Azarian, M. Osterman, S. Mathew, M. Pecht, The use of "canaries" for adaptive health management of electronic systems, in: ADAPTIVE 2010, IARIA Conference, Lisbon, Portugal, November 21-26, 2010.
[55] P. Chauhan, M. Osterman, M. Pecht, Canary approach for monitoring BGA interconnect reliability under temperature cycling, in: The MFPT 2012 Proceedings, 2012.
[56] S. Matthew, M. Osterman, M. Pecht, A canary device based approach for prognosis of ball grid array packages, in: IEEE Conference on Prognostics and Health Management (PHM), Denver, CO, June 18-22, 2012, http://dx.doi.org/10.1109/ICPHM.2012.6299546.
[57] D. Goodman, B. Vermeire, J. Ralston-Good, R. Graves, A board-level prognostic monitor for MOSFET TDDB, in: IEEE Aerospace Conference, 2006, http://dx.doi.org/10.1109/AERO.2006.1656127.
[58] C.Y. Yin, H. Lu, M. Musallam, C. Bailey, C.M. Johnson, A physics-of-failure based prognostic method for power modules, in: 10th Electronics Packaging Technology Conference, 2008, pp. 1190–1195, http://dx.doi.org/10.1109/EPTC.2008.4763591.
[59] N. Anderson, R. Wilcoxon, Framework for prognostics of electronics systems, in: International Military & Aerospace/Avionics Cots Conference, Exhibition & Seminar, 2004.
[60] J.H. Lau, The roles of DNP (distance to neutral point) on solder joint reliability of area array assemblies, Sold. Surf. Mt. Tech. 9 (2) (1997) 58–60, http://dx.doi.org/10.1108/09540919710800674.
[61] J.W. Simons, D.A. Shockey, Prognostics modeling of solder joints in electronic components, in: IEEE Aerospace Conference, 2006, http://dx.doi.org/10.1109/AERO.2006.1656126.

[62] L. Nasser, M. Curtin, Electronics reliability prognosis through material modeling and simulation, in: IEEE Aerospace Conference, 2006, http://dx.doi.org/10.1109/AERO.2006.1656125.

[63] N. Vichare, M. Pecht, Prognostics and health management of electronics, IEEE Trans. Compon. Packag. Technol. 29 (1) (2006) 222–229, http://dx.doi.org/10.1109/TCAPT.2006.

[64] J. Gu, M. Pecht, Prognostics and health management using physics-of-failure, in: 54th Annual Reliability & Maintainability Symposium, Las Vegas, NV, 2008, http://dx.doi.org/10.1109/RAMS.2008.4925843.

[65] M.J. Cushing, D.E. Mortin, J.J. Stadterman, A. Malhotra, Comparison of electronics reliability assessment approaches, IEEE Trans. Reliab. 42 (4) (1993) 542–546.

[66] S. Mishra, M. Pecht, Remaining life prediction of electronic products using life consumption monitoring approach, in: European Microelectronics Packaging and Interconnection Symposium, Cracow, Poland, June 2002.

[67] A. Ramakrishnan, M.G. Pecht, A life consumption monitoring methodology for electronic systems, IEEE Trans. Compon. Packag. Technol. 26 (3) (2003) 625–634, http://dx.doi.org/10.1109/TCAPT.2003.817654.

Modern instruments for characterizing degradation in electrical and electronic equipment

4

P.D. Goodman, R. Skipper, N. Aitken
ERA Technology Ltd, Leatherhead, United Kingdom

4.1 Introduction

4.1.1 Modern instruments

Degradation of electrical and electronic equipment (EEE) occurs for a wide variety of reasons, many of which are described elsewhere in this book, and a variety of techniques are used to study and measure these processes. Degradation can lead to failure of the equipment and many of the techniques described in this chapter are used to determine the cause of failure. Investigations can be simply visual examination and this is often the first stage of any evaluation. Techniques can be destructive or nondestructive where the latter can also be used for quality assurance of EEE during manufacture and assessment of future reliability of used equipment.

The most appropriate technique will depend on what is being investigated and why. For example, surface contamination can cause adhesion problems or poor conductivity, and so surface analysis for organic compounds such as infrared, Raman, or mass spectroscopy might be used. Defects inside integrated circuits require techniques that need very high magnification such as scanning electron microscopy (SEM) or transmission electron microscopy (TEM) and some types of degradation can be detected only by electrical measurements. The choice of analytic techniques will depend on what is being examined and why. All have limitations that need to be understood by the user, and this chapter will describe the most commonly used instruments and their main applications, advantages, and disadvantages.

4.2 Destructive techniques

4.2.1 Cross sections

When microscopy is required to examine the internal construction of a part and where nondestructive methods such a microfocus X-ray examination cannot provide all the information required, it will be necessary to obtain a section through the sample. The methods required to obtain sections through mineral and metallurgical samples are well known and comprise four distinct phases, cutting, potting, grinding, and polishing. These methods have been developed to obtain clean and undistorted surfaces that reflect the true nature of the material under investigation. Sections through whole

Reliability Characterisation of Electrical and Electronic Systems. http://dx.doi.org/10.1016/B978-1-78242-221-1.00004-6

devices and assemblies that are frequently required for failure analysis pose additional problems because of the large difference in the hardness of the materials in the section and the possibility of voids within the samples.

Cutting. Either diamond or silicon carbide abrasive wheels are usually used, but in either case, water-based lubricants are essential to avoid thermal damage. In some cases, contamination due to dust and water during cutting occurs. In such cases, samples can be enclosed with plastic or foil prior to cutting. Cutting small samples without water cooling is possible, but the cutting speed needs to be closely controlled to avoid damage to the sample and blade.

Diamond saws designed for cutting silicon wafers can be used for small samples as these give fine clean cuts, but the diamond-loaded cutting edge is easily clogged by softer material such a plastics and copper.

Potting (encapsulation). Both hot and cold methods are available, but for electronic failure analysis, cold mounting is generally preferred. Cold mounting resins produce clear mounted samples that better allow the examination of the section location, and this can be particularly important if a sequence of sections is required.

Good adhesion between the sample and the resin is important to maintain edge retention during polishing and this is best ensured by thoroughly cleaning the samples prior to potting. A brief wash in isopropanol and then drying are particularly helpful in this respect.

When potting many samples together, maintaining the correct sample identity can pose problems, so it is recommended to indelibly mark each sample by engraving the ID on the outer surface after potting.

All cold mounting resins produce some heat during the curing step, and care should be taken using oven-cured resins. Large samples can produce sufficient heat to damage delicate samples and produce voids (due to gas formation and/or expansion) that can interfere with the polishing step.

For complex samples, vacuum potting is frequently helpful in ensuring all openings are fully impregnated with resin and voids do not appear at the polished surface. Vacuum impregnation with a dye added to the resin can also be helpful in showing possible ingress paths and in distinguishing between potting material and the sample's own plastic parts.

Grinding and polishing. These are closely related steps. The first is the bulk removal stage and the second aims to remove all surface damage caused by the prior grinding. The principle difference is the particle size or grade of the media used.

Grinding with grit coarser than 180 would be unusual; it is generally better to minimize the required grinding by arranging the samples to be cut as close as possible to the required sites. Silicon carbide papers are the most commonly used media for grinding. These are relatively cheap and can be quickly changed allowing a rapid transition through the paper grades and this avoids surface damage.

The disadvantage of these papers is that their performance changes quickly during use, most being useful for only a minute or so. If grinding to a precise location within a mounted sample is required, then judging the rate of loss on papers can be difficult. The use of diamond or silicon carbide-impregnated pads is easier because grinding rates are more consistent.

It is difficult to generalize a preparation method as much depends on the nature of the materials and the quality of the finish required. After the coarse grinding stage, a good general approach is fine polishing using sequential fine grinding on 500, 1000, 2400, and 4000 grit followed by polishing on 1 μm diamond and sometimes further polishing with ¼ μm diamond followed a colloidal alumina finish. For metal samples, further etching processes may be required to provide additional surface relief.

Whichever preparation method is used, it is essential that the samples are thoroughly cleaned between grit stages as this will avoid cross contamination and spurious scratching.

4.2.2 Jet etching and depotting components

The vast majority of modern integrated circuits are used in a fully encapsulated state, the chip itself being enclosed in a plastic resin with fine gold wires conducting input and output from the chip surface to a lead frame that forms the leads extending into the outside world. To conduct failure analysis of the chip surface or the internal lead frame, procedures have to be developed to expose the chip and the bond wires in a clean and undamaged condition.

The best results are usually achieved using a dedicated jet etcher but whole chip depotting methods can also be successful. Jet etchers need to be specially constructed of appropriate acid-resistant materials and with well-protected internal electronics.

The basic principle of jet etching is the delivery of a fine jet or pulses of hot concentrated acids onto a small confined area of the package. The force of each pulse of hot acid will decompose the surface plastic to form a small pit and also removes any undissolved fillers and expose fresh polymer surfaces for the next pulse.

In this manner, a circular patch of the packaging can be removed exposing the chip and bond wires but leaving the bulk package intact allowing easier manipulation. The resistance of the potting material to such treatment varies considerably and a range of acid strengths and mixtures can be used. Concentrated sulfuric and nitric acids are the most common and most systems will allow for the automatic mixing of such acids in some circumstances. For the most resistant polymers, fuming nitric and fuming sulfuric acids need to be used.

Silicon and gold are completely resistant to these acids, but difficulties arise when depotting gallium arsenide chips as these are considerably less acid-resistant.

Whole chip depotting is essentially a simple dissolution step conducted within a fume cupboard in which all of the polymeric components are removed in a strong acid mixture.

Hot nitric acid is perhaps the most common for this purpose, but the so-called piranha mixture composed of nitric acid with 25% hydrogen peroxide solution can be used.

All these methods suffer from the indiscriminate attack of the metal parts and the subsequent collapse of the wire bond system. Some improvement can be made by prior soldering of the leads to a grid of gold wires. These wires can be used to remove the unit from the acid solution at intervals, allowing the dissolution process to be monitored.

4.2.3 Chemical analysis

There are many chemical analytic techniques available. Methods that are most useful for characterizing the degradation of electrical equipment are described below.

4.2.3.1 Ion chromatography

The technique of ion chromatography originated from the 1940s, but became a mainstream technique more recently with the development of styrene/divinylbenzene resins in combination with small and stable conductivity meters. By attaching different functional groups to these resins, the ion-exchange characteristics can be tailored and the chromatography method for separating and quantifying a wide range anions or cations in solution became possible. With modern electronics and IC columns, sensitivity levels for both anions and cations better than 1 mg/l are commonplace.

The technique now finds wide application in many fields from food to pharmaceuticals and electronics. In electronics, it offers the most convenient and sensitive method to assess ionic contamination levels in process solutions and on the surfaces of circuits. With the ever-decreasing size of electronic assemblies with an increasing demand for reliability, ionic cleanliness is of paramount importance in many electronic assembly processes. Ion chromatography offers the only method with the required sensitivity for such investigations.

In its simplest form, ion chromatography comprises an eluant, a low-noise HPLC (high-performance liquid chromatography) pump, a sample loop leading to either an anion or a cation separation column, and finally a detector. The detector is usually a low volume sensitive conductivity detector, but UV-vis detectors are also used for organic species such as carboxylic acids. Even greater sensitivity or selectivity can be achieved with various post-column derivatization methods.

Any species that exists in an ionic form can potentially be analyzed using ion chromatography methods but larger biological species or species that are only weakly ionizing require modifications. Weak acids such as carboxylic acid are commonly analyzed using "ion exclusion chromatography."

IC instrumentation starts with simple isocratic systems (a constant eluant composition) with conductivity detection, but ion suppression is invariably used to reduce the background conductivity. These are methods in which the nonmeasured ion is exchanged for either an OH^- or H^+ to give better signal to noise characteristics. A greater versatility is obtained using dual-pump systems allowing the control of a gradient in the eluant composition for the analysis of mixtures containing ions that show a wide range of retention times on a particular column.

The trend in modern instruments is toward faster and more fully automatic systems including the autogeneration and recycling of eluants.

4.2.3.2 Infrared spectroscopy

Infrared spectroscopy is concerned principally with the energy transitions involved in the vibrational transitions within molecules and molecular functional groups. In general, its use offers a relatively simple method to identify functional groups in complex organic compounds and to a lesser extent in inorganic compounds. The combination of possible vibration modes can provide spectrographic fingerprints, and the comparison with a library of known compounds offers a valuable method to identify unknown materials.

Until the 1970s, almost all infrared spectrometers were based on the light from a broad spectrum incandescent source being passed through a scanning monochromator and measuring the intensity of the transmitted light at each successive wavelength. The results were precise adsorption spectra of the infrared light in either transmission or reflection modes but with the disadvantage that the process was relatively slow.

Today, almost all laboratory infrared spectrometers are of the Fourier transform type. In these, the light from the incandescent source is first passed through a Michelson interferometer, the output being dependent on the instantaneous position of a moving mirror modifying the interference path length. The signal obtained is initially a function of the infrared intensity versus interferometry mirror position. However, a Fourier transformation of this signal can reconstruct the signal frequencies and plot this in the traditional manner. The advantage of such a system is that all the light passes through the sample and provides very rapid spectra with a better energy throughput. The improved energy throughput also allows better sensitivity and offers greater possibilities for specialized sampling accessories.

The extinction coefficient of the various chemical bonds and vibrational modes vary enormously, but in general, organic material shows a strong absorption for infrared radiation and this can pose difficulties in presenting materials for transmission spectroscopy. There are many conventional sampling techniques used and the choice depends mainly on the physical nature of the materials, for example:

- cast films from appropriate solvent,
- dispersions (mulls) in a liquid in which the compound is insoluble,
- pressed disks with an infrared-transparent media such a potassium bromide,
- liquid cells,
- material directly held between potassium bromide windows,
- gas cells.

The greater energy throughput of modern FTIR instruments also leads to the increased use of new sampling accessories, in particular ATR (attenuated total reflection) [1].

ATR depends on the total internal reflection of infrared radiation within an appropriate crystal in contact with the sample. Adsorption occurs because in a material with a high refractive index, the radiation extends a short distance beyond the crystal surface, the so-called evanescent wave.

In use, the sample is clamped in contact with the crystal surface, increased sensitivity being achieved by increasing the number of internal reflections at the contact surface.

The principal advantage of the ATR approach is that the method can be used with a wide range of samples, often with the minimum sample preparation. Liquids, powders, plastics, and rubbers are amenable to this approach, whereas infrared analysis of plastics and rubbers is difficult using most of the more traditional sampling methods.

The use of smaller and harder crystals, such diamond, has allowed the extension of the method into harder inorganic materials and into small sample sizes that may have traditionally required the use of an infrared microscope.

The faster sample throughput and general versatility of this method have replaced other traditional methods in many labs. It is used, for example, to analyze surface contamination and the identification of materials such as polymer types and fluxes on circuitry.

4.2.3.3 Raman spectroscopy

The significant difference between Raman spectroscopy and infrared spectroscopy is that rather than measuring the absorption frequencies within a sample, Raman spectroscopy is concerned with the inelastic scattering of light from the surface of the sample. The scattered photons can show a difference in energy (either positive or negative), which reflects vibrational energy of the bond at which the interaction occurred.

The method has the positive advantages of remote noncontact techniques that can be applied to a very wide range of sample types with the minimum sample preparation. The "selection rules" that govern whether a particular bond vibration transition is detectable are not the same as in infrared, and this results in a method that is much less sensitive to water and hydroxyl groups and finds more application to inorganic species.

Raman spectra are obtained by irradiating the sample with a laser. Ultraviolet, visible, and infrared wavelength lasers are used depending on the type of material being analyzed. Most of the laser radiation is elastically scattered from the substrate and has to be filtered out. A very small proportion of emitted photons are inelastically scattered at wavelengths that are higher and lower than the wavelength of the laser, and these inelastically scattered photons have wavelengths that are characteristic of the molecule composition, crystal structure, and other properties. Obtaining useful Raman spectra is however quite difficult and is sometimes impossible with some dark materials. The main limitation is that the useful inelastically scattered photons are an extremely small proportion of all scattered photons, and so the signal strength is very weak. Also, some materials fluoresce from the laser light and the fluorescence can mask or overwhelm the weak signal. Modern Raman spectrometers use a variety of techniques to enhance spectra such as Fourier transform-Raman spectroscopy, and SEM with Raman analysis is also available.

Infrared spectroscopy, described above, is more sensitive than Raman and so is used more frequently, especially for organic materials, but Raman spectroscopy does have some advantages. Raman works well with hydrated or wet samples as water does not have a large adsorption, unlike in infrared spectra. Raman spectra are obtained with very little or no sample preparation, and no chemicals need to be added to obtain a spectrum. Analysis of components such as the surfaces of worn electrical contacts can be carried out directly on the surface of the part to analyze for surface composition and the presence of any contamination. As very small areas can be analyzed (<1 μm diameter), imaging of areas is possible.

Examples of the uses of Raman spectroscopy include chemical analysis of semiconductors and other electrical materials and measurement of stresses within materials in semiconductor devices, and it can be used to measure temperature variations on a microscopic-scale. Raman can be used to analyze surfaces without causing any damage or contamination so that other techniques can be used to complement the Raman data [2].

4.2.3.4 Mass spectrometric techniques

Mass spectroscopy is a technique that is used to analyze, usually organic, substances. A wide variety of mass spectrometric instruments have been designed for different specific purposes, but the main differences are the way the sample enters the instrument and how different components are separated. Different types of analyzers are also used. Mass spectroscopy most often is used for studying degradation of EEE by analysis of very thin layers of surface contamination that increase contact resistance of electrical connectors, analysis of corrosion products, and analysis of materials down to parts per billion (PPB) concentrations that is useful for semiconductor dopant analysis.

Mass spectrometers (MSs) separate and analyze ionized molecules in the gas phase. Molecules are ionized to form positively charged ions when an electron is ejected, for example, by impact with a high-energy electron. It is common for larger molecules to break up when ionized to form a number of ionized fragments and the primary ion may have lost only an electron and so has the same molecular weight as the original molecule. This all occurs in the gas phase as the interior of MSs is a vacuum so that there is no air to interfere with molecular fragmentation. Each compound produces a characteristic fragmentation spectrum that is useful for the identification of the original molecular species, even if no primary ions remain. Polymers can be analyzed by some mass spectroscopic techniques, but only monomer units and fragments are detected as all larger molecules fragment [3].

The three main parts of MSs are ionizer, mass analyzer, and detector.

Ionizer. This forms the positively charged ions from the material to be analyzed. Ionization can be achieved by bombardment with electrons, high-energy atoms, and molecules from the surface of components or from very small samples of material placed into the ionization chamber either as a solid or sprayed solution.

Mass analyzer. Once the material is in the form of positive ions, these are accelerated out of the ionization chamber by an electric field and into the analyzer region. Most ions have a single charge so are separated by their mass using magnetic or electric fields.

Detector. Modern MSs use various types of detector to detect ionized particles. Typically, the mass spectrum is generated by varying the magnetic or electric field in the mass analyzer to expose the detector to the full range of molecular weight. Time of flight is also used in some instruments where the ion formation is pulsed and the time for ions to reach the detector is measured. Heavy ions take longer to reach the detector than lighter ions.

Of the many types of MS available, a few specific types are useful for investigating the degradation of EEE. Two examples are time-of-flight-secondary ion mass spectrometer (TOF-SIMS) and pulsed glow discharge MS.

TOF-SIMS. This technique is used to analyze very thin surface layers and is often used to analyze surface contaminants that may be only a few molecules thick. TOF-SIMS can also generate images showing spatial surface composition information and also depth profiling by using an ion gun to sputter away surface layers. This technique can be used to analyze and measure the thickness of very thin surface layers and detect any unexpected impurities that might be present between layers and is used with integrated circuits on silicon substrates [4].

Pulsed glow discharge MS. This is a relatively new type of MS that is used to analyze coatings and thin films and is used for examination of hard disks, light-emitting diodes, and photovoltaic materials. Depth profiling is possible so that thickness measurement down to 1 nm is possible and analysis of PPB concentrations in alloys, semiconductors, corrosion products, etc., is possible [5].

4.2.3.5 SEM imaging with energy-dispersive X-ray and wavelength-dispersive X-ray analyses

SEM is one of the most useful techniques for the investigation of degradation of EEE. It has the advantage over imaging with visible light in that much higher magnification is achievable and high depth of field images can be obtained even at very high magnification. SEM imaging is used for detecting wear of electrical contacts, viewing various types of damage and small features such as gold wire bonds on silicon and tracks and other features on integrated circuits, cracks, and surface porosity, and for measuring dimensions such as coating thickness. Chemical analysis using energy-dispersive X-ray (EDX) or wavelength-dispersive X-ray is also possible and is used to assess material oxidation, metal corrosion, and surface contamination.

In an SEM, a beam of electrons is generated and focused on the surface of the specimen where these electrons undergo a variety of effects. Some are adsorbed by atoms very close to the surface, and then, electrons are ejected from the K-shell. These are secondary electrons that are detected to create secondary electron images that have good image resolution that clearly show surface topography. Backscattered electrons are higher-energy electrons that are either reflected or "backscattered" from atoms from a depth of the specimen that depends on the atomic mass and beam energy. Heavy atoms such as lead backscatter more effectively than light atoms such as carbon and so heavy atoms appear "brighter" in backscatter images than lighter elements. Depth of analysis can be varied by changing the beam energy so that at low beam energy, only material at and close to the surface is analyzed, whereas at higher beam energies, analysis of a greater depth is possible. For example, the analysis depth of gold at 10 keV is only 44 nm, whereas at 25 keV, the depth is 1.0 μm.

When the electron beam strikes the specimen, some are adsorbed by atoms, which then emit X-rays of wavelengths that are characteristic of the atom and so generate characteristic spectra that can be used for chemical analysis. The composition of single particles can be analyzed, and elemental mapping is also possible to generate X-ray images.

SEM is a powerful tool for assessing degradation of electrical equipment as small changes can easily be seen. Examples of its uses are given in the following text.

Nonnoble metals such as tin have thin protective air-formed oxides, but frequent sideway movements of one electrical contact across the surface of another disrupt this oxide, exposing the base metal, which then reoxidizes. Gradually, the amount of electrically insulating oxide builds up until there is enough to cause a high resistance. This process is fretting and can easily be seen by SEM in backscatter mode as the oxide appears darker than tin. EDX analysis can be used to measure the amount of surface oxide and give an estimate of whether failure is likely to occur.

Contaminated printed circuit boards can suffer from corrosion and, under the influence of an electric field "dendrites" of metals, can grow from one conductor to another eventually causing a short circuit. These often form as thin filaments of metal that are difficult to see among associated corrosion products but can usually be visualized by element mapping.

Circuitry of integrated circuit packages is too small to visualize with optical microscopy, and SEM is used to look for defects or damage, such as from electrostatic discharge, corrosion after delamination, and wire bond failures.

4.2.3.6 Focused ion beam sample preparation

A common mode for preparing sections of samples is potting the sample in an epoxy or other polymer, followed by grinding and polishing to the plane of interest. A disadvantage with the technique is that relatively soft materials, for example, copper or tin layers, may be smeared that can interfere with thickness measurements or disguise defects such as fine cracks or the onset of intermetallic formation. A useful technique to overcome this is the use of focused ion beam milling to prepare the surface for inspection.

The sample must be placed in a vacuum chamber and a beam of ions (often gallium) is accelerated toward the sample and focused into a beam using electrostatic lenses. The equipment is similar in many ways to an electron microscope. The beam can be scanned across the sample surface sputtering away the surface in a controlled manner in terms of the rate and depth of material removed and the location from where the material is removed.

In some applications, there may be disadvantages associated with this technique as the sputtering process leaves the surface amorphous, and therefore, crystallographic information may be lost especially in metallographic investigations. Ions are also implanted into the sample surface, and the nonvolatile sputtered material will redeposit contaminating other areas of the sample that needs to be considered if making chemical analysis in other regions.

4.2.3.7 Transmission electron microscopy (TEM)

TEM is similar to SEM in that images are created from an electron beam in a high vacuum, but with TEM, a much higher energy beam (up to 1 MeV with TEM compared with up to 40 keV with SEM) passes through the sample before being focused onto the detector. TEM can be used to view real images from the beam that passes through the specimen or to determine crystal structure information from the diffraction patterns generated by crystals in the sample. The main advantage of TEM is its very high magnification that is much higher than with SEM (it is possible to visualize individual atoms), but the biggest drawback is that sample preparation can be very complex as only very thin materials can be imaged.

Examples of the uses of TEM include detection of defects in single-crystal silicon, examination of diffusion barriers, detection of defects in gate oxide layers, and examination and measurement of implanted regions in materials such as p/n junctions. Very small structures in semiconductor devices such as interconnections between layers can be examined from thin cross sections.

4.3 Nondestructive techniques

4.3.1 Visual inspection

The first stage of any investigation is to examine the equipment. Is there anything obvious that points to the cause of deterioration or failure such as burn marks or contamination? This will help to pinpoint the areas for closer examination. A photographic record of the "as-received" condition of parts is normally collected at this stage. A hand lens is sometimes useful but a good low-power binocular microscope will usually be the best first stage of visual examination. In some areas, higher magnification will be needed with different types of microscope.

4.3.2 Optical microscopy

Optical microscopy has been used to investigate materials for over 200 years and is now very well established and an essential technique in almost all fields of failure analysis. Although there are many specialized techniques that use microscopy, it is still the simple optical methods that are the main application in electrical failure investigations.

In the field of failure analysis, three types of microscope would normally be required:

(1) Stereo microscope
(2) Metallurgical microscope
(3) Transmission microscope

4.3.2.1 Stereomicroscopes

Almost all investigations start with an optical examination of some type, and the importance of such steps should not be underestimated. This is often the first time a sample is examined, and this can affect all further investigations. If destructive methods are to be use, it is essential that the as-received condition of the samples is correctly recorded. The primary tool for such examinations is the stereomicroscope that comes in a wide variety of forms.

There is no such thing as the perfect microscope, but there are a range of options to take into consideration. Among the most important of these are the following:

- A sufficient reach that large and complex parts can be correctly positioned in the field of view. It should be noted that such system needs to be robust if optical instability is to be avoided.
- An appropriate magnification range for the intended work. For most applications, the range from ×5 to ×50 is particularly useful. Microscopes with magnification to ×100 are useful for many forensic applications but handling objects and depth of field issues become more difficult at higher magnifications.
- Flexible lighting. No one light arrangement is likely to meet the demands of all sampled and options for ring lighting and movable oblique lighting are recommended. The new white LED (light emitting diode) systems have made the task of microscope light easier in recent times.
- Useful facilities for digital photography and an essential camera system of sufficient quality that can handle both matt and high contrast surfaces. If photography is important, a trinocular microscope designed to securely fit the digital camera will be needed.

4.3.2.2 Metallurgical microscopes

Metallurgical microscopes are ideal for the inspection of prepared polished sections of metal parts or sections though components. These have objectives designed to have good resolution and good light collection to allow high-resolution magnification of the reflected light from the sample. Many metallurgical objective lenses have rather small working distances. Magnification up to ×1000 is possible (with oil immersion).

4.3.2.3 Transmission microscopes

Transmission microscopes of various types are the most commonly encountered type and usually cover the magnification range of ×50 to ×1000, the latter only with the use of oil-immersion objectives. It is this type of microscope that would be used to examine fibers in transmitted light. Examination will be either directly on a glass slide or, for finer particles, mixed with an appropriate solvent below a glass coverslip.

4.3.2.4 Combination systems

The most flexible microscope design will include both substage and vertical illumination sources so that with appropriate objectives, both reflection (metallurgical) and transmission options are available. Zoom options are not normal in such microscopes so a five-head nosepiece is the best to give a full range of magnification options. The most usual would be ×3.5, ×5, ×10, ×20, and ×50. The use of ×100 oil-immersion objectives is only for a limited use but can help bridge the gap between optical and SEM methods for microelectronic investigations.

The objectives are the main factor governing the resolution and brightness of the final image; generally the larger the numerical aperture (NA) of the lens, the greater the resolution power. However, the NA value of the substage condenser will also affect final image resolution.

Most of the specialized optical microscope techniques have been designed for biological systems and find limited use in failure analysis, but the ability to use polarized light for both transmitted and reflected light will be advantageous. The birefringent properties of materials such as plastics can help both in their identification and resolve them from a nonbirefringent background but can also be used in reveal stress in polymers and in eliminating unwanted reflections from metals and similar surfaces. This will help both to characterize materials in transmitted light and to remove unwanted reflections when a vertical light source is used.

With appropriate objectives, Nomarski interference techniques can use polarized light to enhance contrast and find some utility in failure analysis work allowing better discrimination or find surface details.

For failure analysis, the microscope should be a trinocular type to allow easy attachment of a digital camera although some of the more modern cameras may have more integrated camera systems. Such image systems are now frequently linked to image archives and image analysis software that can further expand its usefulness. For many real-world problems, image stacking and image stitching options can be useful and particular in the field of fracture analysis where depth of field issues usually precludes simple imaging.

4.3.3 X-ray imaging techniques

Many defects in electrical equipment and components are hidden because they occur within complex multilayer structures or they are encapsulated to prevent corrosion from moisture and atmospheric contamination. This however makes visualization very difficult as destructive dismantling can destroy any evidence of the original cause of failure. The most commonly used technique to obtain images of the interior of electronic components and printed circuit boards is by X-ray imaging. The techniques used are similar to those used in the hospital to examine the interior of patients, and 2D and 3D images can both be obtained. Unlike medical X-ray imaging, the designs used for electronic examination can magnify the interior, typically up to 5000 times with some models so that a resolution of as little as 1 μm is achievable.

The simplest approach used is to focus a high-intensity X-ray beam onto the part being examined and to measure the intensity of the X-rays that emerge from the other side. The intensity map then generates a 2D image. As no harm comes to the equipment, the part can be examined in the X-ray beam for long periods during which the part can be moved and rotated in all three axes in order to see inside at any viewing angle. The images can be photographed or videoed while the part is moved to provide a permanent record of the interior. Full 3D images can also be generated using computed tomography (CT) X-ray imaging, which is essentially the same as medical CT imaging. With CT, the X-ray source and detector rotate around the part being examined and move along its length. The data from the many slices through the part are then used to generate the 3D image.

X-ray imaging is used to visualize solder bonds to detect voids and cracks, to see inner-layer conductors and vias to detect poor bonding or unusually thin areas, and to examine wire bonds inside semiconductor devices. Cracks in internal conductors are not easy to see with 2D X-ray but can be visualized by CT. An advantage of this technique is that it is quick and nondestructive so that the equipment will function normally after being examined. As a result, X-ray imaging is used for quality control of manufacturing and for failure investigations. Another advantage is that it is not necessary to prepare cross sections if X-ray imaging reveals the cause of a malfunction, and CT can generate "virtual" cross sections [6].

4.3.4 Infrared thermography

Thermography is a technique that is used to measure surface temperatures of a wide variety of objects including electrical equipment, printed circuit boards, and connectors. Elevated temperature is an indication of current flowing through a resistive pathway and so can detect overheated electrical connections that are likely to fail. Measurement of component temperatures can give an indication of potential future reliability problems, for example, due to the formation of intermetallic phases between solder and substrate causing brittle bonds and cyclic stresses due to large temperature fluctuations or components operating at or above their rated temperature.

Infrared thermography detects radiation usually in the 8-14 μm range, which corresponds to fairly warm and hot surface temperatures of up to 2000 °C, but some instruments operate in a wider range such as 0.9 to 14 μm to detect colder temperatures as low as −40 °C and hot. This technique should not be confused with night vision goggles that operate in the near-infrared range of 0.7-1.4 μm, which corresponds to temperatures of around human body temperature or with starlight cameras that enhance visible light.

Obtaining thermographs with modern infrared thermal cameras is relatively straightforward; the camera measures the surface temperature and converts the detected infrared radiation into a color image where temperature is represented by color, typically with red as hot and blue as cold. However, care is needed as small hot areas can be missed by incorrect use of the camera, which requires training and experience. This technique is a line of sight technique, and so only the visible surface temperature is measured. If the enclosure of equipment is imaged, only the outer surface will be analyzed, and detailed information about internal parts will be missing, and so it is essential to be able to see the components and connections whose temperature are to be measured. Another issue is that the outer surfaces of components such as transformers, connectors, and MOSFETs will always be cooler than internal windings of transformers, junctions within MOSFETs and other semiconductor devices and electrical connections within connectors. Heat will be conducted away from hot components by the interconnected circuitry and the component's encapsulation acts as a thermal insulator, and so the actual internal temperature can be considerably hotter than the external surface temperature.

4.3.5 X-ray fluorescence analysis

X-ray fluorescence (XRF) is a rapid nondestructive chemical analysis technique that determines the elemental composition of solids and liquids and is used for characterizing the degradation in EEE. It can be used to analyze metals, plastics, glass, and ceramics and thin coatings. There are many variations in instrument design that affect the performance, such as the smallest area that can be analyzed, analysis sensitivity, and possibility of mapping. However, equally important is the software that controls these instruments and calculates the results.

XRF works by focusing an X-ray beam onto the surface of the material. These X-rays are either scattered by atoms or are adsorbed and expel an inner electron from an atom. This creates an unstable situation that is rectified by transferring electrons from outer shells to inner shells. When this occurs, secondary or fluorescent radiation is emitted with characteristic wavelengths that are used to identify the element. As the incident X-radiation is either adsorbed or scattered by atoms close to the surface, XRF analyzes predominantly the outer surface layers but the analysis depth will depend on the atomic mass of the elements present. The analysis depth for light elements such as in plastic materials can be 10 mm or more depending on the X-ray beam intensity, whereas the analysis depth of heavy elements such as lead is much less at about 10 μm.

Two types of analysis systems are used: the more common is energy-dispersive (ED) where the energy of fluorescent X-rays is measured and less common and more complex wavelength-dispersive (WD) where the energy is passed through a diffraction grating to separate the X-rays into different wavelengths. ED systems can only detect elements that are heavier than neon, whereas WD systems can detect beryllium and all heavier elements. Although carbon, oxygen, and hydrogen are invisible to ED-XRF, some designs of ED-XRF analyze plastics using special software that estimates the carbon content from the intensity of scattered X-radiation so that heavier element concentrations can be calculated.

XRF spectrometers are available with different designs. The simplest are handheld devices that are used for routine analysis of materials and most analyze circular areas with diameter of ca. 10 mm although some models can analyze areas of 1 mm diameter. There are also many designs of desktop instrument. These tend to be more sensitive and some can analyze areas as small as 0.1 mm. Small "spot-size" XRF analyzers are also used for mapping to show the distribution of materials (e.g., by using false-color images) and are useful for identifying contamination. Most instruments analyze two-dimensional but a few can generate 3D images. Multilayer coatings can be analyzed by some models, and it is also possible to measure the thickness of each of several multilayers. This is used to determine the thickness of electroplated metal coatings on components and thin layers on semiconductors. XRF is particularly useful for examination of materials inside circuit board laminates and inside electronic components because the polymeric materials are only weakly adsorbing of X-rays. Components and the conductor tracks in flexible and rigid PCBs are encapsulated by relatively thin polymer coatings, and although SEM cannot see below these polymer layers, as the penetration depth is too small, they can be imaged and analyzed by ED-XRF.

4.3.6 Acoustic microscopy

Acoustic microscopy is used to detect delamination, cracks, and other types of "boundaries" that occur within materials. Cracks and delamination, for example, may occur gradually during the life of electrical equipment and eventually lead to failure. Acoustic microscopy relies on the partial or complete reflection of high-frequency sound waves that occur when the waves propagate through a change in acoustic impedance. The acoustic impedance difference is proportional to the magnitude of the reflection and therefore the magnitude of the reflection can be used as a measure of material change.

Modern acoustic microscopes typically operate at frequencies from 10 to 400 MHz and consist of an ultrasonic transducer that can scan in two dimensions (*X*-*Y*) while emitting ultrasonic pulses in a third dimension (*Z*). The normal configuration is to have a scanning transducer that operates as an emitter and receiver. A pulse of ultrasound is emitted and then the time for a reflected pulse to return to the transducer and the magnitude of that reflection is measured, before the transducer moves in *X*-*Y* and the next pulse is emitted. In this way, an acoustic impedance difference map of the sample is generated. Knowing the acoustic impedance of the materials being scanned allows the depth into the sample at which features of interest are detected to be determined.

Normally, the sample is immersed in water (or other liquid), and the sample must remain in the liquid throughout the scan process, which typically varies between a few tens of seconds up to 1 h depending on the size of the sample and the resolution of the scan. Systems are available where a stream of water is used to couple the transducer to the sample, avoiding complete immersion; however, the sample always ends up wet.

During scanning, the system "triggers" on the first reflection seen by the transducer, and this trigger is used to maintain a consistent reference between features within the sample and the top surface.

The system is particularly sensitive to changes in material and especially to delamination where the technique is sensitive to air or vacuum gaps of nm in depth. High-frequency transducers can be used to achieve scan resolutions of ~10 μm with detection of features as small as 2-3 μm; however, this can only be achieved in dense crystalline materials at depths of <1 mm. For thicker samples or to look at features deeper into the sample, then lower-frequency transducers are required, which have lower spatial resolution; for example, a 50 MHz transducer may have a spatial resolution of approximately 30 μm and can scan at depths of up to ~5 mm. Spatial resolution is determined by the spot size of the focused ultrasound waves combined with the resolution of the *X*-*Y* encoder used to control transducer movement.

The technique is particularly good at looking at delamination in optically opaque materials, for example, in IC packages, where its sensitivity is far greater than X-ray. There are two significant limitations of the technique. One is visualizing features that occur below a delamination layer as all of the sound energy will have been reflected effectively shadowing any features below the delamination. The other is imaging features in materials that scatter the sound energy, for example, samples with very rough top surfaces or materials like FR4-PCB.

4.4 *In situ* measurement techniques

4.4.1 Electrical measurements

Degradation of electrical equipment can occur as a result of changes in the electrical characteristics of electrical components and connections, but measurement of the extent of degradation of these is however complicated by two issues: (1) It is difficult to measure the characteristics of a single component or connection *in situ* because electrical results will be affected by the entire interconnected circuit and (2) small changes may not be detectable if the original characteristics are unknown. Many types of components have rated characteristics such as resistance and capacitance that are within tolerance ranges. Therefore, a 10 μF capacitor with tolerance of ±10% could when new be anything from 9 to 11 μF so a measurement after a period of use of 9.2 μF will not give an indication of the extent of degradation. However, as long as these limitations are accepted, electrical *in situ* measurements are a useful tool.

4.4.1.1 Electrical conductivity/resistivity measurement

In failure analysis work, it is often necessary to measure surface resistance properties or low-level leakage currents, for example, in situations where surface contamination or moisture-related effects are causing problems with a sensitive circuit or where the onset of a problem such as dendritic growth or surface tracking faults are being investigated. In these situations, a high-voltage resistance meter capable of supplying >1000 V test voltages and measuring resistance at >GΩ levels is required (Ω = ohm). A common failure mode of electrical equipment is electrical connections becoming resistive. Accurate measurement of electrical resistance, particularly of connector and contact resistances, should be made using a four-point probe measurement technique, especially for resistance values of less than 100 Ω where the contact resistance of the measurement probes forms a significant portion of the measurement. To characterize these electrical contacts requires accurate measurement of resistances using a four-point resistance measurement technique, which involves a constant current passing through the connection in question with two leads across the contact and the potential difference being measured with the other two leads. As the potential difference is measured using a voltmeter that will have a relatively very high resistance, the current through the test leads and hence also the test lead contact resistance are negligible. These measurements involve passing a constant current through the sample and measuring the voltage drop across a known length of the sample through which the current is passed. Resistance is calculated according to ohms law. As the voltmeter has relatively very high impedance, there is therefore negligible current flow through the voltage measurement contacts and so their resistance does not influence the measurement.

Many digital multimeters have a four-point measuring capability; however, it is important to ensure that the meter used is suitable for the type of electrical contact or connection being assessed. For example, a multimeter will typically use a measuring current of several milliampere, whereas the type of connector that is to be measured may be designed to carry currents of many amps in normal use. Additionally, for measuring very low resistances, a typical digital multimeter may be unstable and so a microohm meter should be used, which is capable of supplying several amps of measuring current. Ideally, the measuring device will also reverse the polarity of the current and produce an average result.

To measure very low resistance (e.g., below ~0.1 Ω), a dedicated microohm meter may be needed as the stability of standard multimeters can be poor as the measurement circuit struggles to maintain a constant current through low-resistance samples. Microohm meters can typically measure the electrical resistance of samples (as the name suggests) at microohm levels and also at significant currents of several amps. Measurement of electrical resistance (particularly for electrical contacts and connectors) at high currents can provide more meaningful results as the effect of any resistive heating in a defective contact is more likely to be observed as opposed to when using a standard multimeter that will typically use a measurement current of milliampere.

4.4.1.2 Passive component measurement

The characteristics of passive components need to be measured accurately. For inductors and capacitors, this requires an LCR bridge that can measure true values. "True" values include the resistive component of capacitors or inductors, and some lower cost instruments are not able to calculate this. The LCR bridge works by tuning internal test circuits to achieve a null current in a test circuit and using the values of the bridge circuit components to derive the *L*, *C*, or *R* values of the component being tested (which also forms part of the circuit).

It is important to have an instrument that can calculate true values including the equivalent series resistance of reactive components in order to properly characterize component failure/deterioration, particularly in capacitors. It is also important to have an instrument that can measure over an appropriate frequency range, typically from a few Hz up to >100 kHz.

A curve tracer is required to properly characterize transistors and enable calculation of transistor gain. A unit that can apply high voltages (up to >1000 V) is useful for assessing the performance of transient suppression devices. Curve tracers are also useful for properly analyzing diodes as a simple digital multimeter will not reveal reverse bias breakdown voltages or the "sharpness" and precise onset of forward bias conduction.

4.4.2 Measurement of physical characteristics

4.4.2.1 Profilometer surface roughness

Stylus profilometry is used to measure small features and surface roughness of materials. Stylus profilometry tools rely on a stylus, normally with a diamond tip, being pulled across the surface of the item being analyzed while measuring the vertical position of the stylus along with the X position to produce a two-dimensional plot of height (Z) versus distance (X). The technique can make measurements over distances X from <1 mm up to several 100 mm with an accuracy of <1 μm and with Z (variations in height) ranges of up to several mm and accuracies of <0.01 μm.

Some systems have motorized X-Y-θ stages that allow automated collection of a series of line scans that can be used to build up a full 3D picture of the sample.

The technique is particularly suited to measurement of step height, form or curvature of a sample, and surface roughness. However, measurements are often made in two dimensions, and there is an uncertainty over whether the scan that is taken is truly representative of the feature that is being measured. The technique is used, for example, to assess wear patterns on surfaces such as on electrical contacts or to calculate the bearing ratio of surfaces.

When measuring surface roughness, care must be taken to ensure that the values being recorded are real and are not affected by vibration or artifacts of the measurement (e.g., if the scan speed is too high, then the tip can bounce off the surface, generating a false reading). It can also be difficult to position the measurement tip in exactly the right position to measure the required feature, especially if there is a radius of curvature measurement to be made.

The size and shape of the measuring tip impose limits on the height or depth of features that can be measured, and typically, sharp steps of more than a few millimeters may risk damage to the tip. The tip force on the surface is typically only a few milligrams; however, some soft surfaces cannot be measured.

4.4.2.2 Atomic force microscopy

Atomic force microscopy (AFM) is a technique that measures extremely fine surface features and depending on the instrument and the measuring mode can measure a variety of surface properties. The basic mode of operation is topography measurement where measurements are taken over a few square micron, and surface topography can even be measured at atomic levels. The instrument makes a series of line scans across the surface and assembles these scans into a 3D image of the surface. A very fine tip (typically made by anisotropic etching of silicon or silicon nitride) located at the end of a silicon cantilever is deflected by atomic forces as it approaches the surface of the sample. As the tip is moved across the sample, the deflection of the cantilever is measured, normally using a laser or interferometry measurement.

AFM provides the highest resolution measurement of a surface at subnanometer accuracy; however, great care must be taken when setting the instrument up and ensuring that it is isolated from any vibration. For example, when measuring surface topography, the way in which the AFM tip is controlled can have a significant influence on the results. The tip position is typically adjusted using a feedback loop that can be tuned to have different responses to the surface.

AFM can be used to measure surface topography, surface chemical attractions, capacitance, and magnetism. In failure analysis work, the technique is useful for comparison of surfaces, such as of integrated circuits and the assessment of fine wear patterns.

There are some situations where AFM can be an alternative to electron microscope inspection with the advantage that sample preparation and an expensive vacuum system are not required but with the disadvantage that scan areas are only a few tens of micron in area and take several minutes to acquire data.

4.5 Conclusions

Electrical equipment degrades and eventually fails in a wide variety of ways, and so many different types of analytical instrument are needed to characterize these changes. Some are more commonly used than others such as optical microscopes, SEM, and electrical measurements, but sometimes the other methods described in this chapter need to be used. All analysis techniques have advantages and disadvantages, and all have limitations that the analyst must understand. Some techniques are relatively easy to use such as optical microscopy, whereas others require extensive training, such as TEM. Usually, degradation characterization will require several techniques to be used, and careful interpretation of the results as determining the degradation mechanism is far from obvious, and there is no alternative to previous extensive experience and careful analysis of the equipment.

4.5.1 Future trends

Most of the techniques described here are fairly mature and so only incremental improvements are likely in the near future. What is changing however is the size of electronic components and circuitry that are relentlessly becoming smaller. For example, CMOS (complementary metaloxide semiconductor) devices are increasingly produced with features smaller than 100 nm. Features such as gate oxide defects in older technology with 0.4 μm can be viewed with SEM or TEM but the oxide and any defects are too small in features of 90 μm circuits and smaller and so different investigation techniques are needed [7]. Printed circuit boards are also changing with smaller components, higher component density, and smaller gaps between conductors.

Chemical analysis of features on very high-density circuits such as mobile phone printed circuit boards needs to use techniques that can analyze materials present in very small areas. The minimum analysis area for XRF analyzers is about 0.1 mm, and most instruments analyze much larger areas and so SEM may be needed that can analyze down to 1 μm^2. Soldering flux chemistries have changed significantly in the past few decades and flux residues can be a common cause of failures.

Another continuous trend is to reduce costs as far as possible. This results in the avoidance of expensive materials such as gold with thinner coatings on connectors or use of cheaper coatings. There is also research into the use of copper bond wires to replace gold although these trends may not need changes in the designs of analysis instruments but can cause different types of failures to occur.

4.5.2 Sources of further information

Microscopes, spectrometers, and other analysis instruments are constantly being improved and manufacturers provide useful information on new models on their websites that describes the capabilities and some also give examples of how they are used.

References

[1] FT-IR spectroscopy—attenuated total reflectance (ATR), Perkin Elmer Technical Note, 2005, Available from: http://www.utsc.utoronto.ca/~traceslab/ATR_FTIR.pdf (accessed 28.03.14).
[2] Horiba Website—Raman Spectroscopy, Available from: http://www.horiba.com/uk/scientific/products/raman-spectroscopy/raman-academy/raman-faqs/when-should-i-use-raman-spectroscopy/ (accessed 28.03.14).
[3] S.D. Hanton, Mass spectrometry of polymers and polymer surfaces, Chem. Rev. 101 (2001) 527–569, Available from: http://www.ch.pw.edu.pl/~maldi/pomiary/Artykuly/Mass%20Spectrometry%20of%20Polymers%20and%20Polymer%20Surfaces.pdf (accessed 28.03.14).
[4] PHI Physical Electronics Website, Available from: http://www.phi.com/surface-analysis-techniques/tof-sims.html (accessed 09.04.14).
[5] Horiba Website—Glow Discharge MS, Available from: http://www.horiba.com/scientific/products/atomic-emission-spectroscopy/glow-discharge/ (accessed 09.04.14).
[6] Nordson CT Brochure, μCT Options for Nordson DAGE X-ray Inspection Systems, Available from: http://www.nordson.com/en-us/divisions/dage/support/Literature/Documents/CT%20brochure_14%208%20mar%202013.pdf (accessed 09.04.14).
[7] R. Mulder, S. Subramanian, T. Chrastecky, Atomic force probe analysis of non-visible defects in sub-100 nm CMOS technologies, in: Proceedings of the 32nd International Symposium for Testing and Failure Analysis, November 12–16, Renaissance Austin Hotel, Austin, TX, 2006.

Reliability building of discrete electronic components

5

T.-M.I. Băjenescu[1], *M.I. Bâzu*[2]
[1]C.F.C., La Conversion, Switzerland; [2]National Institute for R&D in Microtechnologies, Bucharest, Romania

5.1 Introduction

Electronics manufacturers try to get the largest share of the market by developing products with the highest levels of performance, including high reliability. So, they naturally tend to study the reliability of electronic systems in order to prolong the life of the technical systems they produce. But manufacturing reliable electronic equipment begins with reliable electronic components, in order to minimize the failure risks [1]. Consequently, manufacturers must carefully investigate the reliability of the electronic components, creating rules for building reliable components, as well as methods for reliability assessing. Today, most companies understand that the reliability of components and systems must be established at the design phase and then monitored during the whole manufacturing process [2].

In this chapter, we first describe the basic rules of reliability building. Then, we discuss the failure risks associated with some representative types of discrete electronic components (capacitors, diodes, and transistors). In considering these risks, we explain the main characteristics of the device, the typical failure modes (the electrical behavior of failed items), and the methods of failure analysis (FA) used in reliability studies of the device. We then identify the failure mechanisms (FMs) (the physical-chemical processes that produce the failure), before recommending corrective actions for improving reliability. We conclude by considering electrostatic discharge (ESD), a FM that could arise in any electronic component, exploring possible methods for preventing ESD.

5.2 Reliability building

Reliability building includes all techniques and procedures intended to ensure a predetermined reliability level for a given product. This concept is linked with reliability assessing, which consists of the whole evaluation system aimed at recording the reliability of a product batch during and after the manufacturing process (see Figure 5.1). This evaluation system includes tests, electrical measurements, FA, and statistical processing of data [3].

Reliability Characterisation of Electrical and Electronic Systems. http://dx.doi.org/10.1016/B978-1-78242-221-1.00005-8

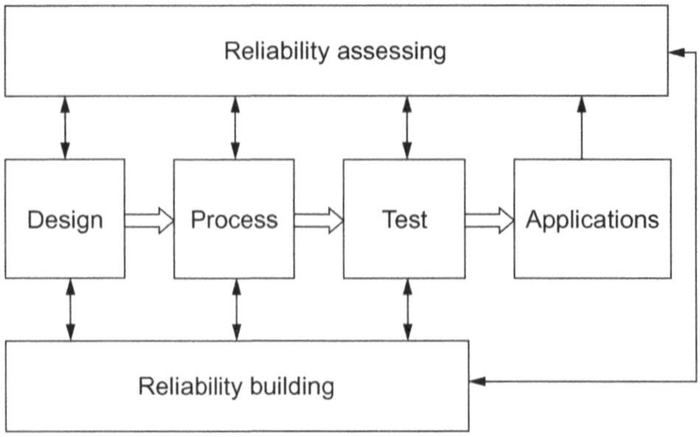

Figure 5.1 Reliability building and reliability assessing during product fabrication and usage.

According to the rules of reliability building, reliability issues must even be taken into account during the design of the process or product, the so-called design for reliability (DfR), and also during manufacturing, by monitoring process reliability. Special attention must be given to the selection of the most reliable items from a batch of products, which could be made by screening or by burn-in.

5.2.1 Design for reliability

DfR aims to understand, identify, and prevent underlying failures before the devices are built. In designing a product, engineers usually miss the following characteristics: (a) key failure modes and failure rate of the product, (b) key FMs that may be present in the service environment, (c) usable life of the product, (d) cost of maintenance required to sustain the inherent reliability, (e) availability, and (f) rigorous testing. DfR is a relatively new concept, and it is an important step in building the reliability of a product or component (in other words, to achieve built-in reliability), which is being linked with the concept of concurrent engineering (CE). CE is a feature that ensures a design is not completed before reliability requirements are identified and dealt with. Basically, DfR consists of the following two elements:

(a) A collection of design rules for making an electronic component reliable, not only electronically, but also mechanically and visually. The design rules have to be continuously updated to reflect best practices, ensuring the maximum component reliability. Robust design and thermal design produce the major part of these rules.
(b) Predictive methods able to asses the reliability of the future device, based on design data and on models describing the time and stress behavior of similar products. Bâzu [4] provides an example of predictive method based on fuzzy logic applied for the manufacturing of electronic components.

The DfR approach begins with capturing the customer's voice and translating it into an engineering function [5]. Then, the engineers create a design immune to the actions of perturbing factors (robust design), possibly using Taguchi methods: (i) developing a metric for capturing the function while anticipating possible deviations downstream and (ii) designing a product that ensures the stability of the metric in the presence of deviation. Finally, the design team must use reliable prediction methods. In principle,

DfR allows the engineer and manufacturer to shift from evaluation and repair to anticipation and design. In recent years, researchers have proposed several approaches for integrating robust design [6,7] and reliability-based design [8,9]. Reliability-based design optimization is a method for achieving confidence in product reliability at a given probabilistic level, and robust design optimization is a method for improving product quality by minimizing the variability of the output performance function. Because both design methods make use of uncertainties in design variables (and other parameters), the two different methodologies have been integrated to develop a method called reliability-based robust design optimization (RBRDO) [10].

The physics of failure (PoF) is a key approach for implementing DfR in a product design and development process. PoF refers to knowledge of how things fail and the root causes of those failures. On the other hand, the PoF approach can be time-intensive and not always definitive, offering limited insight into performance during a device's operating life [11].

Yadav et al. [12] have proposed a multiobjective framework for RBRDO, which captures the degradation behavior of quality characteristics to provide optimal design parameters. The objective function of the multiobjective optimization problem is defined as a quality-loss function considering both desirable and undesirable deviations between target values and the actual results. The degradation behavior is captured by using an empirical model to estimate the amount of degradation accumulated during time t.

5.2.2 Process reliability

The reliability of a product depends directly on the quality of the manufacturing process. Once established, this quality must be kept at the same level during the entire period of product fabrication, and maintaining the quality of the manufacturing process is called process reliability. This is a method for identifying problems that have significant cost-reduction opportunities, if improved. Very often the problems have roots in the operations area [13].

In order to ensure appropriate process reliability, the engineer must consider the following elements:

- Wafer-level reliability, which covers all the activities focused on achieving a reliability goal for the wafer: quality of the equipment, material-environment interactions, synergy of the technological factors, test structures for monitoring the reliability level, and so forth.
- Reliability-driven assembly process: an assembly process that has controls tight enough for the reliability level to be adequately monitored.

The designers must also ensure device traceability and statistical process control during the above-mentioned technological steps.

5.2.3 Screening and burn-in

Last but not least, the finished devices undergo burn-in and screening tests (both are 100% tests) in order to eliminate the weak devices from the batch—those likely to fail in the first hours of functioning. Generally, the stress factors for screening are electrical, mechanical, and climatic ones, followed by a parametric electrical or functional control, with the aim of eliminating the defective items, the marginal items, or the items that will probably have early failures (potentially unreliable items). Burn-in is a short version of screening, which involves a functioning test at high temperature (sometimes, burn-in is the last operation included in screening).

For at least two reasons, defining a cost-effective screening sequence can be difficult. First, screening may activate FMs that would not appear in field operation (and this totally forbidden for a screening test!). Second, it could introduce damages (transients, ESD) that may be the cause of further early failure. Engineers can realize the goal of the screening tests in two ways: (a) Usage of maximum allowed load, because the components predestined to fail during the early failure period are very sensitive to overloading; (b) Usage of efficient physical selection methods that can provide information concerning any potential weaknesses in the components (noise, nonlinearity, etc.). Research has proven that the combination of different stresses for producing early failures in the elements, followed by a 100% electrical test (burn-in), is optimal and efficient, especially if the costs must be taken into account. Establishing the optimal stresses (their sequence and stretching) is a delicate problem, and failures depend on the integration degree, the technology, and the manufacturing methods.

It can be difficult to find the optimum load conditions and burn-in duration so that nearly all potential infant mortality components are eliminated. There must be a substantial difference in the lifetimes of the infant mortality population and the lifetimes of the main (or long-term) wear-out population under the operating and environmental conditions applied to burn-in [14]. The situation may differ depending on the state-of-the-art components, the new technologies, and the custom-designed circuits. The trend is toward monitored burn-in [15]. The temperature should be high, without exceeding 150 °C, for the semiconductor crystal. One may distinguish three types of burn-in, as summarized by Băjenescu and Bâzu [3]. In static burn-in, temperature stresses and electrical voltages are applied, and all the component outputs are connected through resistors to high or low. Dynamic burn-in involves temperature stresses and dynamic operation of components (or groups of components). Power burn-in consists of operation at maximum load and at the ambient temperature (0 to +150 °C), with functioning also monitored under the foreseen limits of the data sheet for 25 °C. It is often difficult to decide whether a static or a dynamic burn-in is more effective. Should surface oxide and metallization problems dominate, a static burn-in is better; a dynamic burn-in activates practically all FMs. Thus, the engineer must choose a burn-in on the basis of practical results. The static burn-in is used by the manufacturers and by the users as a control selection. The static burn-in is particularly adequate for the selection of great quantities of products, and it is economical. Another parameter for dynamic burn-in is the resolution that determines the maximum frequency of the stimuli sent to the components. The best solution involves choosing a stimulus rate in the proximity of the effective operation frequency of the component.

5.3 Failure risks and possible corrective actions

5.3.1 Discrete electronic components

Discrete electronic components could be considered the "pioneers" of electronics, being the first devices invented and used in this field. As a consequence, a huge quantity of published literature addresses reliability building for such components. Taking this body of research into account, we do not intend to provide supplementary information. Rather, we mainly organize the existing material and offer the reader a key for understanding the main issues of the subject. The "long and winding road" of reliability building is detailed, passing from failure modes through FA and typical FMs to the technological procedures for avoiding the failure risks.

One must know that, under the name of discrete electronic components, this chapter could consider a large diversity of devices, including passive components, such as resistors, capacitors, inductors, transformers, relays, varistors, and connectors, as well as active components, including silicon diodes, nonsilicon diodes, thyristors, silicon transistors (bipolar or metal-oxide semiconductor—MOS), and nonsilicon transistors (bipolar or MOS). In this chapter, we do not study the families of discrete electronic components that are already covered by other sections of this book, such as: optoelectronic components, MEMS and nanodevices, respectively.

From the eligible families of discrete electronic components, we have chosen the following families as the most representative: aluminum and tantalum capacitors (the most complex discrete component), silicon and nonsilicon diodes (because the pn junction could be considered "the fundamental brick" of most complex devices), and four families of transistors: bipolar silicon, bipolar nonsilicon, MOS silicon, and MOS nonsilicon. For each family of components, the structure of the text is similar. First, we describe the main applications of the components. Then we consider their typical failure modes and mechanisms and possible corrective actions.

5.3.2 Capacitors

When carrying out the same function, few electronic components differ more from the constitutive materials point of view than does the capacitor. In this section we examine the reliability of two families of capacitors: aluminum electrolytic capacitors and tantalum capacitors (Table 5.1 synthesizes the typical FMs).

5.3.2.1 Aluminum electrolytic capacitors

Half-dry electrolytic wound capacitors (the most used) are formed from an oxidized aluminum foil (anode and dielectric) and a conducting electrolyte (cathode). A second aluminum foil is used as a covering cathode layer. They are available with two formed nonpolarized foils, and they have a large loss factor (frequency and temperature dependent), a limited useful life, and low reliability ($\lambda = 10$–50 FIT; drift, shorts, opens). These capacitors are often the first elements in a system to fail. The probability distribution of

Table 5.1 **Typical failure modes and mechanisms of capacitors**

Families of capacitors	Failure modes	Failure mechanisms
Aluminum electrolytic capacitors	Parameter drift	Degradation of oxide layer produced by material defects Variations in the conductivity of the impregnating electrolyte Poor encapsulation
	Popping	Increase of internal temperature
Tantalum capacitors	Electrical short	–
	Electrical open	–
	Parameter drift	–

failure modes is as follows: electrical short 53%, electrical open 35%, electrolyte leakage 10%, decrease in capacitance 2%. The failures can be attributed to adverse operating conditions, such as high temperatures, voltage surges, and current spikes. Research has proven that reliability depends on the size of the case and the electrolyte quantity: the smaller the capacitor, the shorter the useful life and the higher its failure rate. The growth of the leakage current over a certain limit serves as failure criterion for this capacitor type, because tan δ and the capacity have only unimportant variations [3].

The main FM, degradation in the oxide layer, can be attributed to material defects that occur because of the periodic heating and cooling during the capacitor duty cycle, as well as stress, cracks, and installation-related damage [16]. Another possible FM is the increase in the internal pressure [17] due to an increased rate of chemical reactions, which can again be attributed to the internal temperature increase in the capacitor. This pressure increase can ultimately lead to the capacitor popping.

During the operation time, the electrolytic capacitors are submitted to a multitude of stresses. To evaluate the quality and reliability, we must consider not only the electrical stresses due to the voltage and current, but also the mechanical and microclimatic influences caused mainly by the temperature and humidity of the air [18,19]. The main factors that influence the reliability are oxide layer, impregnation layer, and foil porosity. For example, at oxide forming, various hydrate modifications can appear. The conductivity of the impregnating electrolyte works directly on the loss factor of impedance, the chemical combinations, and the stability of electrical values.

Capacitors with high stability, reduced dimensions, and reduced corrosion sensitivity, and reduced dissipation factors and impedances may be obtained by using electrolytes with high ionic mobility, even in poor water media. The depositing volume of the electrolyte directly influences the lifetime. Poor encapsulation leads to a rapid modification of the electrical parameters: a diminution of the electrolyte quantity by evaporation or a modification of its consistency leads to growth of the loss factor, diminution of the capacity, and growth of the impedance. Generally, ceramic capacitors are not more reliable than aluminum electrolytic capacitors, which may self-heal. Because high-capacitance ceramic capacitors may develop microcracks, aluminum electrolytic capacitors are preferred for high capacitance values.

5.3.2.2 Tantalum capacitors

A tantalum capacitor is a metal-oxide rectifier used in its blocking direction (that explains its polarization). Ta_2O_5 dielectric is inherently thermodynamically unstable, but stabilization can be accomplished by kinetic means. It is characterized by reduced dimensions, good stability of electric parameters, very good high-frequency properties, long lifetime, volumetric efficiency, less derating, no risk of ignition, and a large temperature domain. Concerning its reliability, field data shows that the failure rate of tantalum capacitors is smaller than 10^{-8}/h, with a confidence level of 90%. Two restrictions must be underlined, however: the small value of the reverse-biased voltage (compared with the nominal voltage) and the reduced reliability of solid electrolyte tantalum capacitors in pulse operation, as occurs in circuits with small impedance, where the overvoltage can lead to blackout failures.

The probability distribution of failure modes is as follows: electrical short 57%, electrical open 32%, parameter drift 11%. In comparison with aluminum electrolytic capacitors, tantalum capacitors have very good reliability, a favorable temperature and frequency behavior, a large temperature range, relatively reduced dimensions, and better robustness.

The factors that influence the reliability are the environmental temperature T_E, the operation voltage U_E, the series resistance R_S, and, for plastic encapsulated capacitors, the air humidity. The devices should be used in applications in which the operational voltage would be no more than half of the rated one. This 50% derating factor ensures reasonable failure rates, especially for the power-on conditions experienced in most applications. Long-term failures are never reported to the component manufacturers, but turn-on failures are a very divisive issue [20]. The popularity of solid tantalum capacitors has not been shared for space application, leading to some restrictions in power supply filtering circuits where high current-handling capability and low impedance are prerequisite. The overvoltage spikes severely degrade the reliability of tantalum capacitors. Researchers have investigated deformations of chip tantalum capacitors during reflow soldering simulation to evaluate the possibility of the popcorning effect. Until now, wear-out processes for tantalum capacitors have not been observed. In most cases, a decreasing time to failure has been noticed.

The specifications generally allow a capacitor variation of −10% between the measured values at 20 and −55 °C, and a variation of +12% between 20 and +85 °C, respectively. The product *CU* (capacity × voltage) is an important parameter of tantalum capacitors, because these are the only capacitors characterized by a very great value for the product *CU* at small volume and reduced price. Its reliability is strongly influenced by the capacity value.

5.3.3 Diodes

The parameters of diodes that indicate a satisfactory operation are forward voltage drop, reverse current, and breakdown voltage. Diodes are mainly used as rectifiers: they allow the electric current to pass in one direction (called the forward-biased condition) and to block it in the opposite direction (the reverse-biased condition). But other applications for diodes have also been developed: switching circuits (switching diodes), oscillators (varactor, IMPATT, and Gunn diodes), protection circuits (transient voltage suppression and avalanche diodes), and so on.

Basically, there are two main families of diodes: silicon diodes and nonsilicon diodes (manufactured on other materials than silicon), with the typical FMs for these diode types shown in Tables 5.2 and 5.3, respectively. In both cases, two main manufacturing procedures are used: planar technology (with the planar-epitaxial variant) and MESA technology. As previously stated, we are not discussing optoelectronic diodes, which could be built on silicon, but also on other materials.

5.3.3.1 *Silicon diodes*

Theoretically, the reverse current of pn junctions is proportional with $1/n$ power of the voltage (where $n = 2\text{-}3$). Experiments have shown that, beyond a certain voltage level (100-200 V), but well before the breakdown region, an exponential increase of current

Table 5.2 Typical failure modes and mechanisms of silicon diodes

Families of diodes	Failure modes	Failure mechanisms
pn Junction diodes	Increased leakage current	Defects located at the interface between semiconductor and dielectric
Varactor diodes	Increased leakage current	High-frequency modulation circuit does not work or the modulation performance is damaged
	Parameter drift	The signal of high-frequency modulation circuit has a distortion
Switching diode	Thermal runaway	Surface component flowing at the junction edge produced by damages at junction periphery
Z diodes	Drift of the reverse current	Contamination near the semiconductor junction, or under, in, or on top of passivation layers
		Lack of hermeticity, allowing moisture to reach the semiconductor chip
	Degradation of chip ohmic contacts (also called Z_z drift)	Dislocations in the space-charge region of pn junction
	Short circuit (breakdown of the junction)	Power dissipation is not optimized
		Migration of the dopant impurities
		Time degradation of pressure contact
Avalanche diodes	Short circuit	Avalanche effect (recovery is possible)

Table 5.3 Typical failure modes and mechanisms of nonsilicon diodes

Families of diodes	Failure modes	Failure mechanisms
Schottky diodes	Parameter drift (leading to failure)	Defects in material Electromigration or contact migration
	Low power dissipation (leading to failure)	Fatigue of the die attach
	Degradation of the reverse current and excess forward current	Damage by ionizing radiation, producing surface effects
	Increase in forward and reverse current	Neutron radiation inducing bulk damage by carrier removal and decrease in bulk lifetime
	Destruction after stroked by high-energy particles during the off-state	Local heating caused by avalanche multiplication of ion-generated carriers
IMPATT diodes	Parameter drift (leading to failure)	Diffusion of the contact metal into the semiconductor material Improper metallization and bonding processes Gold contamination of pn junction area
	Electrical short	Interdiffusion of Au through Pt
Tunnel diodes	Appearance of a parallel conductance	Local thermal degradation at the interface
	Shift of the resonance voltage to higher values	Degradation of ohmic contact
	Parameter drift	Soft breakdown
Gunn diodes	Parameter drift (leading to failure)	Metallization damage Thermal issues
PIN diodes	Avalanche breakdown	Basal plane dislocations Metal contamination

occurs as voltage rises, a phenomenon attributed to a significant leakage current flow at the junction periphery because of defects located at semiconductor-dielectric interface [21]. These defects could be diminished or even avoided by appropriate passivation. Also, the packaging variant determines the reliability of the diode.

Varactor diodes use the capacitance change with the applied bias, being employed as voltage-controlled capacitors, rather than rectifiers. Their main applications are in parametric amplifiers and oscillators, or voltage-controlled oscillators. The diodes have low resistance, a high capacitance-change ratio and high reverse voltage, although the capacitance-voltage requirements differ in each application. The two typical FMs are both produced by misuse during operation [22]: (i) leakage increase occurring if the high-frequency modulation circuit does not work or the modulation performance is damaged and (ii) parameter drift initiated when the signal of the high-frequency modulation circuit has a distortion.

The switching diode is a pn junction diode with gold or platinum introduced as dopants acting as recombination centers and providing the fast recombination of minority carriers, increasing the switching speed. Switching diodes are slower than Schottky diodes, but they exhibit lower current leakage. Compared with standard recovery diodes, the *I-V* reverse characteristics of switching diodes in the temperature range 100-150 °C are unstable due to thermal runaway caused by the surface component flowing at the junction edge [23]. The FA performed on failed items led to the conclusion that damages at the junction periphery were responsible for such instabilities.

Z diodes operate in the breakdown zone, being used in electrical circuits, in conjunction with switching diodes, to balance the temperature coefficient to near zero. For Z diodes, the most frequent FMs are:

- Drift of the reverse current (I_R) arising in the following circumstances: (i) *After extended operation*, because of contamination near the semiconductor junction, or under, in, or on top of passivation layers. This FM could be eliminated by using a silicon nitride layer on planar junctions as a barrier against ionic contamination with lithium, sodium, and potassium. (ii) *After a damp heat test*, if there is a lack of hermeticity allowing moisture to reach the semiconductor chip, where possible misalignment in very small insulation paths may favor the development of high leakage currents.
- Degradation of chip ohmic contacts (also called Z_z drift), which could be initiated by the existing dislocations in the space-charge region of the pn junction. A $1/f$ noise screening could be used for identifying the possible failure risk [24].
- Short circuit (breakdown of the junction) mainly arising when the power dissipation is not optimized and excessive heat is produced in junction area, leading to thermal runaway. Other possible causes of short circuits are (i) migration of the dopant impurities [3] and (ii) time degradation of the pressure contact, if mounted by a spring. In the second case, the failure tends to occur when the diode is subjected to variations in ambient temperature or in loading; chip cracks arise, and the short circuit is caused by a loose fragment.

Avalanche diodes are designed to conduct in the reverse direction when the reverse-biased voltage exceeds the breakdown voltage, being almost similar to Z diodes (but the temperature coefficients are of opposite polarity). These diodes break down due to a different FM, which is the avalanche effect, without being totally destroyed.

5.3.3.2 Nonsilicon diodes

Schottky diodes are built using a metal-semiconductor contact (a so-called Schottky barrier, which is a potential barrier with rectifying properties) with a lower forward voltage drop than pn junction diodes (0.15-0.45 V, compared to 0.5-0.7 V) and a much shorter switching time (100 ps, compared to 100 ns), basically because the junction capacitance is lower than for pn diodes. Due to these characteristics, Schottky diodes are used in power-supply switching operations and in RF devices (mixers, detectors, etc.). Since 2001, silicon carbide (SiC) is almost unanimously used as the basic material for the diode because the reverse leakage current is 40 times lower than it is for silicon Schottky diodes, making it an excellent choice for any high-power application requiring fast switching and near-zero reverse recovery losses [25]. Reliability studies published on SiC Schottky diodes indicate lifetimes higher than 50 years [26]. The typical FMs for Schottky diodes are:

- Degradation phenomena can be produced by *defects in material*: higher leakage currents, reduced critical electric field in devices, and degradation in on-state performance (time increasing when they are kept in the forward-biased mode for a long time period) [27].
- *Electromigration* and *contact migration* are another form of aluminum migration, with the physical process being different from that of electromigration, and these FMs could be diminished by appropriate design specifications.
- *Fatigue of the die attach*, which is identified by testing at high temperature combined with repeated power cycling, induces low power dissipation, eventually leading to failure.
- *Wear-out mechanisms*, which often arise when the devices are used in the output of power supplies, normally lead to failures due to a single reverse current effect.
- *Degradation by cosmic radiation* is relevant for Schottky diodes because they are currently used in space applications. With a relatively low sensitivity to a radiation environment (due to radiation hardness intrinsic to the SiC material), Schottky diodes could be damaged by ionizing radiation (e.g., low-energy electrons, gamma rays, and X-rays), which produces surface effects, with a build-up of a positive space charge in the oxide and an increase in the surface velocity responsible for reduced reverse current and excess forward current. Also, the neutron radiation is bulk damaging through carrier removal and decreasing the bulk lifetime. The series resistance of the diode increases because of the carrier removal, and the decrease in bulk lifetime results in an increase in forward and reverse current. Accelerated tests have indicated that cosmic radiation affects the reliability of SiC power devices, as is the case for the silicon counterpart, but the problem can be contained very effectively by device design [28]. Schottky diodes may be destroyed by high-energy particles even during the off-state, however. This FM is produced by local heating caused by avalanche multiplication of ion-generated carriers [29].

The *IMPATT* (*impact ionization avalanche transit-time*) *diode* is a pn junction diode biased beyond the avalanche breakdown voltage, with a depletion region adjacent to the junction allowing the drift of electrons and holes. The basic material in this diode is SiC, which allows for a high breakdown field, but Si, GaAs, and InSb could be used, too. A combination of impact avalanche breakdown and charge-carrier transit-time effects produces the negative characteristic. IMPATT diodes are used in high-frequency electronics and as microwave generators (3-100 GHz, or more), and they are suitable for high values of junction temperature and reverse biases higher than those for other semiconductor devices. The main failure risks of IMPATT diodes are linked to:

- *Diffusion of the contact metal into the semiconductor material*, which is an electrochemical process, described by the Arrhenius equation (with an activation energy of 1.8 eV). This diffusion could be eliminated by the methods described by Băjenescu and Bâzu [3]: (i) proper choice of metals used in the contacting system, (ii) careful control while applying the metals, and (iii) appropriate control of junction temperature.
- *Interdiffusion of Au through Pt* (major degradation mechanism of GaAs IMPATT diodes), by forming metallic spikes that extend into the GaAs, leading to a junction short circuit in the bulk or at the metal-GaAs interface.

- *Improper metallization and bonding processes*, which are generally responsible for a high percentage of semiconductor failures. The weak items could be eliminated, however, by screening tests (thermal resistance and high-temperature reverse bias).
- *Gold contamination* of the pn junction area, observable at operating temperatures of the junction between 230 and +50 °C [30].

Tunnel diodes (mainly made with Ge, but also with Si, GaAs, InP, and other compounds) have a heavily doped pn junction only around 10-nm wide, and they work based on electron tunneling effect, with a negative resistance region of operation caused by quantum tunneling. They are capable of very fast operation into the microwave frequency domain. In reliability studies, transmission line pulses (TLP) were used to degrade the resonant tunneling diodes, with interesting results: the appearance of a parallel conductance was attributed to local thermal degradation at the interface, and the shift of the resonance voltage to higher values was produced by ohmic contact degradation [31]. During constant stress tests, investigators noticed a gradual current change superimposed on the soft-breakdown-related steps, as well as an abrupt decrease in current (predominately under high bias), which was considered to be a consequence of the same soft breakdown event [32].

Gunn diodes (made by GaAs, InP, or GaN) are similar to tunnel diodes, but they have a region of negative differential resistance produced by electrons traversing from a high mobility to a lower-mobility valley and thus producing a lower net electron velocity. They are used in high-frequency microwave oscillators (due to the negative resistance property of bulk semiconductor), with a relatively low efficiency (2-5%), but considerable power is dissipated. Consequently, adequate heat-sinking is needed, and *thermal issues* are the major failure risk. Also, *metallization damage* could lead to failures. In an experiment performed on Gunn diodes, which were submitted to a biased life test at elevated temperature (70 °C) for 8.5×10^5 device hours, researchers concluded that the deterioration of the metallization is the main FM [33].

PIN diode (made by GaAs or SiC) has a wide, lightly doped region sandwiched between p-type and n-type semiconductor regions, both of which are heavily doped in order to be used as ohmic contacts. PIN diodes operate as variable resistors at RF and microwave frequencies, being used as RF switches, attenuators, large-volume ionizing radiation detectors, photodetectors, or components in power electronics, because their central layer can withstand high voltages. The typical FM is avalanche breakdown. In an experiment reported by Ohyanagi et al. [34], the breakdown failure points were analyzed by using the electron-beam-induced current mode (EBIC) of a scanning electron microscope and emission microscopy. Researchers observed basal plane dislocations around the failure point and at temperatures below 200 K, and the dark spots were identified using EBIC. Because the X-ray topography image showed no spots around dislocations, the authors concluded that these spots originated from metal contamination: the electric field was multiplied due to a permittivity change, and this multiplication caused the avalanche breakdown.

5.3.4 Transistors

5.3.4.1 Silicon transistors

The *bipolar junction transistor* (*BJT*) operates by using both electrons and holes (bipolar). There are two possible BJTs, PNP and NPN, depending of the doping types of the three main terminal regions and four fundamental parameters (breakdown voltage, current gain, switching speed, and dissipated power), but other parameters could be significant for some specific applications. The failure risks typical for BJTs and MOS transistors (presented in Table 5.4) can arise because of design flaws, poor quality materials, manufacturing problems, improper conditions during transport or storage, and overstress during operation, and these risks may occur under the following conditions: (i) At the *wafer level* due to crystallographic defects, surface contamination, corrosion, microcracks, electromigration, hot carrier effects, diffusion defects, or insulating oxide breakdown; (ii) At the *packaging level* due to a bad solder joint, chip-mounting errors, use of improper materials for the contact area and connection wire,

Table 5.4 Typical failure modes and mechanisms of silicon transistors

Families of transistors	Failure modes	Failure mechanisms
Bipolar junction transistors	Drift of electrical parameters	Surface contamination with impurities Crystallographic defects, which produce a conduction mechanism based on the Poole-Frenkel effect
	Electrical short	Hot spot phenomena Electromigration Diffusion defects Second breakdown
	Electrical open	Bad solder joint Electromigration Melted conductor as a consequence of excessive current
	Electrical intermittent open	Bad solder joint produced by thermal fatigue
MOS transistors	Increase of threshold voltage together with decrease in drain current and transconductance	Negative bias temperature instability (NBTI): generation of interface traps, which are unsaturated Si dangling bonds
	Parameter drift	Soft errors produced by ionization effect of low energy particles
	Single-event burnout (SEB)	Heavy ions traverse the transistor structure, initiating forward-biased second breakdown
	Electrostatic discharge (ESD)	Explained by the charged device model (CDM)

imperfect sealing, and the presence of contaminants and moisture in the case. In describing the specific FMs of BJTs, one must start with the failure modes, as below:

- *Drift of electrical parameters*: Surface impurities produce an increase in leakage current I_{CBO} (generally, without recovering; complementary advanced techniques [35] identified the origin: structure defects producing a conduction mechanism based on the Poole-Frenkel effect) often accompanied by a decrease of the current gain h_{FE} [36].
- *Short circuit*: This failure may be produced by hot spot phenomena, chip problems, or a second breakdown (SB). Known as an irreversible phenomenon, SB can induce device failure because the current concentrates in a single area of the base-emitter junction, causing local heating and destruction of the transistor [37]. Notably, the onset of SB must be characterized in terms of the energy dissipated in the transistor, and furthermore, the energy threshold (or delay time) must depend on other factors, such as ambient temperature, the biasing base current of the transistor, and the involvement of a current-controlled negative resistance region [38]. Failure after SB can be avoided if appropriate measures are taken, including diversion or "crowbarring" of the collector current [39].
- *Open circuit*: This failure might result from a bad solder joint or melted conductor as a consequence of excessive current.
- *Intermittent open circuit*: This failure is a sign of bad solder joint and can be produced by thermal fatigue (TF), especially when occurring at high temperature or after thermal cycling. The mechanism for TF initiation is as follows: the thermal expansion of various materials used in transistor assembling (e.g., Cu, Al, steel, etc.) differs from the expansion of silicon, producing mechanical stresses that lead to cracks that limit the heat dissipation and modify the thermal resistance (by 25% or more). Crack propagation seems to be accelerated by the presence of voids in the joint region.

MOS transistors are unipolar devices because only one kind of charge carrier is involved in their functioning (the majority carriers). MOS transistors are affected by the same basic FMs that affect the bipolar ones, but three specific FMs also affect MOS transistors (see Table 5.5 for a synthesis). The first two are discussed in the following list, and the third was treated in Table 5.4.

- *Radiation-induced FMs*: (i) Soft errors. Charged particles without enough energy to cause lattice effects induce ionization effects that are generally transient, creating soft errors, but, in conjunction with other damage mechanisms, these errors may lead to device destruction. (ii) Single-event burnout (SEB) is caused by the heavy ions that traverse the transistor structure initiating forward-biased SB [40]. For avionics systems at 40,000 ft, the SEB failure rate for 400-V MOS devices is 9×10^{-5} failures per device-day, and the rate is 2×10^{-3} failures per device-day for 500-V devices [41]. (iii) Single-event gate rupture occurs when a heavy ion traverses the transistor through the gate, eventually damaging the transistor via increased gate leakage or rupturing the gate oxide insulation [42].
- *Negative bias temperature instability* (NBTI) is the generation of interface traps formed by unsaturated Si dangling bonds, producing an increase in the threshold voltage, while decreasing drain current and transconductance [3]. For today's deep submicron technologies, NBTI is exacerbated by the very small distances involved. Until recently, several aspects of NBTI were poorly understood [43]. Now the interface-state generation seems negligible when compared to the oxide positive charge formation at the device's operational conditions. Experiments on NBTI degradation of *p*-MOSFETs with an ultrathin SiON gate oxide identified two NBTI components: a slow component due to interface trap degradation and a fast component resulting from the trapping or detrapping of the oxide charge [44].

Table 5.5 **Typical failure modes and mechanisms of nonsilicon transistors**

Families of transistors	Failure modes	Failure mechanisms
Heterojunction bipolar transistor (HBT)	Collector current change and current gain degradation Sudden and catastrophic DC current gain degradation Breakdown-related instabilities	Hot carriers are initiating stress-induced damage mechanisms Recombination-enhanced defect reaction (REDR) process Technology generation Device geometry Bias configuration Current density
High electron mobility transistors (HEMT)	Drift of drain current Decrease of DC current and RF output power over time	Hot electrons Gate tunneling, determined by the magnitude of electric field at the gate edge

5.3.4.2 Nonsilicon transistors

Heterojunction bipolar transistors (HBT) are formed using different semiconductor materials for the emitter and base regions, thus creating a heterojunction. Used in wireless communication, power amplifiers, mixers, and frequency synthesizer applications, they extend the advantages of silicon bipolar transistors to significantly higher frequencies [45]. The typical FMs (shown in Table 5.5) are described in the following list.

- Hot carriers are responsible for stress-induced damage mechanisms, identified for pnp SiGe HBTs [46], inducing collector current change and current gain degradation [45].
- Sudden and catastrophic DC current gain degradation has been observed in GaAs HTBs, produced by recombination-enhanced defect reaction processes [47].
- Breakdown-related instabilities were found to be related to technology generation, device geometry, bias configuration, and current density [48].

Notably, studies have reported excellent performances for HBT: (i) SiGe power HBTs at RF (6 GHz) have shown good power characteristics at cryogenic (77 K) and high-operation temperatures (junction temperature up to 160 °C) [49] and (ii) InGaP/GaAs HBTs have shown zero device failures after over 3600 h of stress (28-V bias voltage, 5.2-kA/cm^2 current density, and 310 °C junction temperature) [50].

High electron mobility transistors (HEMTs) are field effect transistors with channels formed by a heterojunction between two materials (e.g., AlGaN and GaN), without using dopant impurities for generating mobile electrons, as generally happens for a MOSFET. Consequently, the electrons are not slowed down through collisions with the dopant impurities, and high-frequency or high-power performances are improved. Given that HEMTs are relatively new commercial devices, their reliability is still under discussion. However, the basic failure risks seem to be:

- Hot carrier degradation of the drain current [51], confirmed for AlGaN/GaN HEMTs [52].
- Impact-ionization-dominated reverse gate current [51].

- Decrease of DC current and RF output power over time, which is attributed to gate tunneling (determined by the magnitude of electric field at the gate edge), as discovered for devices fabricated from nitride semiconductors utilizing the AlGaN/GaN [53].

Vitusevich et al. [54] reported an interesting result: small doses of gamma irradiation seem to improve the transport properties of HEMTs, a phenomenon that is irreversible over time.

5.4 Effect of electrostatic discharge on discrete electronic components

5.4.1 Electrostatic discharge (ESD)

ESD is a sudden and unwanted electric current between two materials at different electrical potentials, such as parts of an electronic component, which may have disastrous effects on one or both materials. Generally, ESD events involve high voltage (several kV) and high current stress (1-10 A) on small electric devices. Despite the fact that ESD events are of very short duration (0.2-200 ns), the massive current or voltage pulses can do fatal damage to ICs [55].

Studies have reported that between 25% and 75% of all field returns are due to damage caused by ESD or electrical overstress [56]. Process advances have often had the side effect of degrading the performance of the ESD-protection devices. PCBs contain active and passive components particularly sensitive to damage from static electricity. Static electricity and its companion ESD damage electronic circuits every day. To improve the reliability of products, to ensure their performance, and to reduce maintenance costs, it is imperative to provide system health awareness for digital electronics.

5.4.2 ESD-induced failures

In assessing the threat of ESD-induced failure, the engineer must consider not only the amplitude of the electrostatic voltage, but also: (i) the rise time, duration, and peak current of the discharge; (ii) the source impedance of the discharging body; and (iii) the circuit characteristics and whether the charge enters the circuit via conduction or radiation. The heat associated with an ESD is the major threat to semiconductor devices, especially for high-density RAM or ultra-compact microprocessors. Even relatively low-voltage ESDs within a semiconductor device can produce sufficient heat to vaporize semiconductor material and the metal or polycrystalline silicon traces connecting them. ESD-related failures are not always total or catastrophic, but they can be partial or intermittent. Postmortem studies examining failures via electron microscopy or liquid-crystal thermography have little practical value. Much more important is how to prevent damage from occurring in the first place.

5.4.3 ESD robust systems

ESD damage can be avoided by eliminating potentially dangerous electrostatic buildups, preventing the coupling of ESD-induced energy into sensitive circuits and components, increasing noise immunity through the use of clamping elements between critical points and ground, and judiciously placing traces on PCBs that may act as antennas for ESD-generated fields. To understand and characterize how devices are damaged, researchers often use one of three models: the machine model, the human body model, and the charged device model (CDM). Pierce [57] concluded that 99.9% of ESD damage originates from the CDM. Improved ESD testing methods must include TLP testing [58] not only at the wafer level but also after the device has been packaged for qualification processes such as burn-in.

For systems that integrate advanced technology IC components, designing ESD robust systems can be very challenging. There are three specific ESD issues [59]: (i) test specification requirements for ESD for the system versus IC providers; (ii) understanding of ESD failures (physical failure, system upset) and their causes based on the system- and IC-level constraints; and (iii) fault responsibility between system designers and IC providers with regard to the proper system ESD design level. The great majority of ESD failures are catastrophic. The major physical reason for catastrophic failures is an electrical breakdown, and various thermoelectrical current instabilities in devices also contribute to such failures [60]. Experience has demonstrated that bipolar transistors are resistant or very resistant to ESD, but MOSFETs are inherently sensitive. If the gate rupture voltage is exceeded, catastrophic failure is the consequence. A correct assembly process minimizes the potential for failure, but it always remains a risk.

5.5 Conclusions

In this chapter, we identify and describe the reliability-building issues for electronic components, including DfR, process reliability, CE, screening, and burn-in. DfR is a valuable process for lowering cost, reducing time-to-market, and improving customer satisfaction, a major tool for reliability building currently used for designing electronic components. Here, we analyze the reliability of five families of discrete electronic components, and we identify the typical FMs and corresponding corrective actions. Notably, ensuring the reliability of electronic components has become increasingly difficult because of the industry trend toward scaling down device dimensions coupled with increasing complexity and power requirements and the introduction of new materials and technologies. In order to better answer these challenges, reliability studies increasingly focus on the PoF, which is also known as reliability physics. The root causes of failure (fatigue, fracture, wear, corrosion, etc.) can be modeled with the aid of PoF, which has become a powerful tool for leveraging the value of DfR activities. In the future, PoF based on prognostics and health management seems to be a cost-effective means of predicting the reliability of electronic products and systems because it can help identify the most critical failure under the real application conditions for the products [61].

References

[1] M. Bâzu, T. Băjenescu, Failure Analysis. A Practical Guide for Manufacturers of Electronic Components and Systems, John Wiley & Sons, Chichester, 2011.

[2] Anon. Design for Reliability and Quality, Available from: http://nptel.iitm.ac.in/courses/112101005/downloads/Module_5_Lecture_2_final.pdf, 2013 (accessed 20.05.13).

[3] T. Băjenescu, M. Bâzu, Component Reliability for Electronic Systems, Artech House, Boston and London, 2010.
[4] M. Bâzu, A combined fuzzy logic and physics-of-failure approach to reliability prediction, IEEE Trans. Reliab. 44 (2) (1995) 237–242.
[5] R. Batson, M. Elam, Robust Design: An Experiment-Based Approach to Design for Reliability, Available from: http://ie.eng.ua.edu/research/MRC/Elam-robustdesign.pdf, 2013 (accessed 20.05.13).
[6] M. Kalsi, et al., A comprehensive robust design approach for decision trade-offs in complex systems design, ASME J. Mech. Des. 123 (1) (2001) 1–10.
[7] J. Su, J.E. Renaud, Automatic differentiation in robust optimization, AIAA J. 35 (6) (1997) 1072–1079.
[8] X. Du, A. Sudjianto, W. Chen, An integrated framework for optimization under uncertainty using inverse reliability strategy, ASME J. Mech. Des. 126 (4) (2004) 562–570.
[9] B.D. Youn, K.K. Choi, K. Yi, Performance moment integration (PMI) method for quality assessment in reliability-based robust optimization, Mech. Based Des. Struct. Mach. 33 (2) (2005) 185–213.
[10] I. Lee, K.K. Choi, L. Du, D. Gorsich, Dimension reduction method for reliability-based robust design optimization, Comput. Struct. 86 (13-14) (2008) 1550–1562.
[11] C. Hillman, True design for reliability (DfR): understanding what is and what is not DfR, in: SMTA Lead-Free Academy, Toronto, May 17, 2010.
[12] O.P. Yadav, Sunil S. Bhamare, A framework for capturing degradation behavior in reliability-based robust design optimization. Int. J. Reliab. Qual. Saf. Eng. 18 (6) (2011) 531, http://dx.doi.org/10.1142/S0218539311004238.
[13] P.H. Barringer, Process Reliability and Six Sigma, Available from: http://www.plant-maintenance.com/articles/Process_Reliability_and_Six-Sigma.pdf, 2013 (accessed 20.05.13).
[14] F. Jensen, Electronic Components Reliability, Wiley, Chichester, 1995.
[15] J. Robineau, M. Lazartigues, W. Binroth, Reliability approach in automotive electronics, in: Proceedings of the ESREF, 1992, pp. 133–140.
[16] S. Ikonopisov, et al., Theory of electrical breakdown during formation of barrier anodic films, Electrochim. Acta 22 (10) (1977) 1077–1082.
[17] C. Gomez-Aleixandre, J.M. Albella, J.M. Martnez-duart, Pressure build-up in aluminum electrolytic capacitors under stressed voltage conditions, J. Appl. Electrochem. 16 (1) (1986) 109–115.
[18] R. Mäder, K.-D. Meyer, Zuverlässigkeit diskreter passiver Bauelemente, in: H.G. Schneider (Ed.), Zuverlässigkeit elektronischer Bauelemente, VEB Deutscher Verlag für Grundstoffindustrie, Leipzig, 1974, pp. 400–401.
[19] T. Kormàny, H. Barna, Wege zur Beurteilung der natürlichen Lebensdauer von Elektrolytkondensatoren, Nachrichtentechnik 12 (10) (1984) 391–392.
[20] J. Prymak, Derating differences Ta/Ta-polymer/Al-polymer, in: CARTS 2003: 23rd Capacitor and Resistor Technology Symposium, San Antonio, TX, 31 March-3 April, 2003.
[21] V.V.N. Obreja, The voltage dependence of reverse current of semiconductor pn junctions and its distribution over the device area, in: IEEE International Semiconductor Conference (CAS), 2007, pp. 485–488.
[22] J. Goho, GVD30601-001. Available from: http://www.gideonlabs.com/posts/varactor-diode-gvd30601-001/ (accesed 20.05.13).
[23] V.V.N. Obreja, et al., Reverse leakage current instability of power fast switching diodes operating at high junction temperature, in: 36th IEEE Power Electronics Specialists Conference (PESC '05), June 2005, pp. 537–540.

[24] Y. Zhuang, L. Du, 1/f noise as a reliability indicator for subsurface Zener diodes, Microelectron. Reliab. 42 (3) (2002) 355–360.
[25] J. Henning, A. Ward, P. Kierstead, The new standard for high power semiconductors, Power Electron. Eur. 8 (2008) 18–19.
[26] G. Fischer, Derating of Schottky diodes, White Paper, DfR Solutions, Available from: http://www.emsnow.com/cnt/files/White%20Papers/DFRschottky%20diodes.pdf (accessed 20.05.13).
[27] R. Singh, A.R. Hefner, T.R. McNutt, Reliability concerns in contemporary SiC power devices, in: Proceedings of the International Semiconductor Device Research Symposium, 2003, pp. 368–369.
[28] G. Soelkner, et al., Reliability of SiC power devices against cosmic radiation induced failures, Mater. Sci. Forum 556-557 (2007) 851–856.
[29] A.M. Albadri, et al., Coupled electro-thermal simulation of single event burnout in power diodes, IEEE Trans. Nucl. Sci. 52 (6) (2005) 2194–2199.
[30] P. Staecker, Ka-band IMPATT diode reliability, in: International Electron Devices Meeting, December 4-5, 1973, Washington, DC, vol. 19, 1973, pp. 493–496.
[31] A. Vogt, M. Brandt, A. Sigurdardottir, M. Schussler, D. Pena, A. Simon, H.L. Hartnagel, M. Rodewald, M. Roesner, H. Fuess, S.N.N. Goswami, K. Lal, Characteristics of degradation mechanisms in resonant tunnelling diodes, Microelectron. Reliab. 37 (10) (1997) 1691–1694.
[32] R. Khlil, A. El Hdiy, A.F. Shulekin, S.E. Tyaginov, M.I. Vexler, Soft breakdown of MOS tunnel diodes with a spatially non-uniform oxide thickness, Microelectron. Reliab. 44 (3) (2004) 543–546.
[33] I. Mojzes, et al., Comparative reliability study of $n^{+}-n$ and $n^{+}-n-n^{+}$ Gunn diodes, Microelectron. Reliab. 29 (2) (1989) 131–132.
[34] T. Ohyanagi, Chen Bin, Takashi Sekiguchi, Hirotaka Yamaguchi, Hirofumi Matsuhata, EBIC analysis of breakdown failure point in 4H-SiC PiN diodes, Mater. Sci. Forum 615-617 (2009) 707–710.
[35] B. Domengès, H. Murray, P. Schwindenhammer, G. Imbert, Comprehensive failure analysis of leakage faults in bipolar transistors, Semicond. Sci. Technol. 19 (2) (2004) 191.
[36] M. Goroll, R. Pufall, S. Aresu, W. Gustin, New aspects for lifetime prediction of bipolar transistors in automotive power wafer technologies by using a power law fitting procedure, Microelectron. Reliab. 48 (8-9) (2008) 1509–1512.
[37] H.A. Schafft, J.C. French, Second breakdown in transistors, IRE Trans. Electron Dev. ED-9 (1962) 129–136.
[38] M.E. Snyder, Determination of the Unstable States of the Solid State Plasma in Semiconductor Devices, Ph.D. Dissertation, Texas Tech University, May 1988.
[39] S. Menhart, Development of a second breakdown model for bipolar transistors (Ph.D. thesis), Texas Tech University, December 1988.
[40] R. Koga, S.H. Penzin, K.B. Crawford, W.R. Crain, Single event burnout sensitivity of embedded field effect transistors, IEEE Trans. Nucl. Sci. 46 (6) (1999) 1395–1402.
[41] R. Koga, Single-event effect ground test issues, IEEE Trans. Nucl. Sci. 43 (2 (Part 1)) (1996) 661–670.
[42] D.G. Walker, T.S. Fisher, J. Liu, R.D. Schrimpf, Thermal characterization of single event burnout failure in semiconductor power devices, in: 16th Annual IEEE Semiconductor Thermal Measurement and Management Symposium, March 21-23, 2000, pp. 213–219.
[43] V. Huard, M. Denais, C. Parthasarathy, NBTI degradation: from physical mechanisms to modeling, Microelectron. Reliab. 46 (1) (2006) 1–23.

[44] M.-F. Li, D.M. Huang, C. Shen, T. Yang, W.J. Liu, Z.Y. Liu, Understand NBTI mechanism by developing novel measurement technique, IEEE Trans. Device Mater. Reliab. 8 (1) (2008) 62–71.

[45] H. Livingstone, A Survey of Heterojunction Bipolar Transistor (HBT) Device Reliability, JPL Publication 96-25, California Institute of Technology, Pasadena, CA, 2003.

[46] P.S. Chakraborty, et al., Mixed-mode stress degradation mechanisms in pnp SiGe HBTs, in: Proceedings of the IEEE International Reliability Physics Symposium, 26-30 April, Montreal, Canada, 2009, pp. 83–88.

[47] R.E. Welser, P.M. DeLuca, Exploring physical mechanisms for sudden beta degradation in GaAs-based HBTs, in: Proceedings of GaAs Reliability Workshop, 2001, pp. 135–157.

[48] C.M. Grens, et al., Reliability issues associated with operating voltage constraints in advanced SiGe HBTs, in: Proceedings of the 43rd Annual IEEE International Reliability Physics Symposium, 17-21 April, San Jose, CA, 2005, pp. 409–414.

[49] W. Guogong, Guoxuan Qin, Zhenqiang Ma, Pingxi Ma, M. Racanelli, G.E. Ponchak, Power performance characteristics of SiGe power HBTs at extreme temperatures, in: Proceedings of the 45th Annual IEEE International Reliability Physics Symposium, 15-19 April, Phoenix, AZ, 2007, pp. 584–585.

[50] H.-F. Chau, B.J. Lin, Y. Chen, M. Kretschmar, C. Lee, N.L. Wang, X. Sun, W. Ma, S. Xu, P. Hu, Reliability study of InGaP/GaAs HBT for 28 V operation, in: 28th IEEE Compound Semiconductor IC (CSIC) Symposium, November 12-15, San Antonio, TX, 2006.

[51] A. Mazzanti, Physical investigation of trap-related effects in power HFETs and their reliability implications, IEEE Trans. Device Mater. Reliab. 23 (2002) 65–71.

[52] K. Matsushita, S. Teramoto, H. Sakurai, Y. Takada, J. Shim, H. Kawasaki, K. Tsuda, K. Takagi, Reliability study of AlGaN/GaN HEMTs device, in: Proceedings of the CS MANTECH Conference, 14-17 May, Austin, TX, 2007, pp. 87–89.

[53] R.J. Trew, Y. Liu, W.W. Kuang, G.L. Bilbro, The physics of reliability for high voltage AlGaN/GaN HFETapos's, in: Proceedings of the IEEE Compound Semiconductor Integrated Circuit Symposium (CSIC), November 2006, pp. 103–106.

[54] S.A. Vitusevich, Andrey M. Kurakin, Norbert Klein, Mykhailo V. Petrychuk, Andrey V. Naumov, Alexander E. Belyaev, AlGaN/GaN high electron mobility transistor structures: self-heating effect and performance degradation, IEEE Trans. Device Mater. Reliab. 8 (3) (2008) 543–548.

[55] J.-H. Chun, ESD protection circuits for advanced CMOS technologies (Ph.D. thesis), Stanford University, 2006.

[56] S.T. Ward, Electrostatic discharge (ESD) protection in CMOS (MS thesis), University of Idaho, 2002.

[57] R.J. Pierce, The most common causes of ESD damage, Evaluat. Eng. (2002). Available from: www2.evaluationengineering.com/archives/ArchiveWrapper.aspx?file=1102esd.htm (accessed 20.05.13).

[58] J. Barth, J.E. Barth, Koen Verhaege, Leo G. Henry, John Richner, TLP calibration, correlation, standards and new techniques, IEEE Trans. Electron. Packag. Manuf. 24 (2) (2001) 99–108.

[59] T. Welsher, C. Brett (Eds.), Industry Council on ESD Target Levels: 'White Paper 3', System Level ESD. Part II: Implementation of Effective Robust Designs, 2012. http://www.esda.org/documents/IndustryCouncilWhitePaper3_PII.pdf (accessed 20.05.13).

[60] V.A. Vashchenko, V.F. Sinkevitch, Physical Limitations of Semiconductor Devices, Springer, Berlin, 2008.

[61] M. Pecht, J. Gu, Physics-of-failure-based prognostics for electronic products, Trans. Inst. Meas. Control 31 (3-4) (2009) 309–322.

Reliability of optoelectronics

6

J.-S. Huang
Emcore Corporation, Alhambra, CA, USA

6.1 Introduction

Optoelectronic devices typically use direct bandgap materials such as III-V, with the exception being the recently introduced Si photonics [1,2]. The commercial level of maturity for optoelectronics is largely limited to single devices such as laser diodes (LD) and light emitting diodes (LED). Occasionally, industry has used widely separated arrays or very small-scale monolithic integrations, such as MSM photodiodes (PDs) with Si transimpedance amplifiers. Table 6.1 compares the key differences between optoelectronics and electronics. Electronic devices can be fabricated as single devices but also as integrated circuits (ICs) consisting of millions of transistors and interconnects, and the latter application accounts for most mainstream electronic devices, producing a technology roadmap that follows Moore's law, so that transistor count generally doubles every 18 months [3]. For optoelectronics, device development typically does not follow any universal law. For example, the LD market is often driven by customer demand. On the other hand, LED development sometimes follows Haitz's law, which states that light efficiency and output roughly double every 36 months [4]. Given the significant differences in the materials, device structures, feature sizes, and complexity of electronic and optoelectronic devices, the reliability physics of the devices should be different as well.

Interestingly, reliability studies of electronic devices significantly outnumber those of photonic devices. The scarcity of reliability studies for the latter likely results from the shorter history and smaller number of applications for optoelectronic devices. The transistor was invented in 1947. Three Bell Laboratories scientists, William Shockley, John Bardeen, and Walter Brattain, demonstrated the first transistor based on a point-contact germanium (Ge) device [5]. On the other hand, the semiconductor laser was discovered 15 years later in 1962. Robert Hall and colleagues at General Electric (GE) and Marshall Nathan at IBM's T.J. Watson Research Center demonstrated the first semiconductor injection laser based on GaAs [6,7].

The reliabilities of electronic devices in terms of electromigration (EM), oxide breakdown, hot carrier, and plasma damage have been well documented in the past several decades [8–13]. This chapter focuses on the reliability of optoelectronic devices. In the following sections, we review current findings on optoelectronics, discussing recent developments, reliability modeling, and failure mechanisms, as well as electrostatic discharge (ESD). The reliability of LD provides case studies and examples.

Reliability Characterisation of Electrical and Electronic Systems. http://dx.doi.org/10.1016/B978-1-78242-221-1.00006-X

Table 6.1 Comparison of optoelectronic and electronic devices in terms of device characteristics, materials, and structures

	Optoelectronic	Electronic
Light emission/ absorption	Yes	No
Material	Varieties of material platforms (>10): • *LD*: InGaN, ALGaInP, GaAlAs, InGaAs, InGaAsP, GaInAsSb, etc. • *LED*: AlN, InGaN, ZnSe, InGaN, GaP, InGaN, GaP, AlGaAs, GaAs, YAG, etc.	Si-based material: Si, SiGe
Device structure	Single device usually	Integrated circuits, system-on-a-chip consisting of millions of transistors and interconnects

After Ref. [2].

6.2 Overview of optoelectronics reliability

Before we discuss optoelectronics reliability in depth, we should compare it with the reliability of electronics. The two technologies demonstrate varying levels of similarity. However, optoelectronic devices are also associated with unique failures. Failures related to buried heterostructure (BH) interface, crystal quality, metal diffusion, and optical coating have been also observed in emitters, and studies have documented failures related to interface/surface, dielectrics, and diffusion in receivers [14–17].

Table 6.2 summarizes the major reliability differences between electronic and photonic devices. When studying both technologies, researchers use accelerated aging

Table 6.2 Comparison of photonic and electronic devices in terms of reliability aspects

	Photonic	Electronic
Aging acceleration factor	Current, temperature, and photons	Current and temperature
Measurement parameter of reliability test	• Optical power (LED, LD) • Threshold current (LED, LD) • Breakdown voltage (PD)	• Resistance (for interconnect) • Breakdown voltage (for gate oxide) • Substrate current (for hot carrier)
Test structure	Real device	Dedicated test structure
Reliability failure time	Extrapolation involved	Experimentally determined

After Ref. [2].

in order to accumulate data in a reasonable time frame. Device aging is often performed by accelerated life testing (ALT) during which sample devices are stressed by means of elevated current and temperature. One obvious difference between ALT for the two types of semiconductor devices is the measurement of optical power that is often required for photonic devices such as LDs. Although *in situ* optical power measurement during aging provides a valuable measure of the health of the device under test conditions, it can also significantly increase the cost of testing. Some of this cost may be mitigated by a simplified wafer-level test design in the case of vertical cavity surface emitting lasers [18,19]. However, *in situ* optical power measurements at the wafer level are not possible with edge emitting lasers. The high cost of testing has been an impediment to large-population ALT studies of edge emitting lasers.

Other differences in measurement parameters for photonic and electronic devices depend upon the specific nature of the devices in question. For example, testers use a plot to monitor an LD threshold current extracted from a light versus forward drive current (LI) during life-test aging, but they monitor breakdown voltage for PDs. Examples of parameters monitored during the life testing of electronic devices include forward-voltage drop or differential resistance changes for the purpose of studying EM effects on semiconductor-to-metal interconnects, breakdown voltage of transistor gate oxides, and substrate current for gauging hot carrier reliability effects.

Test structures for reliability stresses are also different for optical devices and electronic ICs. For the ICs, standardized test structures are commonly used to evaluate the reliability performance of specific components. For example, the EM behaviors of metal lines and vias are often studied by using lines connected to bond pads [9] and lines terminated by vias [20–23]. Specialized test structures allow the tester to isolate one variable from the others, however. For optical devices, dedicated test structures are usually not available, so complete functional devices are stressed for reliability assessment. Although the reliability data obtained can be more relevant to the actual application of the device, the root cause of reliability failures is often more difficult to identify.

Extrapolation based on mathematical fitting to the aging curves may be needed for photonic devices because, in some cases, failures cannot be induced using ALT techniques in a reasonable time frame. In contrast, time to failure is often readily determined by ALT experiments on electronic devices. The typical experimental time-to-failure periods for IC devices under accelerated stress conditions are between 10 and 1000 h, but many photonic devices do not manifest failures until several thousand hours have passed.

6.3 Approaches and recent developments

There are two major structures for semiconductor lasers. One is the ridge waveguide (RWG), and the other is the BH. The latter typically involves etch and regrowth, but the former does not require regrowth on the etched interface. RWG lasers are common in the fabrication of uncooled digital lasers. Figure 6.1(a) shows the schematic of an

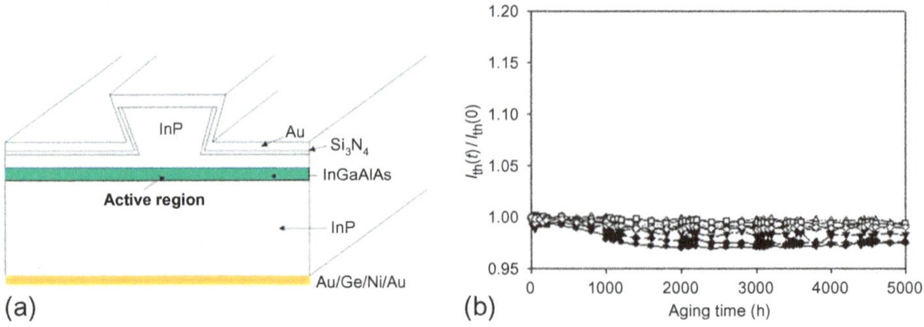

Figure 6.1 (a) Schematic of a ridge waveguide laser and (b) the relative threshold current change versus life-test aging time.

RWG laser [24–27]. In this structure, only the top layers are etched to form a ridge in order to provide carrier and optical confinement. The active layer underneath remains intact, and hence, no defects are generated in the active region. Due to the absence of regrowth defects, the threshold current typically does not show any increase during life-test aging, as shown in Figure 6.1(b). Random failure is occasionally observed, and the sudden failure mode is likely related to the mechanical stress.

The BH structure has been widely used to achieve low threshold current, high slope efficiency (SE), and good linearity for analog applications. The BH laser is used in an uncooled design for gigabit passive optical network applications, and in a gain chip design for telecommunication tunable lasers. The etched sidewall of the mesa structure is often a prolific source of crystalline defects, however. The defect density at the BH interface controls the device reliability of lasers during long-term operation. Common remedies for this problem have included the addition of a subsequent wet etch to refine the sidewall surface, called melt-back process [15,28], and the insertion of a buffer layer to prevent Zn diffusion [29]. Figure 6.2(a) shows a schematic diagram of the BH laser. For the BH, a double heterostructure is grown by metal organic

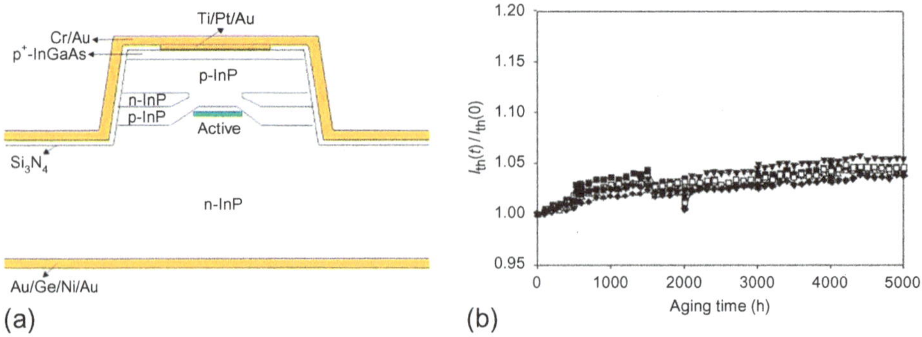

Figure 6.2 (a) Schematic of a buried heterostructure laser and (b) the relative threshold current change versus life-test aging time.

chemical vapor deposition (MOCVD) on an n-type InP substrate. Typically, the layers of the double heterostructure are an n-type InP cladding, an InGaAsP active layer, and a p-type InP cladding layer. The epitaxial layers are subsequently wet-etched to form the mesa structure. After the mesa etching process, the wafer is returned to the MOCVD chamber where the mesa is buried with a regrowth of p-InP and n-InP blocking layers. Finally, the heavily doped p-type InGaAs layer is often used to form low contact resistance. Figure 6.2(b) shows the relative threshold currents, $I_{th}(t)/I_{th}(0)$, as a function of aging time (t) for the BH lasers. The BH devices exhibit gradual increases in threshold current. The faster performance degradation in the BH laser could be attributed to the defects created during the mesa etch and subsequent regrowth. The sidewall of the etched InGaAsP active layer and InP cladding layer could potentially be damaged during the regrowth process.

In order to enhance the bandwidth, wavelength division multiplexing (WDM) network has become increasingly popular in fiber optic deployment [30]. The modern WDM network requires a high density of channels in order to support broadband applications. Due to the increasing channel density, the reliability of each channel has become more stringent than ever. The reliability requirement expected by the end-users or network operators typically consists of two aspects: wavelength stability and threshold current robustness. Although these two aspects are seemingly two different specifications, they are correlated in many cases. For example, as the threshold current increases during aging, it requires higher operating bias current to maintain the same level of optical power. The increase in the operating current consequently results in a higher junction temperature and a longer wavelength. In the past, most wavelength stability required no more than 0.1 nm for 100G dense wavelength division multiplexing (DWDM) spacing. Recently, many network and/or cable operators have begun requesting that LD component manufacturers show wavelength stabilities of 0.03-0.09 nm [31].

The other important factor for a WDM network is energy conversion efficiency. The energy efficiency is especially critical for long-distance transmission using high-power lasers. One common way to measure the energy efficiency is by means of wall plug efficiency, which is defined as the ratio of the total optical output power to the input electrical power.

For high-power BH lasers, the input electrical power is given as follows:

$$P_s = IV \tag{6.1}$$

where I is the bias current and V is the forward voltage. At the bias current of 500 mA, the typical forward voltage is about 2.3 V. Consequently, the input electrical power is 1150 mW.

The optical output power is given by:

$$P_o = P_{o-f} + P_{o-r} \tag{6.2}$$

The optical output power from the front laser facet (P_{o-f}) is typically about 125 mW. Assuming 20% power from the rear facet (P_{o-r}), the total output power is 150 mW. The ratio of the front to rear facet depends upon the reflectivity of the optical coating.

Hence, the wall plug efficiency can be calculated as follows:

$$E = \frac{P_o}{P_s} \tag{6.3}$$

In the case of C-band high-power lasers, the wall plug efficiency is estimated to be about 13%.

Table 6.3 shows the thermal characteristics of high-power BH and low-power RWG lasers. The wall plug efficiency for the RWG lasers is higher due to the lower input power. The temperature rise of the LD chip is related to the heat generation ($P_s - P_o$). The thermal impedance for the laser chip and AlN submount is inversely proportional to the cavity length. Using the thermal impedances of 0.02 and 0.10 °C/mW, the temperature rises in the chip are estimated to be 20 and 18 °C for high-power BH and low-power RWG lasers, respectively. The temperature rise data are consistent with the results determined by the wavelength shift.

Figure 6.3(a) and (b) shows the wavelength changes versus aging time data based on the stress current of 450 mA at 75 and 85 °C, respectively. The wavelength was measured *in situ* in the laser module butterfly package where the fiber coupling was fixed in the module. To extrapolate the failure time, the wavelength shift of tradition DWDM specification of 0.1 nm was used as the criterion for end-of-life. The median value of the extrapolated failure times at the test condition of 75 °C was about 67,700 h.

The activation energy (E_a) of the wavelength drift can be determined by life-test aging at different temperatures. Figure 6.4 shows the failure time versus the reciprocal of the test temperature at three aging temperatures: 65, 75, and 85 °C. The activation energy based on linear regression is estimated to be 0.96 eV.

In order to correlate the experimental wavelength shift with the theoretical value based on the thermal model, the power change is measured during the life-test aging. Figure 6.5(a) shows the relative power change versus aging time of the laser submounts subjected to the stress current of 400 mA at 100 °C. The power change is about 10% after 3000 h of aging at 100 °C, corresponding to 1.7% (or about 1.7 mW) at

Table 6.3 Thermal characteristics of BH and RWG lasers

	Formula	High-power BH laser (C-Band)	Lower-power RWG laser (O-Band)
Operating bias, I		500 mA	110 mA
Forward voltage, V_f		2.3 V	2.0 V
Input power, P_s	$P_s = IV$	1150 mW	220 mW
Output power, P_o	$P_o = P_{o-f} + P_{o-r}$	150 mW	40 mW
Wall plug efficiency	$E = P_o/P_s$	13%	18%
Thermal impedance, R_{th}		0.02 °C/mW	0.10 °C/mW
Temperature rise, dT	$dT = \kappa(P_s - P_o)$	20 °C	18 °C

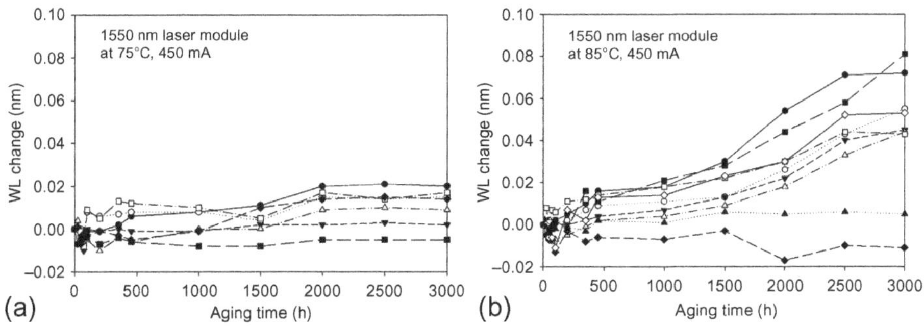

Figure 6.3 Wavelength shift data for high-power lasers aged with a stress current of 450 mA at the case temperatures of (a) 75 °C and (b) 85 °C.
After Ref. [32].

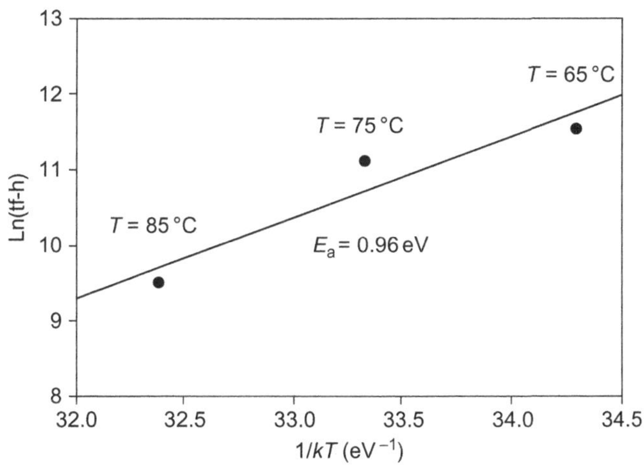

Figure 6.4 Failure time versus $1/kT$ plot. The E_a of 0.96 eV was determined based on the case temperature.
After Ref. [32].

75 °C. Based on the thermal impedance of 0.10 °C/mW, the temperature increase due to the power drop is 0.17 °C [33]. The wavelength shift induced by the power drop is 0.015 nm, which is close to the experimental observation of 0.014 nm. Figure 6.5(b) shows the wavelength drift of experimental versus thermal modeling at 65, 75, and 85 °C. The comparison shows good agreement between the experimental measured value and the theoretical estimate based on thermal modeling.

Tables 6.4 and 6.5 show the extrapolated device lifetimes based on the conventional wavelength-drift specification of 0.100 nm and the most stringent specification of 0.025 nm, respectively. The data shows that our DWDM lasers meet the 0.025-nm specification with a great margin in device lifetime over the 20-year requirement.

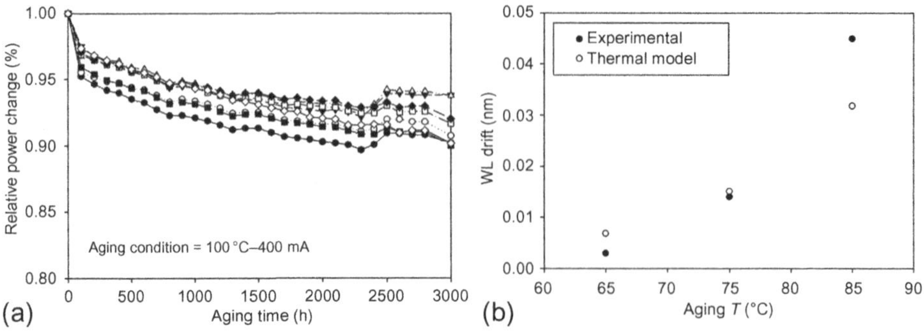

Figure 6.5 (a) Power change data for high-power lasers aged with a stress current of 400 mA at a case temperature of 100 °C. (b) Wavelength (WL)-drift data of experimental versus thermal modeling. After Ref. [32].

Table 6.4 Extrapolated device lifetimes at use condition based on the wavelength-drift specification of 0.1 nm

Aging *T* (°C)	tf(test) (h)	Acceleration factor	tf(25 °C) (h)	tf(25 °C) (years)
65	101,400	83.1	8.4×10^6	961.9
75	67,700	214.1	1.5×10^7	1654.4
85	13,400	523.1	7.0×10^6	799.5

After Ref. [32].

Table 6.5 Extrapolated device lifetimes at use condition based on a wavelength-drift specification of 0.025 nm

Aging *T* (°C)	tf(test) (h)	Acceleration factor	tf(25 °C) (h)	tf(25 °C) (years)
65	10,100	83.1	8.4×10^5	95.4
75	6700	214.1	1.4×10^6	164.1
85	1300	523.1	6.9×10^5	79.3

After Ref. [32].

6.4 Case study: reliability of buried heterostructure (BH) InP semiconductor lasers

Semiconductor lasers are the major light sources in the optoelectronics component market, and the reliability of these devices is of critical importance in many applications [34–36]. The InP BH lasers exhibit two major degradation patterns [37]. One degradation pattern (A) involves only the increase in the threshold current with little change in the external quantum efficiency, as illustrated in Figure 6.6(a) and (b). The degradation is typically related to the regrowth interface. The other degradation

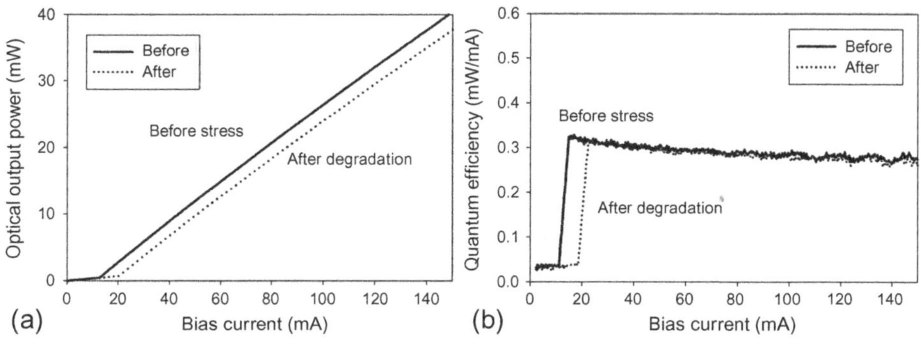

Figure 6.6 Degradation pattern A of BH lasers: (a) power change and (b) quantum efficiency change plots before and after stress.

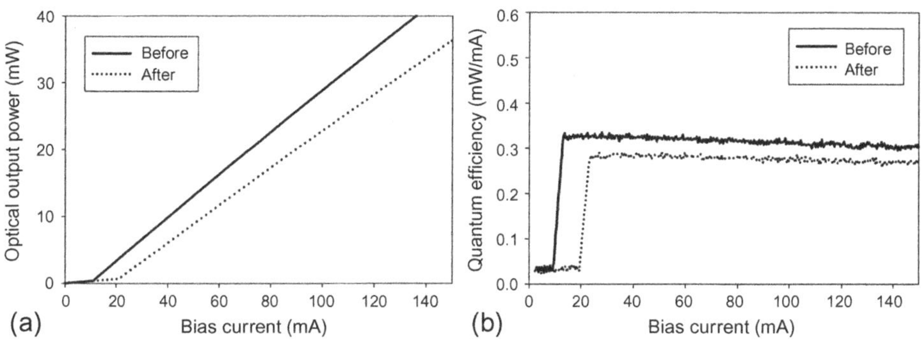

Figure 6.7 Degradation pattern B of BH lasers: (a) power change and (b) quantum efficiency change plots before and after stress.

pattern (B) involves changes in both threshold current and external quantum efficiency. After degradation, the threshold current increases, and the external quantum efficiency decreases, as shown in Figure 6.7. The degradation is typically related to the active region and the regrowth interface.

Several potential factors may be responsible for reliability degradation in semiconductor lasers such as BH interfaces, crystal quality, metal contacts, metal bonding, and facets [14]. The dominant factor typically depends upon the device design, structure, and materials. In this case study, we focus on the effects of p-metal contact, regrowth interface, substrate quality, and cavity length on the reliability degradation of BH lasers. The BH lasers are based on distributed feedback (DFB) multiquantum well InP lasers.

6.4.1 Effects of p-metal contact

6.4.1.1 p-metallization

Alloy and Schottky-type metals are widely used to make ohmic contacts to p-type semiconductors. The early generation devices often incorporate alloy-type ohmic contacts in order to achieve low contact resistance. The alloy contact is devised by alloying a metal with a semiconductor. Metal atoms, such as Au and Zn, are thermally diffused into the semiconductor layer, resulting in a reduction of potential barrier height and resistance. The modern optoelectronic devices typically employ Schottky-type contacts in order to establish robust reliability with good control. The Schottky-type contacts are formed by metal deposition on a heavily doped semiconductor. The width of the energy barrier at the metal/semiconductor interface decreases with increasing doping concentration.

Although electrical performance is comparable in alloy and Schottky contacts, the interfacial reactions in the two systems are dramatically different. For the alloy-type, significant interfacial diffusion occurs after alloying. Figure 6.8 shows an example of alloy morphology for a Au/Zn/Au/Cr/Au contact to p-InGaAs/p-InP taken by scanning transmission electron microscopy (STEM). The chemical composition is characterized by energy dispersive X-ray spectroscopy (EDS). The sample is subjected to the alloy condition at 430 °C for 30 s. The morphology of the compound is nonuniform. There are two types of compounds: one is large, dark, and Au-rich, and the other is small, white, and GaAs-rich [38–40]. When the Cr layer is reduced to the critical thickness, the Au-rich compound diffuses and penetrates to the InGaAs/InP interface [40]. For the Schottky-type, Pt acts as a diffusion barrier between the metal and the semiconductor.

Figure 6.9 shows the corresponding EDS spectra of an alloyed p-InP/InGaAs/Au/Zn/Au/Cr/Au sample. A line-scan across the large, dark, Au-rich compound is shown. There is considerable interdiffusion between AuZnAu and InGaAs. From the top (a distance from 0 to 0.3 μm), some amount of Zn appears to out diffuse to the top Au layer. The Cr atoms out diffuse to the bottom portion of the top Au layer. A significant amount of As out diffuses into the Cr layer to form a Cr-As compound. Immediately below the Cr-As layer, another region consists of Ga and As. This corresponds to the GaAs-rich compound, which appears small and white in the bright field image. Below the GaAs-rich compound, there is an Au-rich compound and an InGaAs layer.

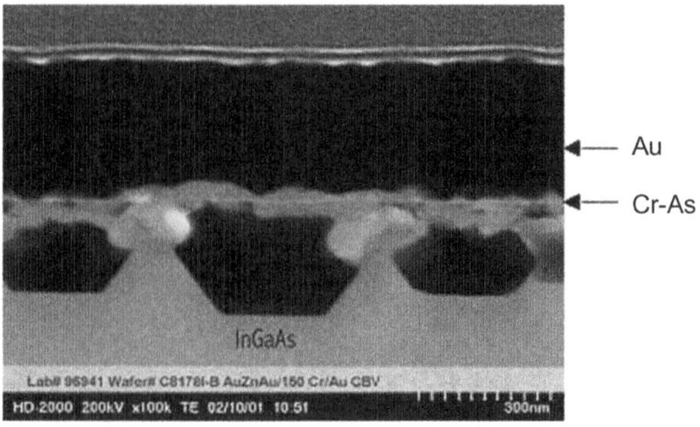

Figure 6.8 STEM image showing the p-metal alloy morphology of an Au/Zn/Au/Cr/Au contact to a p-InGaAs/p-InP semiconductor. After Ref. [38].

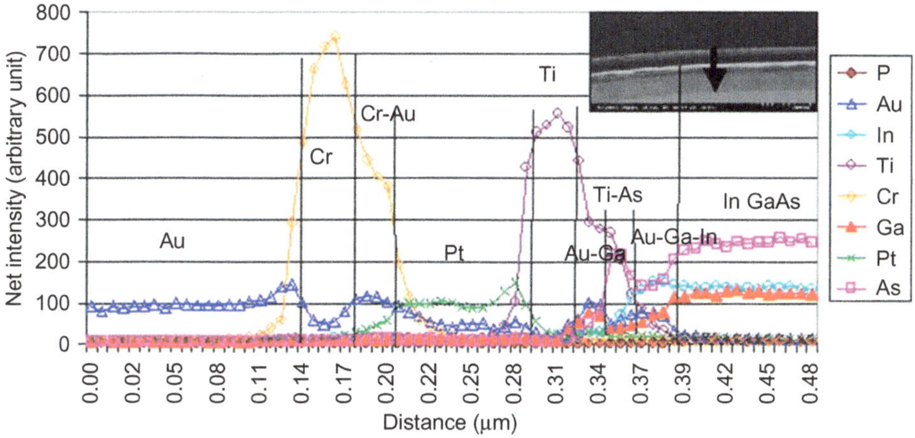

Figure 6.9 STEM spectra of an alloyed Au/Zn/Au/Cr/Au contact to a p-InGaAs/p-InP semiconductor. The STEM line scan is performed across the Au-rich compound. After Ref. [38].

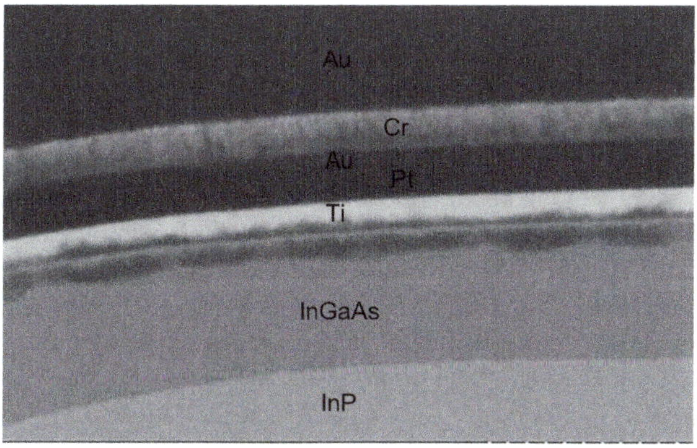

Figure 6.10 STEM image showing the p-metal alloy morphology of an Au/Ti/Pt/Au/Cr/Au contact to p-InGaAs/p-InP semiconductor. After Ref. [38].

Figure 6.10 shows a STEM image of an Au/Ti/Pt/Au/Cr/Au contact to p-InGaAs/p-InP. The samples were annealed at 430 °C for 30 s. In contrast to the Au/Zn case, the compound formation is more uniform and confined to the metal/InGaAs interface. The majority of the InGaAs layer remains unreacted. The Pt and the metal layers above it are also intact. Figure 6.11 shows the corresponding EDS spectra of the annealed p-InP/p-InGaAs/Au/Ti/Pt/Au/Cr/Au sample. The Ti and Au atoms diffuse downward, while the Ga and As diffuse upward. The interdiffusion between the metal and the semiconductor leads to the formation of three compounds. These compounds are identified as Au-Ga, Ti-As, and Au-Ga-In [38]. The Pt layer remains intact after

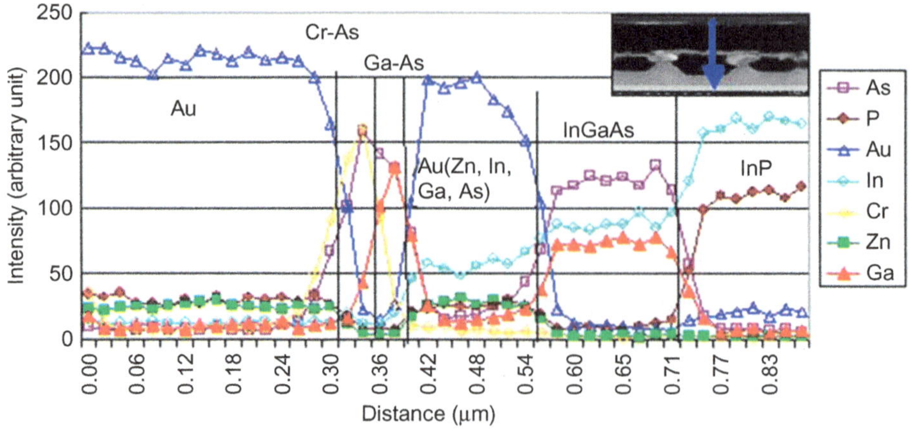

Figure 6.11 STEM spectra of an alloyed Au/Ti/Pt/Au/Cr/Au contact to a p-InGaAs/p-InP semiconductor.
After Ref. [38].

the annealing, indicating that Pt provides an excellent diffusion barrier preventing interaction below and above the Pt. The Cr shoulder overlaps the middle Au layer to form a Cr-Au mixture. Phase analysis by Ivey et al. shows that the interfacial reaction in the p-InGaAs/Ti/Pt/Au system results in the formation of Ti-As, metallic In, and Ga-rich InGaAs [41]. The cross-section transmission electron microscopy (TEM), selective area diffraction, and EDS show that Ti is the major diffusing species in the formation of Ti-As.

Figure 6.12 shows the I_{th} and power change box plots of the lasers of the Au/Zn/Cr/Au (top, thick) and Au/Ti/Pt/Au/Cr/Au (top, thick) groups. The changes in threshold current and output power of the Au/Ti/Pt/Au are significantly less than those of the Au/Zn. The reliability improvement of the Au/Ti/Pt/Au group is attributed to the diffusion barrier by the Pt layer. The Pt barrier prevents the thick Au from interacting with the InGaAs and underlying InP layers, thus limiting the Au interfacial reaction.

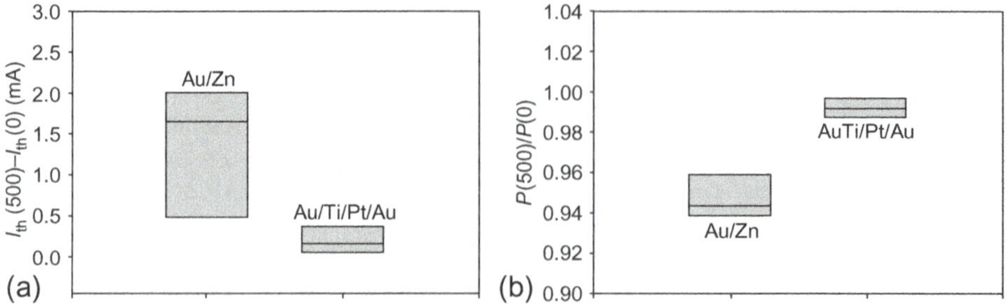

Figure 6.12 Reliability comparison of Au/Zn versus At/Ti/Pt/Au p-metal devices. The box plots show the changes in (a) threshold current and (b) optical output power.
After Ref. [42].

For Au/Zn, the faster degradation is attributed to the defect formation resulting from the migration of Au [43,44]. Due to the low resistivity of the Au-rich compound, the current flow tends to concentrate on the alloy spike and accelerate the migration of Au. When the Au reaches the active region, dark spot defects (DSDs) or dark line defects can be formed, reducing the lasing efficiency.

6.4.1.2 Plasma damage

The other important factor regarding the p-contact is the plasma damage induced by reactive ion etching (RIE). Figure 6.13 shows the distributions of threshold current change for two experimental groups with different contact etch conditions. The lasers of both contact etch groups were subjected to the same aging condition, with stress current of 175 mA at the case temperature of 100 °C for 48 h. The lasers in one group were processed with 240 s of contact etch by RIE, and the lasers in the other were etched with 80 s of etch. The I_{th} change in the former was very small (around 0.8-1.2 mA, median). On the other hand, the I_{th} change in the 240-s RIE was 12 mA, larger by a factor of 12.

Because interfacial diffusion is driven by defects, the quality of the epitaxial contact layer should influence the defect formation and propagation toward the active region. When the contact etch time is excessive, more defects tends to be generated by plasma charging damage during the dry etch. Those defects that originate from the p-contact could subsequently propagate toward the active region during aging. The defects could also enhance the Au diffusion at the p-contact.

In the discussed study, the damaged region induced by RIE was likely a superficial layer that behaved as a collection of deep traps that captured conduction electrons. Foad et al. have proposed the conduction technique to estimate the extent of surface etching damage [45]. Morello et al. reported that the damage depth induced by RIE was about 21-50 nm, as measured by the conduction technique [46]. The damage generation rate depended upon the energy of impinging ions. The RF power, reactant, carrier gas flow, and chamber pressure are also important for the contact etch in order to minimize the plasma damage [47–50].

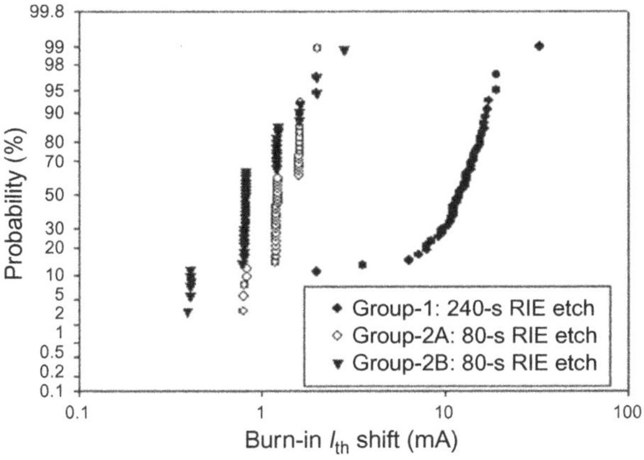

Figure 6.13 Distributions of threshold current changes in the two contact etch groups. Group 1 is etched with 240-s RIE, and Group 2 is etched with 80-s RIE. After Ref. [42].

6.4.1.3 p-InGaAs contact layer thickness

The contact layer thickness also appears to influence laser degradation. Figure 6.14 shows the I_{th} change as a function of contact layer thickness. The figure compares LDs with InGaAs thicknesses ranging from 0.1 to 0.4 μm. The contact layer thickness is verified by the scanning electron microscopy and TEM. The group of the 0.1-μm InGaAs lasers shows the largest I_{th} change, suggesting that the lasers with the 0.1-μm InGaAs are most vulnerable to laser degradation. In addition, the range of the I_{th} changes of the 0.1-μm InGaAs group is larger than the others, implying that the process window is less tolerant. As the InGaAs thickness increases, the I_{th} change becomes smaller, indicating that the lasers with the thicker InGaAs are more reliable.

As the InGaAs thickness decreases, it becomes easier for defects to propagate through the InGaAs layer into the InP. In addition, the thinner InGaAs also makes the underlying InP more susceptible to the plasma damage generated during the RIE etch. Finally, the Au diffusion in the Au/Ti/Pt/Au to p+-InGaAs contact is typically 0.1 μm in depth and confined to the InGaAs layer [38]. The Au diffusion into the InP becomes increasingly likely when the InGaAs thickness is 0.1-0.2 μm.

6.4.2 Effects of BH interfaces

Another major laser degradation mechanism has been associated with epi-regrowth interface defects [51]. The quality of the BH interface between the etched mesa sidewall and burying layer was found to be critical in reliability performance. The reliability degradation was found to correlate with the residual damage at the sidewalls of the mesa [52]. The probable causes of the residual damage included inadequate removal of process-induced defects and defect generation during the epitaxial regrowth. For the latter, the III-V ratio and Zn-dopant were typically important factors. The common remedies included the addition of a subsequent wet etch to refine the sidewall surface and the insertion of a buffer layer to prevent the Zn diffusion [53]. For the former, defects could originate from the RIE-induced steps. The defects served as the nonradiative recombination sites that were responsible for laser degradation.

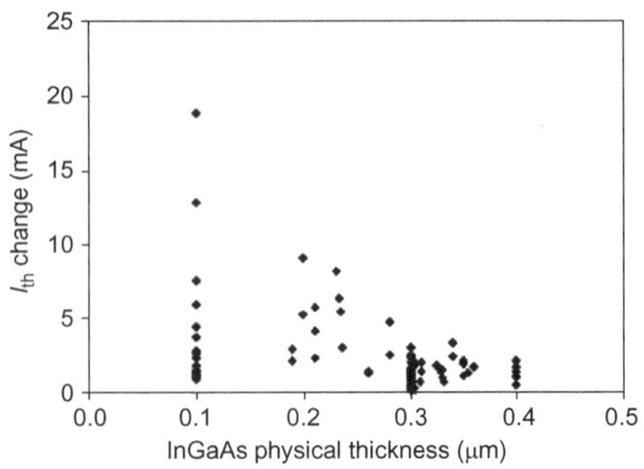

Figure 6.14 Threshold current change as a function of InGaAs contact layer thickness. The lasers with thinner InGaAs layers have higher tendencies to show larger I_{th} increases during aging.

6.4.3 Effects of substrate quality

It has been reported that generation and growth of dislocations may also cause performance degradation in semiconductor lasers [28]. The propagation of the dislocations can follow one of two crystallographic orientations, ⟨1 0 0⟩ and ⟨1 1 0⟩ and the rate of growth can be enhanced by nonradiative recombination at the dislocation [54]. The energy emitted by the nonradiative recombination is transformed into lattice vibrations, leading to low temperature defect migration. Two models have been proposed for the defect migration. Petroff and Kimerling suggested that the dislocation climbing resulted from the absorption of interstitials [55]. O'Hara et al. proposed that the climbing resulted from the emission of vacancies [56].

By absorbing interstitials or emitting vacancies, the dislocations grow and form a complicated network structure [57,58]. Threading dislocations have appeared as DSDs in electroluminescence, photoluminescence, and electron-beam-induced current images of the active regions of photonic devices. The formation of observable DSDs was shown to correlate well with the increase in threshold current [52].

A threading dislocation may originate from defects in the substrate. In practice, substrate defects can be controlled by selecting only high quality material or by employing improved crystal growth techniques. For the latter, buffer layer deposition on the substrate has been widely used [59–61]. For the former, etch pit density, doping concentration, and crystal orientation must be carefully controlled and monitored to obtain the substrate material quality required for high device reliability.

The buffer layer is designed to accommodate the lattice misfit and confine the threading dislocations from propagating to the active region. For example, studies of $In_xGa_{1-x}As$ epilayer/(001) GaAs substrate systems have shown that the confinement of the threading dislocations is determined by the composition and thickness of the epilayer [62,63]. Figure 6.15(a) and (b) shows the box plots of the relative threshold current changes of InGaAsP/InP lasers built using two different n-type buffer designs. With a thick buffer layer, the reliability degradation rate is significantly reduced.

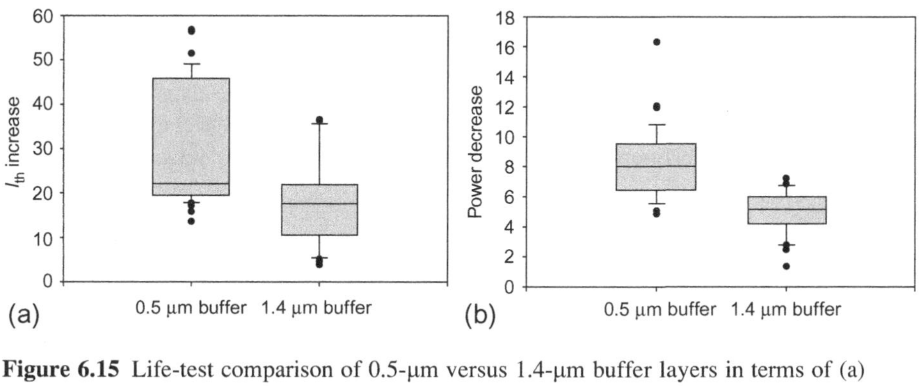

Figure 6.15 Life-test comparison of 0.5-μm versus 1.4-μm buffer layers in terms of (a) threshold current and (b) power changes.
After Ref. [42].

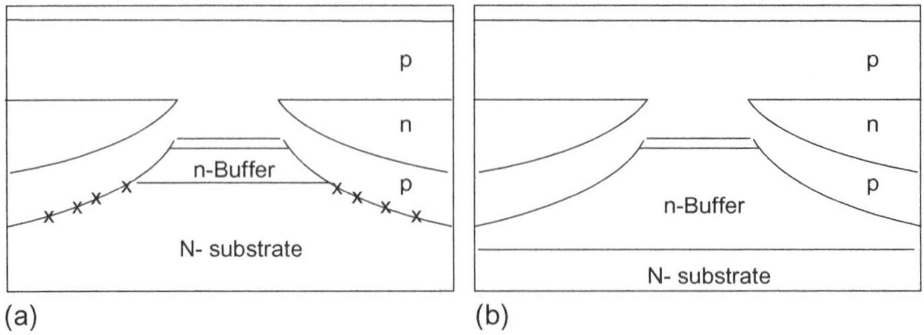

Figure 6.16 (a) BH laser with a 0.5-μm buffer layer and (b) BH laser with a 1.4-μm buffer layer.
After Ref. [42].

To understand the reliability improvement in the thick buffer structure, researchers proposed the mesa etch and regrowth defect mechanism, as shown in Figure 6.16(a) and (b). In the thin buffer structure, the mesa etch punched through the substrate and the epi-regrowth was interacting with the defects in the substrate. The mesa etch/ substrate interface became defective in the region marked by "x" in Figure 6.16(a). During the aging, the defects were likely migrating toward the active region, degrading the lasing efficiency. In the thick buffer structure, there was no interaction between mesa etch and the substrate. Consequently, the active region was less susceptible to the defect formation during aging.

Alternatively, the reduction of laser lifetime in the thin buffer structure could be related to the growth and propagation of dislocation networks arising from the substrate crystal due to recombination-enhanced defect motion [64]. The defect motion could occur via climb and glide motion.

6.5 Reliability extrapolation and modeling

6.5.1 Sublinear model extrapolation

For analysis of semiconductor laser life tests of finite duration, researchers must often fit an extrapolation curve to the experimental data. A sublinear model usually provides the best fit to the experimental life-test curves. In the sublinear model, an experimental curve is fitted by

$$y = 1 + at^m \quad (6.4)$$

where y is the ratio of the threshold current after aging to the initial threshold current, t is the aging time, and the constant a and the exponent m are the free parameters for the curve fitting [65]. The constant a is then used to determine the failure time. For low threshold lasers that have initial threshold currents below 20 mA, the failure criterion for the life test is defined as a 50% increase in the initial threshold current [66].

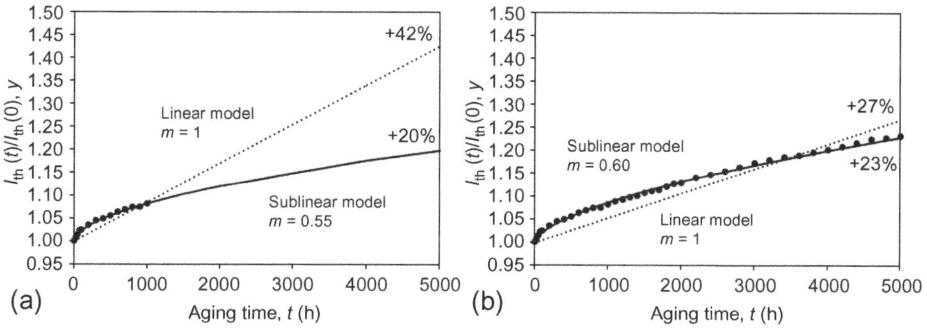

Figure 6.17 Life-test curve fittings to experimental data from (a) a 1000-h test and (b) a 5000-h test. The linear and sublinear models are shown for comparison.

Figure 6.17(a) and (b) shows examples of life-test curve fittings of 1000- and 5000-h data, using linear and sublinear models. For 1000-h data, the linear model projects to an increase of 42% in the threshold current, while the sublinear model extrapolates to +20%. For 5000-h data, the linear model projects to an increase of 27% in the threshold current, and the sublinear model extrapolates to +23%. The sublinear model extrapolation is consistent with the experimental results. The estimated threshold current increase based on the sublinear model is consistent and independent of the accumulation time taken for the fitting analysis. On the other hand, the linear model strongly depends on the aging time and tends to generate overly pessimistic results.

Table 6.6 shows the sublinear fitting extrapolation results from laser devices with Au/Zn metallization stressed with a constant current of 150 mA at 85 °C. The fitting correlation factor in this case is maximized when the exponent $m = 0.65$. It is noted that the projected failure time increases with decreasing exponent. This is expected because a smaller exponent corresponds to a greater saturation of the performance degradation rate. The physical meaning of the sublinear model ($m < 1$) is that the rate

Table 6.6 Life-test curve fitting results based on a group of laser devices with Au/Zn p-metallization stressed with a constant current of 150 mA at 85 °C

Exponent (m)	0.30	0.40	0.5	0.55	0.60	0.65	0.70	0.75	0.80
Correlation factor (f)	0.844	0.935	0.980	0.992	0.998	0.999	0.995	0.987	0.977
Projected failure time (h) at 85 °C	31418.5	15930.6	10886.2	9543.8	8581.4	7864.2	7313.1	6878.6	6528.9
Projected device lifetime (year) at 25 °C ($E_a = 0.8$ eV)	661.0	335.2	229.0	200.8	180.5	165.5	153.9	144.7	137.4

After Ref. [2].

of degradation will diminish after a certain point in time during aging. An analytical simulation by Lam et al. showed that the sublinear model is accurate in describing the saturation of degradation in aging curves [67]. The Lam analysis assumed that the change in threshold current is given by the change in the nonradiative recombination current that is proportional to the defect density. It further assumed that the source supplying the creation and growth of defects is finite.

The results shown in Table 6.6 are based on the average values of the total population. Ideally, the life-test curve of each device needs to be fitted separately because life-test behavior may be dissimilar among chips. In practice, however, fitting to the whole population is often employed for simplicity. The fitting accuracy can also be affected by tester instability, test stops and restarts, and other external factors that may cause false values in the data set. Under these circumstances, artificial curve smoothing may be required to obtain accurate results [68,69].

6.5.2 Temperature and current accelerations

The activation energy can be determined based on Black's equation for EM [70], as follows:

$$\mathrm{MTTF} = AJ^{-n}\exp\left(\frac{E_\mathrm{a}}{kT}\right) \tag{6.5}$$

where median time-to-failure (MTTF) is the time for 50% cumulative failure, A is a constant, J is the current density, n is the current exponent, E_a is the activation energy, k is the Boltzmann's constant, and T is the temperature in Kelvin. Theoretically, two temperatures and two stress currents would allow for the determination of the activation energy and current exponent. In practice, estimates based on three temperatures and three currents are more reliable and accurate because the confidence level is higher.

$$\frac{\mathrm{MTTF}(T_2)}{\mathrm{MTTF}(T_1)} = \exp\left[\frac{E_\mathrm{a}}{k(1/T_2 - 1/T_1)}\right] \tag{6.6}$$

Figure 6.18(a) shows the MTTF versus stress temperature of the Au/Zn devices from two wafers. The failure time is determined by extrapolation to 50% threshold current change, using a sublinear model [65]. The temperature acceleration factor is determined by the life-test data at stress temperatures of 40-100 °C. The value of the E_a for both wafers is estimated to be about 0.8 eV. For the current density acceleration factor, life-test data based on three stress currents ranging from 90 to 150 mA was collected. Because the device geometry is the same, the current is analyzed for simplicity, rather than the current density. Figure 6.18(b) shows the MTTF versus stress current of the Au/Zn devices. The current exponent determined from the plot is 1.4 ± 0.1.

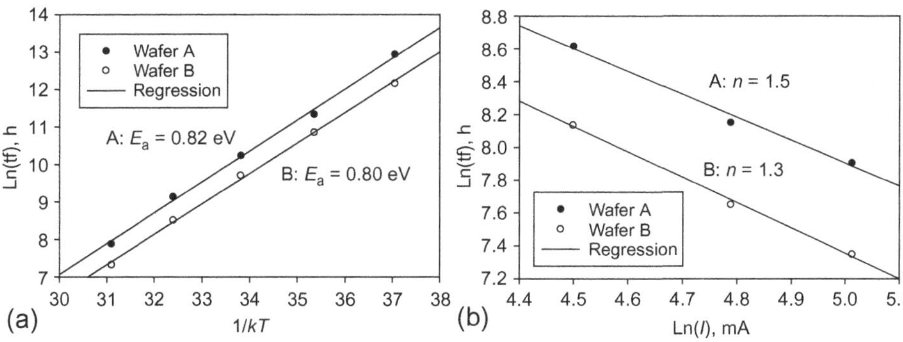

Figure 6.18 Temperature and current accelerations of Au/Zn lasers. (a) MTTF versus stress temperature and (b) MTTF versus stress current.
After Ref. [2].

6.6 Electrostatic discharge (ESD) and electrical overstress (EOS)

Besides the long-term aging degradation discussed in the previous sections, LDs and PDs could also exhibit catastrophic random failures. One common cause for the catastrophic sudden failures is ESD. ESD damage has been widely studied for both electronic ICs and optoelectronic devices [71–79]. Characterization of ESD thresholds and understanding of ESD failure mechanisms are critical for the development of microelectronic and nanoelectronic devices [80]. The ESD damage can result from human operators, machine transients, and so on. The other related failure mechanism is electrical overstress (EOS). EOS can result from lightning strikes and power surges [81]. EOS can cause damage failure by itself, and it can also interact with ESD, accelerating the ESD degradation.

Typically, ESD and EOS can cause changes in a variety of device performance parameters. For Fabry-Perot and DFB semiconductor lasers, ESD can cause changes in device performance parameters such as threshold current, optical power, and optical spectrum. For DFB lasers, ESD damage can cause changes in single-mode lasing, leading to side-mode-suppression-ratio (SMSR) failure. Additionally, the kink, often associated with the ESD damage, could also adversely affect the device linearity during modulation in analog applications. For applications such as DWDM, wavelength stability with respect to the specified channel spacing could also be affected by the ESD damage. For example, the specification limit of the wavelength drift for 50 GHz is 0.1 nm, which is very important for the DWDM application of 0.4-nm channel spacing. Hence, it is critical to ensure that laser components achieve ESD robustness in order to meet the performance requirement of the fiber optic system.

The Telcordia technical reference (TR-NWT-000870) has highlighted the ESD sensitivity classifications, as shown in Table 6.7 [81]. Class 1 is defined as the minimum human-body-model ESD failure threshold (V_{th}) less than 100 V, and it includes certain types of small-area PDs. Class 2 covers the ESD failure threshold from 100 to 500 V, typically including metal-on-semiconductor (MOS) ICs. Class 3 covers the

Table 6.7 ESD sensitivity classifications based on the human body model (HBM)

ESD classification	Minimum HBM failure threshold	Device type
Class 1	$V_{th} < 0.1$ kV	Optoelectronic detectors
Class 2	$0.1 \text{ kV} < V_{th} < 0.5$ kV	MOS ICs
Class 3	$0.5 \text{ kV} < V_{th} < 2.0$ kV	MOS ICs, bipolar ICs
Class 4	$2.0 \text{ kV} < V_{th} < 4.0$ kV	Semiconductor LDs and LEDs
Class 5	$V_{th} > 4.0$ kV	Passive components

ESD failure threshold between 500 and 2000 V, including devices such as well-designed MOS and bipolar ICs. Perhaps for this historical reason, most of the IC and optoelectronic device manufacturers set 500 V as the ESD requirement. Class 4 covers ESD failure threshold between 2000 and 4000 V. Many well-designed and processed semiconductor LDs and LEDs are expected to fall into this category. Finally, Class 5 contains the ESD failure threshold above 4000 V. Typically, passive components such as resistors meet Class 5 criterion.

Besides the ESD classifications, JEDEC also documents the standards of ESD qualification, apparatus calibration, and waveform verification [82].

6.6.1 ESD damage characteristics

Figure 6.19 shows an example of the LI curves of laser devices before and after ESD damage. In this case, the initial threshold current at room temperature was 8.0 mA prior to the ESD stress. The optical output power above the threshold current was linearly proportional to the bias current. After the ESD stress of 4.0 kV in the reverse polarity, the threshold current increased to about 28.0 mA. The SE, defined as the slope of power versus current, decreased from 0.43 to 0.31 mW/mA. Because the laser exhibited high threshold current and low optical power but still lased after the ESD stress, we define this type of ESD damage as an intermediate ESD damage.

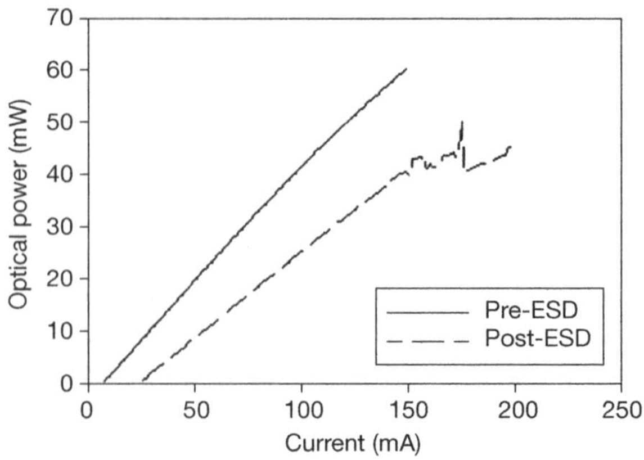

Figure 6.19 LI plot of pre-ESD and post-ESD BH laser devices. The damage is induced by the reverse ESD stress at −4.0 kV. After Ref. [83].

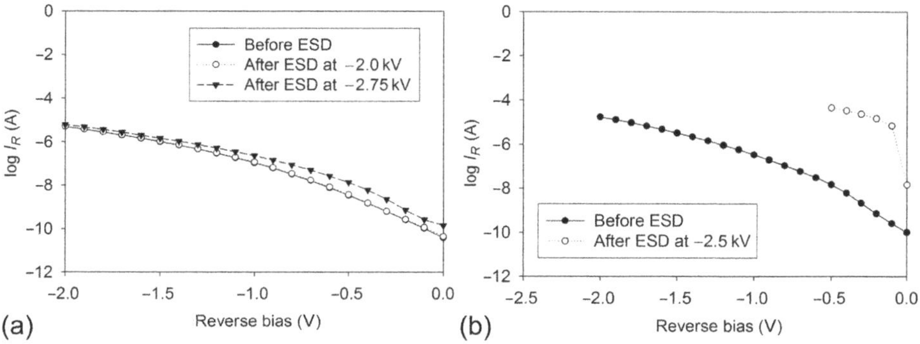

Figure 6.20 Reverse-bias current of lasers before ESD and after ESD. (a) The threshold currents before and after −2.0-kV ESD and after −2.75-kV ESD were 10.2, 10.0, and 10.3 mA, respectively. (b) The threshold currents before and after −2.0-kV ESD were 8.7 and 12.6 mA, respectively.
After Ref. [83].

In addition to the threshold current change, the change in reverse-bias current also appears to be a sensitive parameter for the ESD damage. Figure 6.20(a) shows an example of reverse current before and after ESD stress where the threshold current does not change. The reverse current is plotted logarithmically. The threshold current shows little change after the ESD stress: the threshold currents before and after −2.0-kV ESD stress and after −2.75-kV ESD stress were 10.2, 10.0, and 10.3 mA, respectively. On the other hand, the reverse-bias current at −0.5 V increases from -3.4×10^{-9} A before ESD to -1.31×10^{-8} A after −2.75 kV ESD, representing a 285% change in reverse current. Figure 6.20(b) shows another example of reverse current before and after ESD stress, with the threshold current greater than 3 mA. After the ESD at −2.5 kV, the threshold current of the laser increased from 8.7 to 12.6 mA. This represented an absolute change of 3.9 mA or a relative change of 46% in the threshold current. On the other hand, the reverse current at −0.5 V increased from -1.4×10^{-8} A before ESD to -4.4×10^{-5} A, representing a change of 3097%. In both cases (Figures 6.19 and 6.20), the percentage change in the reverse-bias current was more dramatic than that in the threshold current.

Although the reverse-current change may be correlated with the threshold-current degradation in some cases, the reverse current itself is often not an indicator of long-term reliability. The devices that show high reverse leakage current do not necessarily show early reliability degradation. The devices that show low reverse leakage are also not guaranteed to exhibit robust reliability. The most reliable and direct way of characterizing the reliability is by means of burn-in screening and life-test aging, and testers should carefully interpret the reverse current data.

6.6.2 *ESD polarity effect*

The ESD failure threshold as a function of polarity was also investigated. Figure 6.21 shows the box plot of the ESD damage threshold of forward (F), reverse (R), and

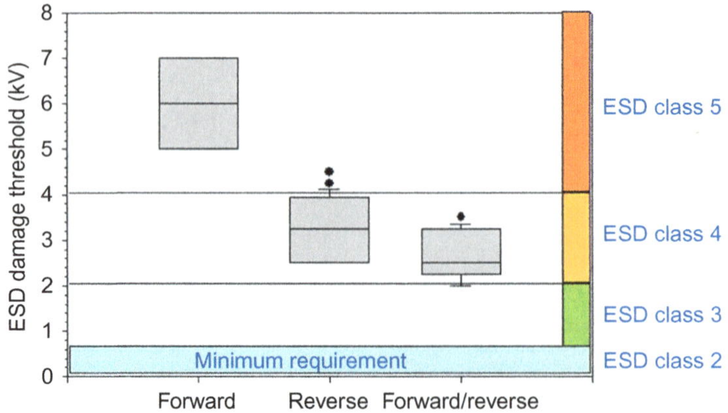

Figure 6.21 Box plot of the ESD damage thresholds of forward, reverse, and forward/reverse polarities.

forward/reverse (F/R) polarities. The box contains the middle 50% of the data, with the upper and lower edges of the box indicating the 75% and 25% the data points, respectively. The line in the box indicates the median value of the data. The ends of the vertical line indicate the maximum and minimum data values, and the points outside the ends of the line represent outliers. For each polarity, the statistical box plot consists of chip-based data, with each data point representing one chip. There were three to six wafers tested for each polarity. For the forward polarity, the chips showed no failure at 5 kV. After being exposed to a 5-kV forward bias, the devices showed no change in the threshold current and SE. To verify the functionality, devices were sampled to test the optical spectrum, and the results showed good DFB peak and SMSR. For graphing purposes, 5-7 kV was assigned to the forward-bias group in Figure 6.21. For the reverse polarity, the majority of the laser devices exhibited shifts in the threshold current and optical power in the range of 2.5-4.0 kV. The most exciting observation was that ESD damage threshold was the lowest for the combination of forward and reverse polarities, ranging from 2.2 to 3.2 kV.

Because both Au/Zn and Ti/Pt/Au p-metal systems have been used in industry, we investigated the correlation of ESD performance and p-metal. Table 6.8 shows the

Table 6.8 Median values (kV) of ESD failure thresholds from Au/Zn wafers

ESD polarity	Wafer B-1	Wafer B-2	Wafer B-3	Wafer B-4	Wafer B-5
Forward	>5	>5	>5	>5	>5
Reverse	2.62	3.98	3.60	3.03	2.41
R/F	2.18	2.79	2.70	2.94	2.32

After Ref. [83].

Table 6.9 Median values (kV) of ESD failure thresholds from Au/Ti/Pt/Au wafers

ESD polarity	Wafer A-1	Wafer A-2	Wafer A-3	Wafer A-4	Wafer A-5
Forward	>5	>5	>5	>5	>5
Reverse	3.13	3.25	3.25	3.13	2.88
R/F	3.06	–	3.25	2.50	2

After Ref. [83].

ESD damage thresholds of Au/Zn wafers under F, R, and F/R bias. The ESD threshold represents the median value of each wafer. Table 6.9 shows ESD damage thresholds of Au/Ti/Pt/Au wafers under F, R, and F/R polarities. On the wafer basis, the ESD threshold is the highest for the forward polarity and the lowest for the F/R polarity. Again, similar dependence of polarity was observed. For each polarity, Au/Ti/Pt/Au and Au/Zn show comparable ESD thresholds.

The lower reverse ESD threshold is attributable to the avalanche breakdown effect. Figure 6.22 shows the example of forward and reverse current as a function of applied voltage (V) in logarithmic scale. The I-V curve is based on the 360-μm cavity BH laser. The Joule heating, known as ohmic heating, is given by Equation (6.7) according to Joule's first law.

$$Q = AI^2R \tag{6.7}$$

where A is a constant, Q is the heat release through a device by the passage of an electric current, I is the electric current, and R is the resistance. The reverse current is roughly four to five orders of magnitude smaller than the forward current.

For the same applied voltage, the reverse current at the forward bias is higher than that at the reverse bias, but the resistance is lower at the forward bias. As a result, the

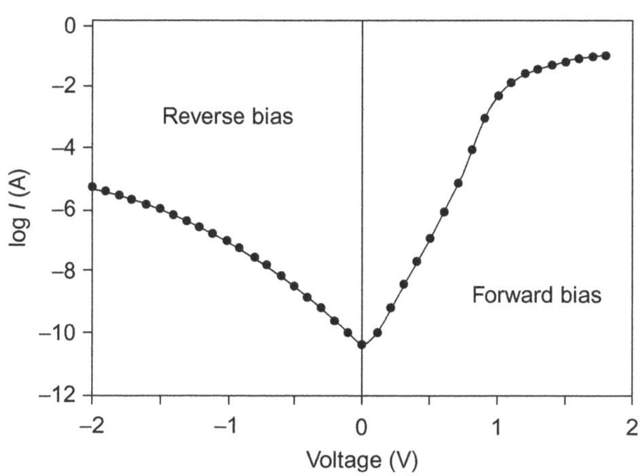

Figure 6.22 Forward and reverse I-V curves of BH lasers shown in logarithmic scale. The I-V is based on the 360-μm cavity BH laser. After Ref. [83].

Table 6.10 Example of joule heating of forward and reverse bias based on the performance of BH InP semiconductor lasers (360 μm in cavity length).

	Applied voltage, V (V)	Current, I (A)	Resistance, R (Ω)	Q (W)
Forward	1.0	5.4×10^{-3}	9.0	2.6×10^{-4}
bias	1.8	9.4×10^{-2}	9.0	1.4×10^{-1}
Reverse	−1.0	-1.1×10^{-7}	3.0×10^{-5}	3.6×10^{-9}
bias	−1.8	-2.8×10^{-6}	3.0×10^{-5}	4.1×10^{-6}

Joule heating, proportional to the square of electric current, is lower at the reverse bias by about four to five orders of magnitude, as shown in Table 6.10. Hence, the avalanche breakdown resulting from the tunneling mechanism may be responsible for the reverse ESD failure. For the F/R polarity, the ESD threshold does not appear to correlate to the order of forward and reverse bias. The lower ESD threshold of F/R polarity, compared to the forward and reverse bias alone, could be related to the memory/cumulative effect. The cumulative ESD effect has been observed in ICs [84–86].

6.6.3 ESD soft and hard degradation behaviors

Two types of ESD degradations have been reported [83]. Soft degradation is characterized by a gradual increase in the threshold current as a function of ESD voltage. Hard degradation shows a sudden jump in the threshold current. Figure 6.23(a) and (b) illustrates examples of soft and hard ESD degradation in threshold current and external quantum efficiencies, respectively. In both cases, lasers exhibit increases in the threshold current and corresponding decreases in the quantum efficiencies. For soft degradation, the laser does not exhibit any threshold current increase until 2.0 kV. Above

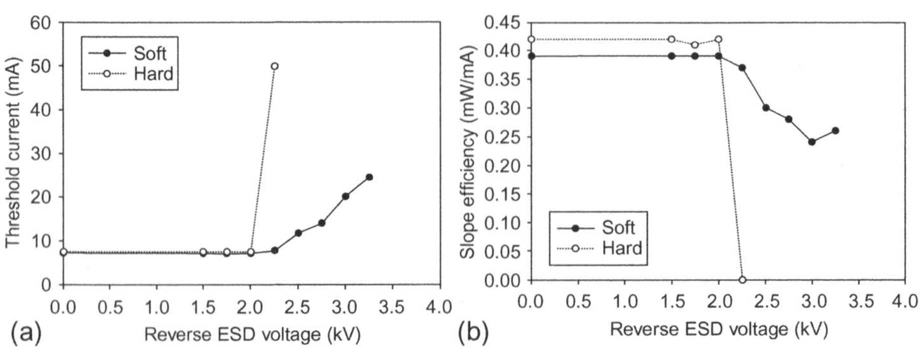

Figure 6.23 Soft and hard ESD degradation characteristics of BH semiconductor lasers shown in terms of (a) threshold current and (b) slope efficiency.
After Ref. [83].

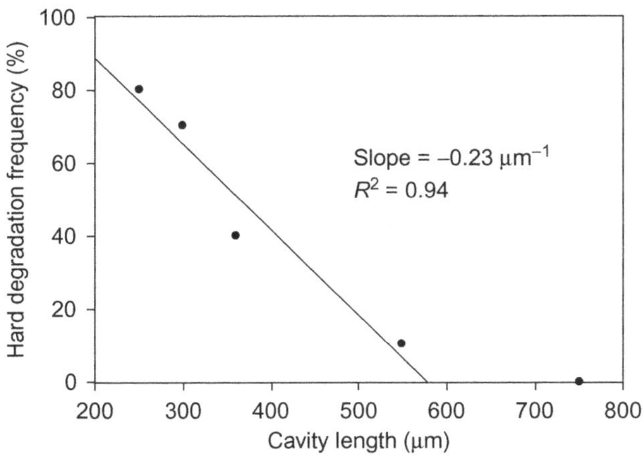

Figure 6.24 Occurrence frequency of hard ESD degradation as a function of cavity length. R^2 is the correlation factor with the higher value representing better correlation.
After Ref. [83].

2.0 kV, the threshold current increases gradually from 7.2 mA at 2.0 kV to 24.6 mA at 3.25 kV. In contrast, the threshold current of the hard-degradation device shows a dramatic jump from 7.6 mA at 2.0 kV to no lasing at 2.25 kV. For graphing purposes, a threshold current of 50 mA is assigned to show the data point.

The degradation behavior depends on the cavity length of the LDs. Figure 6.24 shows the occurrence frequency of hard degradation as a function of cavity length. The occurrence frequency of hard ESD degradation generally increases with decreasing cavity length. The occurrence frequency of hard ESD degradation decreases with increasing cavity length. For example, 80% of the 250-μm lasers fail by hard degradation, but only 10% of the 550-μm lasers fail by hard degradation. No hard degradation occurs for the 750-μm lasers. The length dependence is likely due to the difference in energy flux (energy injection per area). For the shorter cavity length, the energy flux injected into the laser cavity is greater at a given ESD level. Hence, the possibility for a catastrophic degradation (i.e., hard degradation) is likely to increase due to the higher instantaneous power dissipation at the junction. For the longer cavity, the energy flux is less, so the laser tends to degrade more slowly.

6.6.4 *Size effect*

The ESD threshold depends on the length and width of the LDs [87]. Figure 6.25 shows the reverse ESD threshold of BH lasers as a function of cavity length. The contact width of the BH laser is about 6.0 μm. The correlation plot indicates that the reverse ESD threshold increases linearly with the cavity length. The error bars represent the standard deviation. As the laser cavity increases from 250 to 750 μm, the median value of reverse ESD threshold increases linearly from 2.25 to 6.75 kV. There is a performance tradeoff between short and long laser cavities. The long cavity favors high-power operation, and the short cavity favors the speed [88]. Engineers should try to incorporate ESD robustness when designing the high-speed laser.

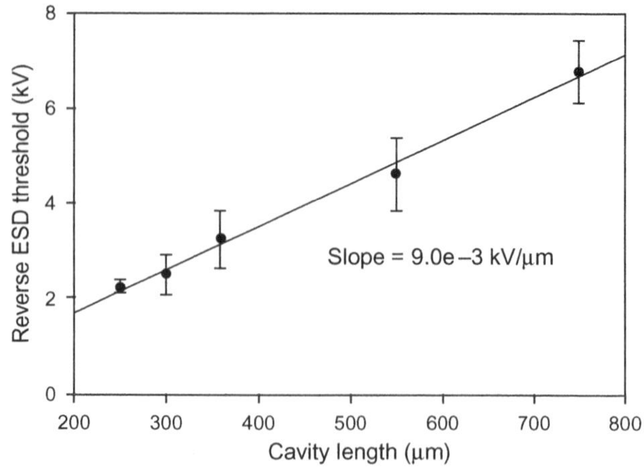

Figure 6.25 Reverse ESD threshold of BH lasers as a function of cavity length. The contact width is about 6 μm.
After Ref. [87].

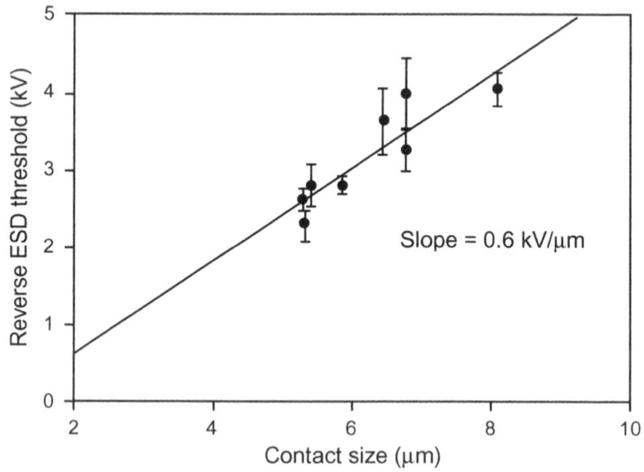

Figure 6.26 Reverse ESD threshold of BH lasers as a function of contact width. The contact length is about 360 μm.
After Ref. [87].

Figure 6.26 shows the reverse ESD threshold of BH lasers as a function of contact width. The cavity length of the BH laser is 360 μm. The correlation plot indicates that the reverse ESD threshold increases linearly with the cavity length. As the contact width increases from 5.3 to 8.1 μm, the median value of the reverse ESD threshold increases from 2.31 to 4.05 kV. Based on the linear scaling from the fitting in Figure 6.25, the ESD threshold of the 8.1-μm contact is estimated to be 3.51 kV, which is lower than the observed value of 4.05 kV. Such deviation could be attributed to the dielectric undercut and current crowding.

Engineers typically employ small contact width to improve the current confinement and reduce the parasitic capacitance [89]. However, the ESD sensitivity is likely to be higher based on the data in Figure 6.26. Therefore, the laser designer may need to carefully evaluate the ESD requirement and incorporate other design-in ESD reliability parameters as necessary.

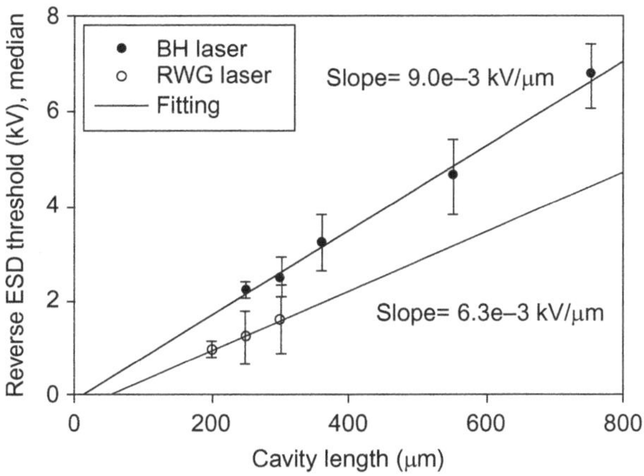

Figure 6.27 Comparison of ESD thresholds of BH and RWG lasers.
After Ref. [87].

6.6.5 BH versus RWG lasers

Figure 6.27 compares the reverse ESD threshold of BH and RWG lasers. The data of long-cavity RWG lasers are not available because RWG is typically designed for high-speed applications that require short cavities. For the same cavity length, the RWG lasers exhibit a lower ESD threshold. We attribute the lower threshold to the narrow ridge width. The width-dependence also supports the data shown in Figure 6.26 showing that the reverse ESD threshold increases with the contact width. However, the ESD threshold does not seem to scale linearly with the contact width. For example, the ESD thresholds of the 300-μm BH and RWG lasers are 2.5 and 1.65 kV, respectively. The contact widths of the BH and RWG lasers are about 5.5 and 2.0 μm, respectively. The ratio of the BH ESD threshold to the RWG ESD threshold is 1.54, and the ratio of contact width is 2.75.

The nonlinear relationship is likely due to current crowding and other processing variables. Notably, the ridge laser exhibited an ESD threshold higher than the expected value based on linear width scaling. The higher-than-expected ESD threshold of the ridge laser is likely related to less current crowding. Detailed simulation may be needed to further quantify the effect of each factor. RWG design is employed for low cost applications for which the current confinement is achieved by ridge etch without involving any epitaxial regrowth. However, the data in Figure 6.26 suggests that there may be an increase in ESD sensitivity for the ridge structure as a tradeoff.

6.7 Conclusions

In this chapter, we review the effects of metal contact, BH interfaces, and substrate quality on laser reliability degradation. For p-metal contacts, the Au/Ti/Pt/Au system exhibits significant reliability improvement over Au/Zn/Au. The times-to-failure for lasers with Au/Ti/Pt/Au contacts are significantly longer than the times for devices with Au/Zn/Au contacts. Chip-to-chip variation in threshold current and aging-related power changes are also smaller in the Au/Ti/Pt/Au system. Studies attribute the improved reliability of Au/Ti/Pt/Au contacts to the Pt diffusion barrier that prevents the formation of device-killing Au alloy spikes. For the Au/Zn/Au system, we estimate the temperature and current density acceleration factors of the Au diffusion by means of life-test degradation data. The activation energy determined based on stress temperatures of 40-100 °C is about 0.8 eV. The current exponent determined based on stress currents of 90-150 mA is 1.4. For the Au/Ti/Pt/Au system, the activation energy is difficult to determine due to the slow degradation. For the BH interfaces, the reliability degradation of BH semiconductor lasers may be related to damage on the sidewall generated during etching and regrowth processes. Process remedies such as melt-back have been attempted to avoid direct interaction between the buried p/n junction and the damage. We attribute the slower degradation rate in RWG lasers to the elimination of the defects in the active region. In terms of substrate quality, researchers have also reported that DSDs originating at dislocations in the substrate material can adversely impact device reliability. Several approaches, such as the incorporation of a buffer layer between the active region and the substrate, have been employed to improve reliability performance. We show that the laser reliability can, in fact, be improved with an increased buffer-layer thickness. Often, engineers cannot determine the failure times of robust optical devices in a reasonable time frame. In this circumstance, testers typically apply a fitting extrapolation using a sublinear model to estimate device failure time. Occasionally, test artifacts lead to irregularities in the life-test curve, however, and curve smoothing may produce the most accurate estimate.

References

[1] C.Y. Chang, S.M. Sze, ULSI Technology, McGraw-Hill, New York, NY, 1996.

[2] J.S. Huang, T. Nguyen, W. Hsin, I. Aeby, R. Ceballo, J. Krogen, Reliability of etched-mesa buried-heterostructure semiconductor lasers, IEEE Trans. Device Mater. Reliab. 5 (4) (2005) 665–674.

[3] G.E. Moore, Cramming more components onto integrated circuits, Electron. Mag. 38 (8) (1965) 4.

[4] R. Haitz, Haitz's law, Nat. Photonics 1 (1) (2007) 23.

[5] American Physical Society, November 17–December 23, 1947: Invention of the First Transistor, American Physical Society and College Park, MD, 1947.

[6] R.N. Hall, G.E. Fenner, J.D. Kingsley, T.J. Soltys, R.O. Carlson, Coherent light emission from GaAs junctions, Phys. Rev. Lett. 9 (9) (1962) 366–368.

[7] M.I. Nathan, W.P. Dumke, G.D. Burns, F.H. Dill, G. Lasher, Stimulated emission of radiation from GaAs p-n junctions, Appl. Phys. Lett. 1 (3) (1962) 62.

[8] S. Wolfe, Si Processing for VLSI Era: The Submicron MOSFET, vol. 3, Lattice Press, Sunset Beach, CA, 1995.

[9] P.S. Ho, T. Kwok, Electromigration in metals, Rep. Prog. Phys. 52 (1989) 301–348.

[10] J.C. Lee, I.C. Chen, C. Hu, Modeling and characterization of gate oxide reliability, IEEE Trans. Electron Dev. 35 (1988) 2268–2278.

[11] G.B. Alers, A.S. Oates, D. Monroe, K.S. Kisch, B.E. Weir, Effect of electronic corrections on the thickness dependence of thin oxide reliability, Appl. Phys. Lett. 71 (1997) 2478–2480.

[12] M.A. Alam, R.K. Smith, B.E. Weir, P.J. Silverman, Thin dielectric films: uncorrelated breakdown of integrated circuits, Nature 420 (2002) 378.
[13] K.P. Cheung, C.P. Chang, Plasma-charging damage: a physical model, J. Appl. Phys. 75 (1994) 4415–4426.
[14] M. Fukuda, Optical Semiconductor Devices, John Wiley & Sons, New York, NY, 1999 (Chapter 7).
[15] M. Fukuda, Y. Noguchi, G. Motosugi, Y. Nakano, N. Tsuzuki, O. Fujita, Suppression of interface degradation in InGaAsP/InP buried heterostructure lasers, J. Lightwave Technol. LT-5 (1987) 1778–1781.
[16] M. Fukuda, K. Wakita, G. Iwane, Dark defects in InGaAsP/InP double heterostructure lasers under accelerated aging, J. Appl. Phys. 54 (1983) 1246–1250.
[17] C.H. Henry, P.M. Petroff, R.A. Logan, F.R. Merritt, Catastrophic damage of $Al_xGa_{1-x}As$ double-heterostructure laser material, J. Appl. Phys. 50 (1979) 3721–3732.
[18] E. Kapon, Semiconductor Lasers II: Materials and Structures (Optics and Photonics), Academic Press, San Diego, CA, 1999.
[19] J. Muller, Semiconductor Laser, US Patent 7, 177, 339 (2007) 47–49.
[20] A.S. Oates, F. Nlkansah, S. Chittipeddi, Electromigration-induced drift failure of via contacts in multilevel metallization, J. Appl. Phys. 72 (6) (1992) 2227–2231.
[21] J.S. Huang, A.S. Oates, Y.S. Obeng, W.L. Brown, Asymmetrical critical current density and its influence on electromigration of two-level W-plug interconnection, J. Electrochem. Soc. 147 (10) (2000) 3840–3844.
[22] J.S. Huang, T.L. Shofner, J. Zhao, Direct observation of void morphology in step-like electromigration resistance behavior and its correlation with critical current density, J. Appl. Phys. 89 (4) (2001) 2130–2133.
[23] J. Proost, K. Maex, L. Delaey, Electromigration-induced drift in damascene and plasma-etched Al(Cu). II. Mass transport mechanisms in bamboo interconnects, J. Appl. Phys. 87 (2000) 99–109.
[24] I.P. Kaminow, L.W. Stulz, J.S. Ko, B.I. Miller, R.D. Feldman, J.C. Dewinter, M. A. Pollack, Low threshold ridge waveguide laser at 1.55 micron, Electron. Lett. 19 (21) (1983) 877–879.
[25] W.T. Tsang, R.A. Logan, A new high-power, narrow-beam transverse-mode stabilized semiconductor laser at 1.5 micron: the heteroepitaxial ridge overgrown laser, Appl. Phys. Lett. 45 (10) (1984) 1025–1027.
[26] H. Temkin, R.A. Logan, M.B. Panish, J.P. Van der Ziel, WL ~1.5 micron InGaAsP ridge lasers grown by gas source molecular-beam epitaxy, Appl. Phys. Lett. 45 (4) (1984) 330–332.
[27] M. Aoki, M. Komori, T. Tsuchiya, H. Sato, K. Nakahara, K. Uomi, InP-based reversed-mesa ridge-waveguide structure for high-performance long-wavelength laser diodes, IEEE J. Sel. Top. Quantum Electron. 3 (2) (1997) 672–683.
[28] M. Fukuda, Laser and LED reliability update, J. Lightwave Technol. 6 (10) (1988) 1488–1495.
[29] H. Oohashi, M. Fukuda, Y. Kondo, M. Wada, Y. Tohmori, Y. Sakai, H. Toba, Y. Itaya, Reliability of 1300-nm spot-size converter integrated laser diodes for low-cost optical modules in access networks, J. Lightwave Technol. 16 (7) (1998) 1302–1307.
[30] B.E.A. Saleh, M.C. Teich, Fundamentals of Photonics, second ed., Wiley, Hoboken, NJ, 2007 (Chapter 22).
[31] J.S. Huang, Design-in reliability of modern distributed feedback (DFB) InP lasers: can we meet up the stringent wavelength-division multiplex (WDM) requirement? in: IEEE Photonic Society Summer Topical, July 18–20, 2011, Montreal, Quebec, Canada, 2011, pp. 89–90.

[32] J.S. Huang, E. Isip, R. Carson, In-situ wavelength aging study and reliability thermal model of C-band 100 mW high-power DWDM lasers, in: IEEE Photonics Conference, Burlingam, CA, September 23–27, 2012, pp. 538–539.
[33] J.S. Huang, R. Miao, H. Lu, C. Wang, C. Helms, H. Demirci, Theoretical and experimental thermal analysis of InP ridge lasers on submounts and TO packages, IEEE Trans. Device Mater. Reliab. 7 (2) (2007) 363–368.
[34] R.V. Steele, Review and forecast of the laser markets, part II: diode lasers, Laser Focus World 39 (2) (2003) 63–76.
[35] P.M. Powell, Laser diodes look for long life, Photonics 37 (7) (2003) 54.
[36] L.A. Johnson, Reliability counts for laser diodes, Photonics 37 (7) (2003) 55–56.
[37] M. Fukuda, G. Iwane, Degradation of action region in InGaAsP/InP buried heterostructure lasers, J. Appl. Phys. 58 (8) (1985) 2932–2936.
[38] J.S. Huang, C.B. Vartuli, Scanning transmission electron microscopy study of Au/Zn/Au/Cr/Au and Au/Ti/Pt/Au/Cr/Au contacts to p-type InGaAs/InP, J. Appl. Phys. 93 (2003) 5196–5200.
[39] J.S. Huang, C.B. Vartuli, STEM provides the answers to some difficult questions, Comp. Semiconduct. Mag. 9 (2) (2003) 37–39.
[40] J.S. Huang, C.B. Vartuli, The effect of Cr barrier on interfacial reaction of Au/Zn/Au/Cr/Au contacts to p-type InGaAs/InP, Thin Solid Films 446 (2004) 132–137.
[41] D.G. Ivey, S. Ingrey, J.-P. Noel, W.M. Lau, Microstructural study of Ti/Pt/Au contacts to p-InGaAs, Mater. Sci. Eng. B49 (1997) 66–73.
[42] J.S. Huang, Design-in reliability of modern wavelength-division-multiplex (WDM) distributed feedback (DFB) InP lasers, Appl. Phys. Res. 4 (2) (2012) 15–28.
[43] K. Endo, S. Matsumoto, H. Kawano, I. Sakuma, T. Kamejima, Rapid degradation of InGaAsP/InP double heterostructure lasers due to $\langle 110 \rangle$ dark line defect formation, Appl. Phys. Lett. 40 (11) (1982) 921–923.
[44] A.K. Chin, C.L. Zipfel, M. Geva, I. Gamlibel, D. Skeath, B.K. Chin, Appl. Phys. Lett. 45 (1) (1984) 37–39.
[45] M.A. Foad, S. Thoms, C.D.W. Wilkinson, New technique for dry etch damage assessment of semiconductors, J. Vac. Sci. Technol. B 11 (1) (1993) 20–25.
[46] G. Morello, M. Quaglic, G. Meneghini, C. Papuzza, C. Kompocholis, Reactive ion etching induced damage evaluation for optoelectronic device fabrication, J. Vac. Sci. Technol. B 24 (2) (2006) 756–761.
[47] S.J. Pearton, F. Ren, C.R. Abernathy, W.S. Hobson, T.R. Fullowan, R. Esagui, J.R. Lothian, Damage introduction in InP and InGaAs during Ar and H_2 plasma exposure, Appl. Phys. Lett. 61 (5) (1992) 586–588.
[48] S.H. Yang, P.R. Bandaru, An experimental study of the reactive ion etching (RIE) of GaP using BCl_3 plasma processing, Mater. Sci. Eng. B143 (1–3) (2007) 27–30.
[49] R. Dylewicz, R.M. De La Rue, R. Wasielewski, P. Mazur, G. Mezosi, A.C. Bryce, Fabrication of submicron-sized features in InP/InGaAsP/AlGaInAs quantum well heterostructures by optimized inductively coupled plasma etching with $Cl_2/Ar/N_2$ chemistry, J. Vac. Sci. Technol. B 28 (4) (2010) 882–890.
[50] K.P. Cheung, C.P. Chang, Plasma-charging damage: a physical model, J. Appl. Phys. 75 (9) (1994) 4415–4426.
[51] M. Fukuda, Laser and LED reliability update, J. Lightwave Technol. 6 (10) (1988) 1488–1495.
[52] T. Kallstenius, J. Backstrom, U. Smith, B. Stoltz, On the degradation of InGaAsP/InP-based bulk lasers, J. Lightwave Technol. 17 (12) (1999) 2397–2406.
[53] H. Oohashi, M. Fukuda, Y. Kondo, M. Wada, Y. Tohmori, Y. Sakai, H. Toba, Y. Itaya, Reliability of 1300-nm spot-size converter integrated laser diodes for low-cost optical modules in access networks, J. Lightwave Technol. 16 (7) (1998) 1302–1307.

[54] C.H. Henry, D.V. Lang, Nonradiative capture and recombination by multi-phonon emission in GaAs and GaP, Phys. Rev. B 15 (1977) 989–1016.

[55] P.M. Petroff, L.C. Kimerling, Dislocation climb model in compound semiconductors with zinc blende structure, Appl. Phys. Lett. 29 (8) (1976) 461–463.

[56] S. O'Hara, P.W. Hutchinson, P.S. Dobson, The origin of dislocation climb during laser operation, Appl. Phys. Lett. 30 (8) (1977) 368–371.

[57] M. Fukuda, Reliability and Degradation of Semiconductor Lasers and LEDs, Artech House, Norwood, MA, 1991 (Chapter 4).

[58] P.W. Hutchinson, P.S. Dobson, Defect structure of degraded GaAlAs-GaAs double heterojunction lasers, Philos. Mag. 32 (1975) 745–754.

[59] A.R. Calawa, F.W. Smith, M.J. Manfra, C.L. Chen, Method of making buffer layers for III-V devices using solid phase epitaxy, US Patent 4, 952, 527 (1990).

[60] I. Akasaki, S. Nobuhiko, Process for growing III-V compound semiconductors on sapphire using a buffer layer, US Patent 4, 855, 249 (1989).

[61] H. Kawai, T. Asatsuma, K. Funato, Method for growing single crystal III-V compound semiconductor layers on non single crystal III-V compound semiconductor buffer layers, US Patent 5, 863, 811 (1999).

[62] V. Krishnamoorthy, P. Ribas, R.M. Park, Strain relief study concerning the $In_xGa_{1-x}As$/GaAs ($0.07 < x < 0.5$) material system, Appl. Phys. Lett. 58 (18) (1991) 2000–2002.

[63] M. Tamura, A. Hashimoto, Y. Nakatsugawa, Threading dislocations in $In_xGa_{1-x}As$/GaAs heterostructures, J. Appl. Phys. 72 (8) (1992) 3398–3405.

[64] M. Fukuda, Historical overview and future of optoelectronics reliability for optical fiber communication systems, Microelectron. Reliab. 40 (2000) 27–35.

[65] J.S. Huang, Temperature and current dependencies of reliability degradation of buried heterostructure semiconductor lasers, IEEE Trans. Device Mater. Reliab. 5 (1) (2005) 150–154.

[66] Generic Reliability Assurance Requirements for Optoelectronic Devices Used in Telecommunications Equipment, GR-468-CORE, Bellcore, Telcordia, Piscataway, NJ, 1998.

[67] S.K.K. Lam, R.E. Mallard, D.T. Cassidy, Analytical model for saturable aging in semiconductor lasers, J. Appl. Phys. 94 (3) (2003) 1803–1809.

[68] J.S. Huang, Reliability extrapolation methodology of semiconductor laser diodes: is quick life test feasible? IEEE Trans. Device Mater. Reliab. 6 (1) (2006) 46–51.

[69] J.S. Huang, Reliability extrapolation methodology for determination of activation energy of buried heterostructure semiconductor lasers, in: CS-MAX, October 26–28, Institute of Physics, Monterey, CA, 2004, pp. 52–54.

[70] J.R. Black, Electromigration—a brief survey and some recent results, IEEE Trans. Electron Dev. 16 (4) (1969) 338–347.

[71] Y. Twu, L.S. Cheng, S.N.G. Chu, F.R. Nash, K.W. Wang, P. Parayanthal, Semiconductor laser damage due to human-body-model electrostatic discharge, J. Appl. Phys. 74 (3) (1993) 1510–1520.

[72] J. Jeong, K.H. Park, H.M. Park, Wavelength shifts of 1.5-μm DFB lasers due to human-body-model electrostatic discharge followed by accelerated aging experiments, J. Lightwave Technol. 13 (2) (1995) 186–190.

[73] L.F. DeChiaro, C.J. Sandroff, Improvements in electrostatic discharge performance of InGaAsP semiconductor lasers by facet passivation, IEEE Trans. Electron Dev. 39 (3) (1992) 561–565.

[74] S.P. Sim, M.J. Robertson, R.G. Plumb, Catastrophic and latent damage in GaAlAs lasers caused by electrical transients, J. Appl. Phys. 55 (11) (1984) 3950–3955.

[75] M. Fukuda, F. Kano, T. Kurosaki, J. Yoshida, Behavior of lasing wavelength change duding degradation in buried-heterostructure (BH)-type DFB laser, Qual. Reliab. Eng. Int. 11 (1) (1995) 295.

[76] H.C. Neitzert, A. Piccirillo, Sensitivity of multimode bi-directional optoelectronic modeules to electrostatic discharges, Microelectron. Reliab. 39 (1999) 1863–1871.
[77] J.S. Huang, ESD breakdown characteristics of buried heterostructure semiconductor lasers, in: Compound Semiconductor Manufacturing Expo 2005, October 30–November 3, Institute of Physics, Palm Springs, CA, 2005, pp. 63–65.
[78] L.F. DeChiaro, B.A. Ungar, Degradation in InGaAsP semiconductor lasers resulting from human body model ESD, J. Electrostat. 29 (1993) 227–250.
[79] J. Wallon, G. Terol, B. Bauduin, P. Devoldere, Sensitivity to electrostatic discharges of "low-cost" 1.3 micron laser diodes: a comparative study, Mater. Sci. Eng. B 28 (1994) 314–318.
[80] S.H. Voldman, Lightning rods for nanoelectronics, Sci. Am. (2002) 91–97.
[81] Electrostatic Discharge Control in the Manufacture of Telecommunications Equipment, TR-NWT-000870, Bellcore, Telcordia, Piscataway, NJ, 1991.
[82] Electrostatic discharge (ESD) sensitivity testing human body model (HBM), JEDEC Standard JESD22-A114D, JEDEC, Arlington, VA, 2006.
[83] J.S. Huang, T. Olson, E. Isip, Human-body-model electrostatic-discharge and electrical-overstress studies of buried-heterostructure semiconductor lasers, IEEE Trans. Device Mater. Reliab. 7 (3) (2007) 453–461.
[84] F. Essely, C. Bestory, N. Guitard, A. Bafleur, A. Wislez, E. Doche, P. Perdu, A. Touboul, D. Lewis, Study of the ESD defects impact on ICs reliability, Microelectron. Reliab. 44 (9–11) (2004) 1811–1815.
[85] A. Kerber, M. Rohner, C. Wallace, L. O'Rian, M. Kerber, From wafer-level gate-oxide reliability towards ESD failures in advanced CMOS technologies, IEEE Trans. Electron Dev. 53 (4) (2006) 917–920.
[86] W.D. Greason, Z. Kucerovsky, K.W.K. Chuam, Latent effects due to ESD in CMOS integrated-circuits: review and experiments, IEEE Trans. Ind. Appl. 20 (1) (1993) 88–97.
[87] J.S. Huang, H. Lu, Size effect on ESD threshold and degradation behavior of InP buried heterostructure semiconductor lasers, Open Appl. Phys. 2 (2009) 5–10.
[88] T.R. Chen, J. Ungar, X.L. Yeh, N. Bar-Chaim, Very large bandwidth strained MQW DFB laser at 1.3 mm, IEEE Photon. Technol. Lett. 7 (5) (1995) 458–460.
[89] H. Lu, C. Blaauw, B. Benyon, G.P. Li, T. Makino, High-power and high-speed performance of 1.3-μm strained MQW gain-coupled DFB lasers, IEEE J. Sel. Top. Quantum Electron. 1 (2) (1995) 375–381.

Reliability of silicon integrated circuits

7

A.S. Oates
Taiwan Semiconductor Manufacturing Company Ltd, Hsinchu, Taiwan

7.1 Introduction

The manufacture of increasing complex integrated circuits (ICs) has been the engine that has driven the growth of the global electronic industry for the past 50 years. Looking ahead, it is clear the vital role of IC will continue: further improvements in IC technology, by either increasing performance and density with lower cost and power, as required by Moore's law, or increasing system functionality, will enable ever more ubiquitous electronics. Throughout the history of IC development, the reliability of circuits has been a primary concern. Ensuring that IC perform their function correctly over the intended lifetime has been achieved by systematically addressing degradation mechanisms as they are uncovered, either by testing prior to use or during field use. The failure rate of circuits has shown a rapid decrease with technology progression, as shown in Figure 7.1, which plots the long-term failure rate of circuits determined from accelerated testing measurements. The exponential decrease in failure rate through the first 30 or so years of IC development can be attributed to the multidisciplinary activities of technology development, manufacturing, circuit design, failure analysis, and reliability engineering professionals. While the industry achieved an acceptable level (from the system designer's perspective) of long-term reliability in the 1990s, it is by no means certain that this high level of reliability can be maintained as we continue to develop new IC technologies. The most advanced IC technologies require new materials and transistor and interconnect structures to achieve performance/power/cost goals. Reliability continues to be a major constraint on the development on these new IC technologies.

In this chapter, we will discuss the reliability of silicon ICs, paying particular attention to the failure mechanisms that generically impact all circuit types, be they digital, analog, RF, or memory. We consider only those issues that affect the IC itself; failure mechanisms associated with IC packages, or due to the interaction between the IC and the package, although an important aspect of overall reliability, are beyond the scope of this chapter. Our emphasis will be on understanding the physical processes that are responsible for degradation phenomena and the abatement techniques that are necessary. We begin in the next section by outlining the various approaches to characterizing reliability over the intended device life. Following that, we discuss the failure mechanisms that are of most significance to advanced IC technologies. Here, we will briefly enumerate mechanisms that impact the fundamental building blocks of circuits: namely, transistors, conductors, and isolating dielectrics that comprise the

Reliability Characterisation of Electrical and Electronic Systems. http://dx.doi.org/10.1016/B978-1-78242-221-1.00007-1

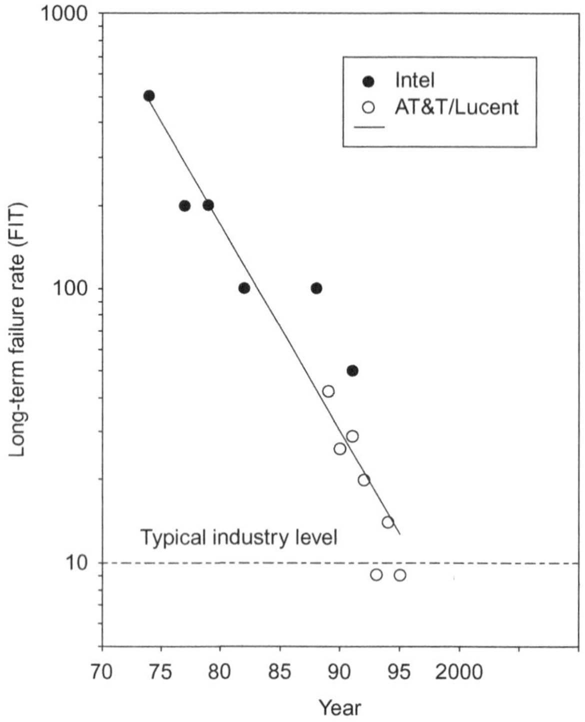

Figure 7.1 Long-term failure rates of IC estimated from accelerated test data obtained from high-temperature, high-voltage stress. Failure rates decreased exponentially with technology progression until the mid-1990s when the 10 FIT level was achieved. The challenge has been and will be to maintain this level of reliability.

circuit interconnects. We include in this section several examples that elucidate the complex relationship between the reliability of these building blocks and that of large circuits. This discussion will also include radiation-induced soft errors (SER) for both memory and logic circuits.

7.2 Reliability characterization approaches

Two distinct approaches have been developed to assess the reliability of silicon ICs. The first is termed probabilistic and is focused on determining failure rates while not attempting to specify the detailed mechanisms involved. With this approach, the goal is primarily to describe complete electronic system failure rates and to achieve reliability improvement by accommodating failure rates of individual components. The probabilistic approach is used extensively in high-reliability applications for military, aerospace, or telecommunications systems. The other approach, which we will use in this chapter, is termed deterministic and is used to describe reliability of the vast majority of circuits used in industrial, consumer, computing, and communication applications. This approach aims to understand the mechanisms responsible for failure and to model their physical and statistical properties. Reliability improvements are achieved by root-cause elimination of these failure mechanisms during circuit manufacture and design.

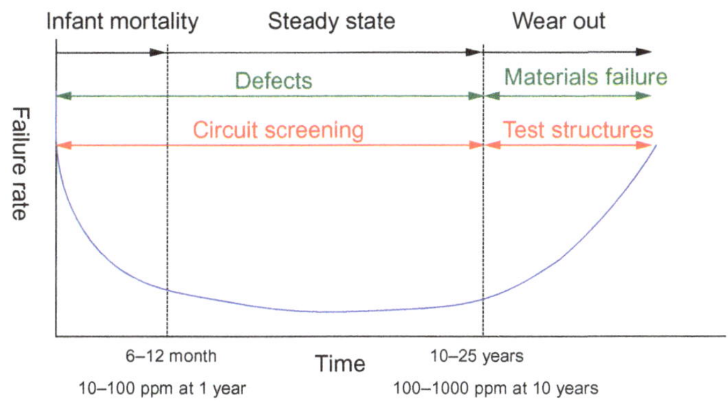

Figure 7.2 The bathtub curve for IC failure rates.

The failure rate of an electronic circuit may be generally thought of as exhibiting three distinct phases over the course of its use, as shown in Figure 7.2. Directly after manufacture, the circuit is in the infant mortality phase, where the failure rate is relatively high due to defects that are introduced during manufacture. As these defects are weeded out of the population, the failure rate decreases and eventually reaches a steady-state regime where the failure rate is constant with time. In this phase, failures are due to residual defects or, as is the case for advanced silicon technologies, from SER (i.e., intermittent) caused by the interaction of energetic atomic or subatomic particles with the circuit. Finally, the failure rate exhibits an increase with time as the device enters the wear-out phase, where the materials of the circuit degrade leading to eventual total failure of the population.

Specialized techniques are used to assess reliability in each of these three phases. Defects introduced by the manufacturing process can induce failures, but with a multitude of signatures, making the prediction of the magnitude of defect failures difficult. Consequently, screening techniques are frequently used to identify and remove defective circuits. The most important of the screening techniques is burn-in, whereby a high voltage and temperature are simultaneously applied to package circuits [1]. Other techniques to screen for defects involve voltage or temperature stress that are applied to circuits directly after manufacture at the wafer level, that is, before individual circuits are packaged. Additionally, since the defects responsible for failures are frequently similar to those that cause yield (or time-zero) failures, yield improvement activities can lead to improved early failure rates [2]. These same approaches are effective in reducing defect-related failures in the steady-state regime. However, wear-out failures, as well as SER in the steady-state regime, are an intrinsic feature of the materials of the circuit or of the transistor architecture. Therefore, the focus of activities to reduce wear-out failure rates is during the technology development phase. Since these failure mechanisms are related to specific materials and structures, it is possible to effectively design small test structures to focus on individual mechanisms, thereby allowing detailed characterization to be carried for each mechanism. Accelerated test conditions need to be applied to force failures in these test structures to occur in a practical time frame, necessitating the development of physics-based models to relate failures observed at such conditions to the lower-stress conditions of circuit operation.

7.3 Integrated circuit (IC) wear-out failure mechanisms

In this section, we highlight detailed failure mechanisms involved in Si circuit wear-out. The issues discussed here are generic in that they occur for all silicon circuits, although their relative importance varies with technology node and also circuit architecture. Each subsection begins with a brief historical perspective to provide a sense how the relative importance of mechanisms has changed with the evolution of silicon IC technologies.

7.3.1 Transistor degradation

During transistor operation, charge trapping occurs in the gate dielectric, either at pre-existing defects or at defects that are generated by the applied voltages. This trapped charge has a profound impact on the reliability of the transistor. Degradation can occur either by electrical breakdown of the dielectric, whereby the insulating properties of the gate are lost, or by shifts in the transistor's electrical parameters, such as threshold voltage or on current. Figure 7.3 provides a historical perspective that outlines the timeline of the discovery of the major transistor reliability issues. One of the earliest transistor degradation mechanisms, first observed in the early 1960s, involved changes in threshold voltage due to the presence of ionic contaminants such as Na, which was introduced during processing [3]. During the 1970s as transistor scaling

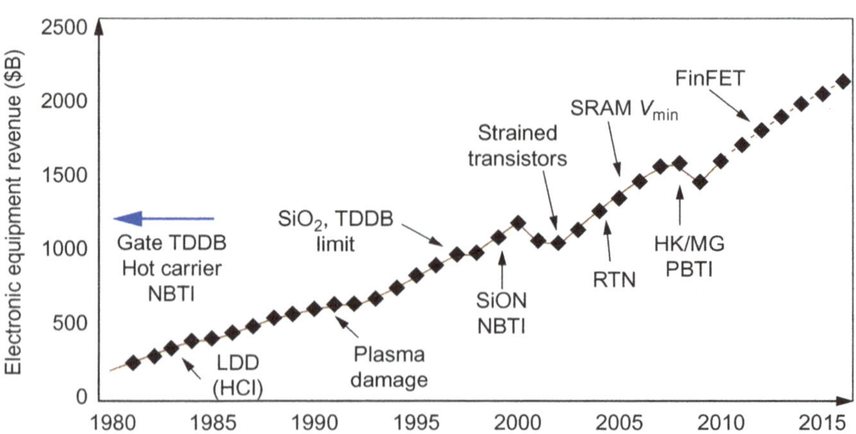

Figure 7.3 Historical progression of transistor and gate dielectric failure mechanisms versus the economic scale of the electronic industry.

was implemented to increase circuit performance and density, electrical breakdown of the gate dielectric was identified as a key reliability issue [4]. Subsequently, charge trapping and trap generation manifested itself as hot carrier injection in NMOS [5] and negative bias temperature instabilities (NBTIs) in PMOS [6]. With the advent of dry plasma cleaning, film deposition, and pattern definition techniques in the late 1980s, trap formation during processing (plasma processing-induced damage) became important as an exacerbating influence on gate dielectric and transistor degradation [7]. Fortunately, the severity of plasma damage diminished with decreasing dielectric thickness, but it remains as an important issue for relatively thick gate dielectrics (~50 Å), as, for example, used in the I/O transistors of digital logic circuits. As SiO_2-based dielectric thickness was reduced close to the practical gate leakage limit of ~14 Å, increasing electric fields aggravated NBTI degradation to the point where it became, and still remains, the dominant source of concern for long-term transistor reliability. The transition to high-*k*/metal gate dielectric stacks provided an opportunity for charge accumulation in the additional high-*k* layer, which became evident as a new failure mechanism of positive bias instability in NMOS (PBTI) [8]. Despite the increased physical complexity of the FinFET structure compared to planar transistors, this transition has apparently not obviously resulted in the introduction of new transistor degradation mechanisms.

7.3.1.1 Time-dependent dielectric breakdown of gate dielectrics

Time-dependent dielectric breakdown (TDDB) describes the increase in the gate dielectric leakage that occurs as a result of the gate voltage stress, which generates electrically active defects to form an irreversible, conductive percolation path between the cathode and the anode. Experimentally, TDDB is studied by using relatively small-area (compared with typical areas present on circuits) transistor or capacitor test structures [9]. Breakdown events are characterized as "hard" if a short is produced or "soft" if the leakage current increase is limited so that transistor operation can be preserved [10–14]. Frequently, as shown in Figure 7.4, soft breakdown is observed initially leading to an eventual hard breakdown as the stress is continued. Constant voltage (TDDB) or ramped voltages (V_{bd}) are applied to capacitor or transistor test

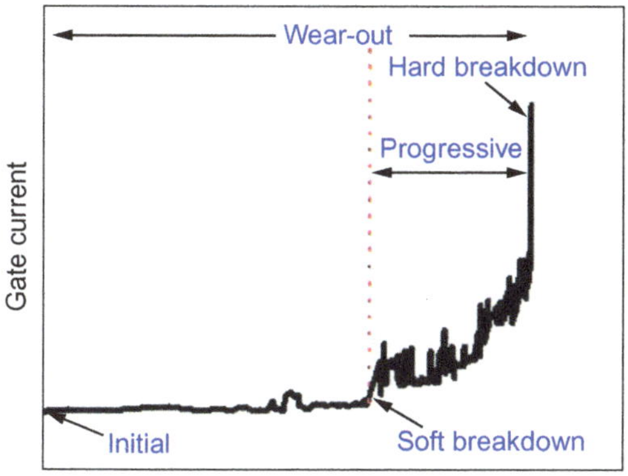

Figure 7.4 Gate leakage as a function of stress time for a SiO_2 gate dielectric of thickness 14 Å. The electrical stress leads initially to a soft breakdown where the leakage current increases as a result of the formation of a percolation path spanning the dielectric thickness. As the size of the percolation path grows, the leakage current increases, leading to eventual catastrophic breakdown.

structures, usually at elevated temperatures up to about 125 °C, to gather failure time data [9]. Failure time distributions are usually well approximated by the Weibull statistics.

The failure time of the dielectric may be expressed as [15]

$$t_f = N_{bd}/P_g J \tag{7.1}$$

where N_{bd} is the defect density for failure, P_g is the defect generation rate, and J is the gate leakage current density. A major challenge for reliability assessment is to determine the mechanism of the defect generation process, which determines the functional form of the failure time versus gate voltage relationship; several models have been proposed. The anode hole injection (the "1/E" model where $t_f \propto A \exp(G/E)$, where E is the gate dielectric field and G is a constant) model [16] that is applicable at larger electric fields (although a low-voltage modification was developed) assumes that electrons are injected from the cathode and generate holes at the anode, which then diffuse into the dielectric film causing defects. The electrochemical (the "E" model where $t_f \propto B \exp(\gamma E)$, where B and γ are constants) model relies on the process of bond polarization and breakage driven by the oxide electric field (it may be assisted by hole trapping in the oxide) [17–19]. The hydrogen release model explores widely observed strong correlations between the hydrogen content in SiO_2 and gate leakage/breakdown [20–22]. This model suggests that the hydrogen is released from the anode/oxide interface by the injected electrons and generates electrically active defects. Failure times for this process are determined by a power law dependence on the gate voltage ($t_f \propto CV_g^{-n}$ where C and n are constants). Each model has limitations in ability to describe the entire range of breakdown characteristics observed over the range of materials and structures that are of current technological importance. Typically, the E model is used when gate dielectric thickness is above about 40 Å, while below this, the power law gate voltage model is predominantly used. The transition between the E and power law models marks the change in gate conduction mechanism from the Fowler-Nordheim to direct tunneling regime. Figure 7.5 shows the large difference in failure times that are anticipated at fields typical of circuit operation from the various models. The defect density, N_{bd}, has been successfully described by percolation theory whereby defects are randomly generated during stress within the dielectric. Increased gate leakage leading to breakdown is associated with conduction through percolation paths that extend across the dielectric between the anode and the cathode electrodes. Failure statistics are, therefore, described in terms of a weak-link approximation, explaining the success of the Weibull statistics in describing failure distributions. As a result, it is relatively straightforward to extrapolate failure times measured on small-area test structures to the large areas that are typical for circuits [23].

The circuit impact of dielectric breakdown depends on the functionality of the circuit. While hard breakdowns are usually catastrophic irrespective of the circuit type, digital logic circuits are mostly impervious to soft breakdowns [24]. However, soft breakdowns can lead to functional failures of memories, particularly SRAM since these circuits are very sensitive to changes in leakage currents of transistors in the bit cell. In addition, gate leakage increase can significantly impact SRAM V_{min}, the

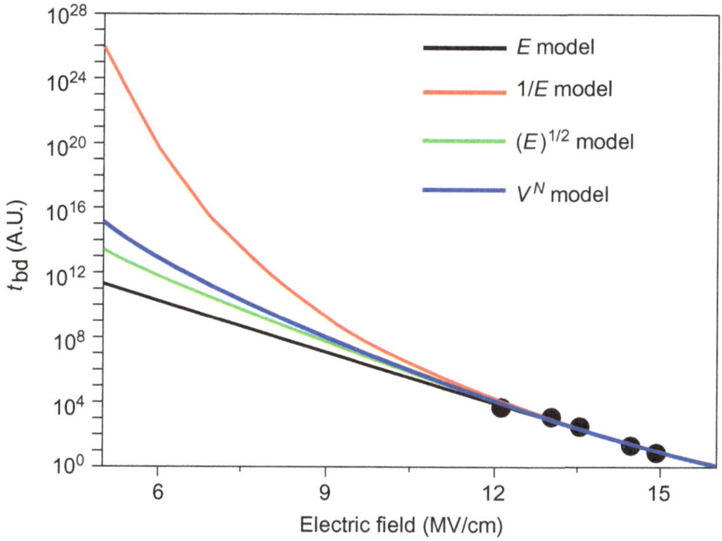

Figure 7.5 The dependence of the gate dielectric failure time on the electric field. As can be seen, experimentally obtainable data (black dots) are consistent with multiple models of failure, leading to difficulty in establishing the correct model of the failure time. The choice of the failure time model has an enormous impact at the lower fields typical of circuit use.

minimum voltage to read data, causing V_{min} to increase to unacceptably large values [25]. Soft breakdown is also a concern for RF and analog circuits such as voltage-controlled oscillators and operational amplifiers, whose performance can be significantly impacted and requires design techniques to mitigate degradation [26].

7.3.1.2 Bias temperature instabilities

Negative bias temperature instability

Shifts of the threshold voltage, V_{th}, of PMOS in the off-state with temperature and inversion gate voltage stress (NBTI) were first observed in the late 1970s [27–29]. Subsequently, it was discovered that recovery of a fraction of the threshold voltage change occurred once the bias was removed [30–32] (Figure 7.6). The V_{th} change during stressing exhibits power law dependence with time and is exponentially dependent on the electric field in the gate dielectric. The V_{th} shift, ΔV_{th}, is accurately given by the reaction-diffusion (RD) model as

$$\Delta V_{th} = A t_{ox} E_c^{2/3} \exp(2\gamma E_{ox}/3) \exp(-E_a/kT) t^n \tag{7.2}$$

where A is a constant, t_{ox} is the gate dielectric effective thickness, E_{ox} is the field across the gate dielectric, E_c is the electric field due to mobile carriers, γ is a field acceleration factor, E_a is an activation energy, T is the temperature, and k is Boltzmann's constant. The power law exponent, n, is characteristic of the degradation

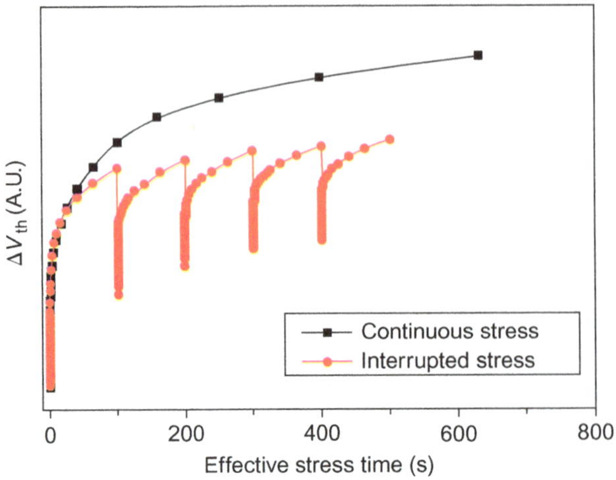

Figure 7.6 Examples of the change in threshold voltage, V_{th}, due to NBTI and the recovery of the V_{th} due to recovery. The recovery in this case is made obvious by interrupting the stress voltage by turning it off the bias and then resuming the stress.

mechanism. As a practical consideration, characterization of NBTI in transistors is problematic because V_{th} recovery occurs over a very wide time range. Care must be taken to accurately account for recovery to characterize NBTI that will be experienced during circuit operation. NBTI measured under AC bias conditions more typical of digital circuit operation is of significantly lower magnitude than under DC bias as a result of recovery effects [33–35]. As Figure 7.7 shows, failure times due to NBTI have decreased rapidly with technology scaling, posing a significant challenge to the development of the most advanced Si process technologies.

The mechanisms responsible for NBTI have been of intense interest for over 40 years, and they are still debated [30,36]. NBTI is proposed to involve positive charge generation at the gate dielectric interface with the Si channel and is observed

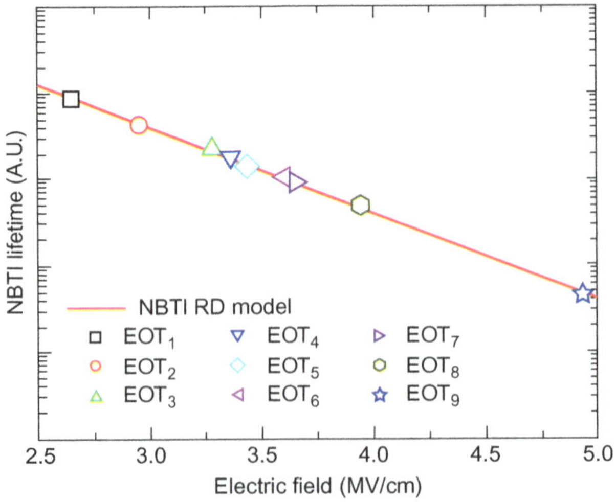

Figure 7.7 The dependence of the NBTI failure time, or lifetime, as a function of the gate dielectric field. The failure time is accurately described by the reaction-diffusion (RD) model. The data points correspond to technology nodes between 90 and 20 nm. The rapid decrease of failure time with technology progression has resulted in NBTI being one of the most serious reliability concerns for technology development.

irrespective of the choice of transistor architecture, that is, planar and FinFET, or the use of a high-k gate dielectric, since the latter usually incorporates a thin SiO_2 layer in contact with the Si substrate. The mechanism has been successfully described by the RD model [30,37–39], which suggests that interface states are formed by hydrogen release from passivated Si dangling bonds at the SiO_2/Si interface, following capture of a hole from the Si channel. Recovery involves detrapping of hydrogen when the gate bias is removed and its subsequent diffusion away from the interface in molecular form. These considerations explain why there is no noticeable difference in NBTI phenomena with high-k/SiO_2 gate stacks, since NBTI is primarily associated with the near-interface region of Si and SiO_2. However, the RD model experiences difficulties in explaining the extremely wide range of threshold voltage (V_{th}) relaxation times [36]. Alternative models have recently been proposed where the injected holes tunnel directly to and are trapped at defects in the bulk oxide. Including both interface state generation and hole trap generation in the dielectric can explain the full range of phenomena associated with NBTI [36,40]. The magnitude of NBTI observed in transistors is strongly modulated by processing conditions; in particular, the presence of nitrogen in gate dielectrics results in additional degradation. Incorporation of nitrogen is frequently used to increase the effective dielectric constant of the dielectric, thereby lowering the gate leakage current [35,41]. The presence of large quantities of nitrogen close to the Si/SiO_2 interface results in hole trap defects that strongly affect the magnitude of V_{th} shift.

Positive bias temperature instability

NMOS transistors with SiO_2 gate dielectrics are much less vulnerable to positive bias temperature instability (PBTI), where a positive bias is applied. However, with the introduction of high-k gate stacks, it was found that PBTI in NMOS was greatly enhanced and could be comparable to that of NBTI in PMOS [42]. The PBTI effect arises from trapping of inversion channel electrons in the high-k layer, which results in increased bulk trapped charge in the dielectric, while the interface layer in contact with the Si channel is unaffected [43]. PBTI also exhibits a power law dependence on V_{th} drift on time [44,45], as well as recovery of V_{th} drift when the bias is removed [46]. Since the mechanism involves bulk electron trapping, the magnitude of V_{th} shift due to PBTI decreases with the decreasing thicknesses associated with technology scaling, so that PBTI is usually less of a problem for circuit reliability than NBTI [47].

Impact of BTI on digital circuit reliability

Effective mitigation of bias temperature instabilities (BTIs) effects on reliability requires optimization of manufacturing processes and circuit design approaches. SRAM circuits are particularly vulnerable to BTI effects, since differing amounts of V_{th} drift in the PMOS pull-down (NBTI) and NMOS pull-up transistors (PBTI) of the bit cell lead to a reduction of the static noise margin of the SRAM cell [48,49]. As Figure 7.8 shows, the minimum voltage required to read the memory (V_{min}) increases, while that to write the memory decreases. The solution to this issue

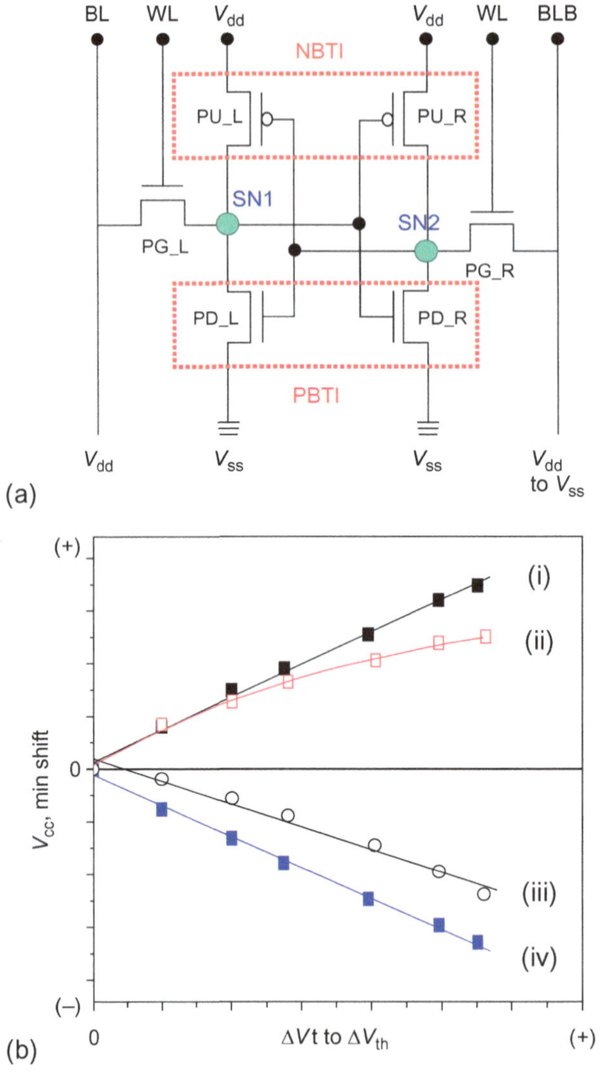

Figure 7.8 (a) Schematic layout of the SRAM cell indicating that the pull-up PMOS transistors suffer from NBTI, while the pull-down NMOS transistors are susceptible to PBTI. (b) Both mechanisms combine to cause shifts in the read margin (RM) and write margin (WM) of the bit cell, which leads to a reduction in the static noise margin and an increase in the minimum voltage to read the cell, V_{min}. (i) The increase in read margin due to NBTI in the pull up transistors; (ii) The increase in read margin due to PBTI in the pull down transistors; (iii) The decrease in the write margin due to PBTI in the pull down transistors; (iv) the decrease in write margin due to NBTI in the pull up transistors.

lies not only in decreasing transistor susceptibility to BTI but also in the design of the bit cell to reduce the read voltage distribution to a sufficiently low level that the V_{min} is rendered relatively insensitive to the increase in the read voltage. On the other hand, random digital logic circuits appear to be relatively insensitive to BTI effects, with degradation in operating frequency appearing to be much smaller than the initial post-manufacture spread in frequency [50]. Finally, we note that NBTI can pose problems for analog circuits; care must be taken to avoid application of DC bias to transistors, and the impact of BTI on transistor mismatch must be carefully assessed with specific test structures [51].

7.3.1.3 Hot carrier aging

When the transistor is turned on with appropriate gate and drain bias, degradation of the transistors parameters such as V_{th}, transconductance (g_m), and linear drain current (Id_{lin}) occurs [52–59]. In this case, the degradation is caused by charge trapping in the drain region of the transistor [52,53]. The trapping arises because of high lateral electric fields in the saturation region of the transistor that accelerate the channel carriers to the point where impact ionization occurs [54]. As a result of the vertical field in the gate, the holes and electrons are separated and either injected into the gate dielectric or ejected into the bulk of the transistor. The driving force for hot carrier aging is effectively reduced by the insertion of lightly doped drain structures into transistors to lower the lateral electric fields [60]. Models of hot carrier aging indicate a strong dependence of failure time (time to reach a threshold V_{th}, g_m, or Id_{lin} change) on the operational voltage and a strong degradation as the channel length is decreased. As a result of the relatively rapid decrease in operational voltages, hot carrier aging has not posed significant limitations for CMOS circuits. It remains, however, an important if not dominant concern for analog and RF circuits as well as power circuits because of the higher voltages involved and/or the more frequent use of NMOS circuitry.

7.3.2 Interconnect degradation

Some of the earliest failure mechanisms observed in silicon circuits involved interconnects that attach the transistors to the outside world. Figure 7.9 provides a historical perspective on the discovery of interconnect reliability issues. Formation of AuAl intermetallic compounds between Al wires and Au package interconnects (the "purple

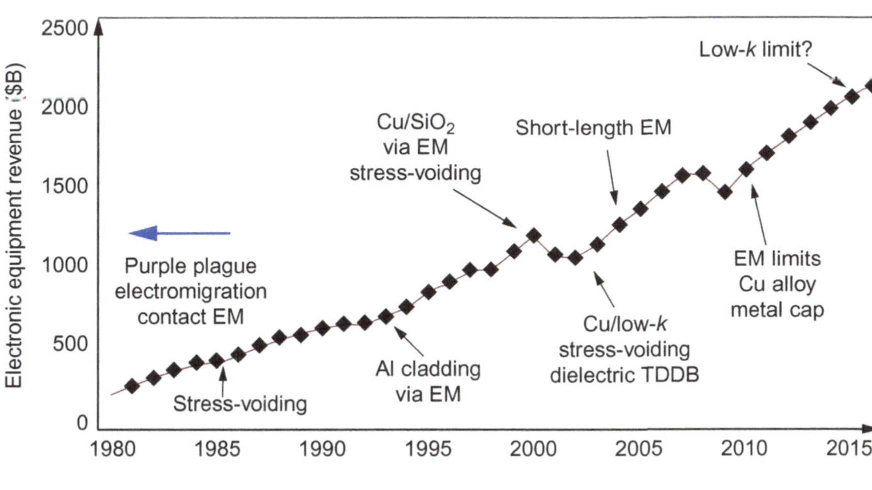

Figure 7.9 Historical progression of circuit interconnect failure mechanisms versus the economic scale of the electronic industry.

plague") was observed in the early 1960s [61]. By the mid-1960s, electromigration failure of on-chip Al interconnects was identified [62]; despite intense effort over the intervening 50 years, it still remains a critical issue for the development of the most advanced Si process technologies. Subsequently, in the 1970s, electromigration failure of the Al contacts to silicon was recognized as an important issue [63]. As Si technology approached 1 μm feature sizes during the 1980s, a new interconnect failure mechanism became evident: stress voiding [64]. This mechanism was particularly insidious since it did not require a current flow to induce open-circuit failure and arose because of the increasing mechanical stresses and stress gradients in scaled interconnects. By the late 1990s, it was widely recognized that interconnects were limiting circuit performance. In response, Cu was chosen to replace the Al conductor, and the damascene Cu trench replaced subtractive etched Al. The SiO_2 insulating layers between metal conductors also had to be replaced with lower dielectric constant materials (low *k*). The transition to Cu/low-*k* interconnects suffered from two major reliability issues. The first involved resurgence in stress voiding at vias between metal levels since there was now no refractory cladding on top of the Cu damascene trench. Limiting how vias could be placed within circuits with placement design rules solved this problem. The second issue, which had not been observed with Al/SiO_2 interconnects, was the breakdown of the insulating dielectric layers as a result of inevitable Cu contamination during processing and operation [65]. The geometry reduction associated with technology progression exacerbated this issue to the point where it now is also a critical problem for development of advanced Si process technologies.

7.3.2.1 Electromigration

Thin-film IC interconnects carry relatively high current densities $\sim 10^5$-10^6 A/cm^2, which leads to a large flux of atoms in the direction of electron flow—this is termed electromigration [66]. In regions of a conductor where this flux experiences a divergence, in particular at grain boundary triple-point intersections and at the interfaces between different materials in contacts and vias, open and short failures can occur. Following the discovery of electromigration failures in Al interconnects in the mid-1960s, it was realized that the only feasible means of prevention was to limit the current density. Since smaller currents negatively impacted circuit performance, process solutions were vigorously sought. The initial process solution to the problem involved reducing the rate of transport along grain boundaries by doping the Al layer with Cu, which segregated at grain boundaries [67,68]. Subsequently, systematic control of the grain structure was implemented to further improve lifetimes [69]. With increasing circuit integration and density, multiple levels of metal began to be employed (below the 1 μm node), requiring W-plugs as contacts between the metal levels, which exacerbated the electromigration problem since the plug interface is an inherent site of flux divergence. This led to the introduction of refractory cladding layers (e.g., TiN above and below the AlCu layer) to prevent open circuit at vias connecting metal levels. Al interconnects were industry standard until the 0.13 μm node, which saw the introduction of damascene Cu/SiO_2 and ultimately Cu/low-*k* structures for cost and performance reasons where typically the low *k* is a porous

SiO_2-based material. From a reliability perspective, Cu offered larger current-carrying capability $>10^6$ A/cm^2 due to the slower self-diffusion of Cu compared to Al.

Electromigration testing is usually carried out on test structures that consist of several hundreds of microns of Cu conductor, or small numbers of vias, by applying a constant current at and elevated temperature in the range 250-350 °C. Failures occur primarily as a result of void formation, which gives rise to resistance increases, although under some circumstances extrusions of the Cu conductor can also occur, giving rise to short circuit failures. Cu exhibits several failure modes due to electromigration [70], as shown in Figure 7.10. Voids form either in the metal trench or under vias. In a nonoptimized interconnect, it is also possible to observe voids form inside vias. Optimizations of interconnect processing to eliminate the formation of voids inside vias are important to ensure high current-carrying capability.

Failure time distributions of electromigration test structures are usually modeled with lognormal failure statistics. Failure times for long conductor lengths (>100 μm) when failures are determined by the grain structure of the conductor are described by equation [71]:

$$t_f = \frac{A}{j^n} \exp\left(\frac{E_a}{kT}\right) \quad (7.3)$$

where A is a constant that depends on the processing of the conductor, j is the current density, E_a is the activation energy for failure, T is the temperature of the conductor, and k is Boltzmann's constant. The parameter n is a constant that depends on the mechanism that limits the rate of void formation: $n=2$ when nucleation is the rate limiting step, while $n=1$ for void growth [72]. Figure 7.11 shows examples of the current density dependence of the median time to fail for microstructure-related void failures in Al interconnects [73]. However, for conductors that are terminated by vias, where failure is predominantly associated with the vias, to a good approximation, void growth kinetics dominate the failure process. For short conductor lengths (<10 μm), failure times are influenced by the existence of an electromigration-induced stress gradient within the Cu [74], which induces a backflow of atoms to counteract the electromigration flux. This leads to the existence of a critical current density below which electromigration failure does not occur. Now, the failure time is given by

$$t_f = \left(\frac{B}{j - j_c}\right) \exp\left(\frac{E_a}{kT}\right) \quad (7.4)$$

where B is a constant and j_c is the critical current density. As a result, short-length conductors can carry considerably more current than long lengths [75], where the

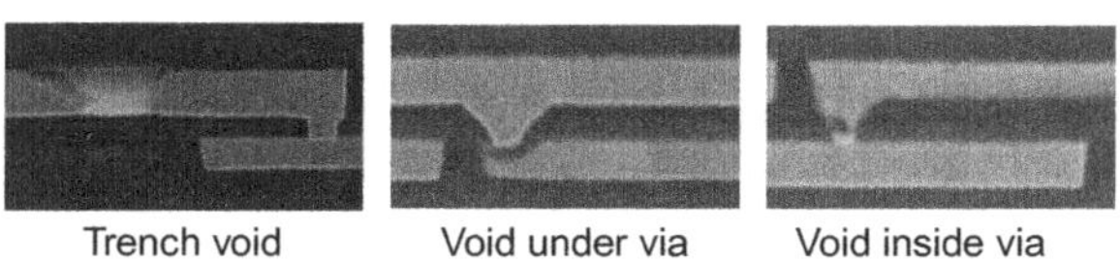

Figure 7.10 Examples of void formation associated with electromigration for Cu/low-k interconnect.

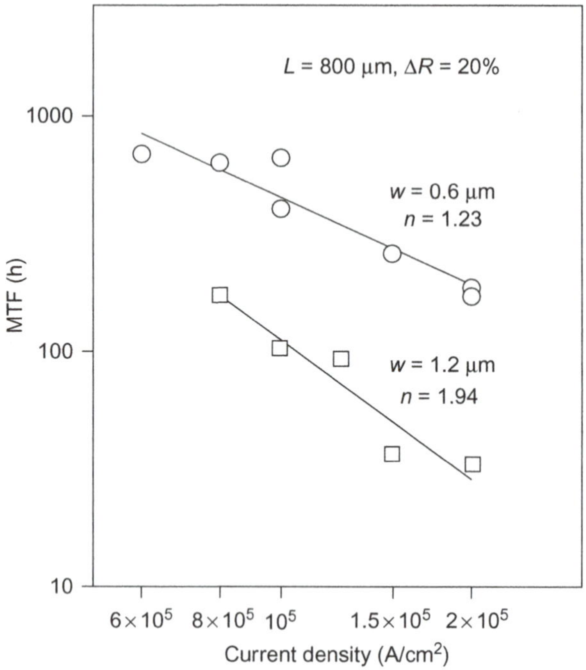

Figure 7.11 The current density dependence of the electromigration failure time for microstructure-related voiding in AlCu interconnect. The data indicate that current density exponent can vary between 1 and 2 depending on the relative importance of void nucleation and growth to the failure time.

stress gradient-induced backflow is negligible. The existence of the critical current density leads to deviations in the statistics of failure away from lognormal for short-length conductors at current densities close to the critical value, as shown in Figure 7.12.

Due to the decrease of critical feature size with scaling, as well as changes in the microstructure of the Cu, failure times have decreased significantly with technology scaling [76–78]. This reduction poses a serious challenge to attaining the increasing current densities required for continued performance enhancement and density increase. To compensate, analogs of the techniques used to improve Al

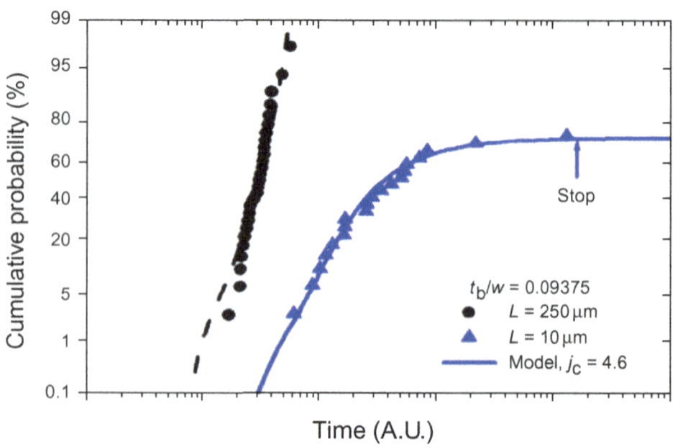

Figure 7.12 Electromigration failure distributions for Cu/low-*k* interconnect of two lengths. The longer length exhibits a lognormal failure distribution, while the shorter-length distribution deviates significantly from lognormal at high percentiles, in this case saturating at about the 70th percentile.

electromigration performance have been introduced for Cu: namely, Cu alloys to limit fast grain boundary diffusion [79–84], which controls the rate of electromigration in highly scaled Cu, and the incorporation of metallic cap layers at the Cu surface to impede diffusion along the top surface of the Cu [85,86].

7.3.2.2 Stress voiding

Stress voiding refers to the formation of voids within IC conductors as a result of the tensile mechanical stresses that are present following thin-film deposition and patterning. It became an important technological issue as feature sizes approached 1 μm in the late 1980s. The growth of stress voids is driven by two competing mechanisms: thermal stresses arising from mismatch in the coefficient of thermal expansion between the conductor and the Si substrate and vacancy diffusion that allows void growth [87]. The interplay of these mechanisms leads to the rate of void growth showing a peak typically in the temperature range 150-250 °C. Al interconnects exhibit stress voids at grain boundary intersections with the film surface, and so refractory metal cladding (as described earlier for electromigration improvement) also serves to provide electrical redundancy to limit the electrical impact of the voids. For Cu damascene interconnects, voiding occurs preferentially under vias attached to large Cu plates, where stress gradients are coupled with a large atomic sink [88]. Multiple via placement is required to prevent open-circuit failure when vias attach wider Cu plates, necessitating the use of via placement design rules. However, at the most advanced technology nodes (28 nm and below), voids can also form where narrow trenches are attached to wider plates due to large stress gradients. Voids in these regions can only be controlled by limiting stresses in the interconnect materials during processing [89]. Stress voids may also exacerbate electromigration failure by acting as preferential sites for electromigration void formation [90,91].

The development of predictive, mechanistic models of stress voiding has proved to be problematic, and so it is difficult to estimate failure times due to stress voiding from accelerated test data in a manner similar to that for electromigration failure. Most commonly, accelerated testing is performed for a fixed length of time (usually 1000 h or greater at temperatures in the range 150-250 °C) with a variety of test structures that contain a wide range of via and line geometries of interest in circuit designs. The goal of the testing is to confirm that processing conditions and allowable via configurations are optimized so that no resistance increase failures are observed during the test.

7.3.2.3 Time-dependent breakdown of interlevel dielectrics

With the introduction of Cu, breakdown of interlevel dielectric (ILD) was observed to occur at relatively low voltages, in contrast to the situation for Al interconnects. While the precise details of the breakdown mechanism still remain to be clarified, it is clearly driven by the increase in electric fields that result from feature size reduction and by Cu impurities in the dielectric that either are residual from processing or are introduced from the trench during voltage stress [92]. Unlike gate dielectrics, where soft breakdown frequently is observed, failure of ILD occurs by hard breakdown. ILD breakdown typically occurs at the locations where the electric field between conductors within the same horizontal level is a maximum and dielectric thickness is a minimum. Dielectric thickness reductions are most pronounced where intermetal via contacts are in close proximity to adjacent metal lines because of misalignment either between the metal conductors and vias during manufacture or between adjacent metal lines within the same level as a result of line edge roughness that arises from the pattern definition and etching processes. Breakdown occurs along the interface between the low-*k* dielectric and its dielectric capping layer, as shown schematically in Figure 7.13, since this location is frequently defective as a result of CMP polishing and cleaning processes. As a consequence, failure times are highly sensitive to the processing of the dielectric and in particular the interface region.

Failure time distributions are accurately described by the Weibull statistics, consistent with the weak-link nature of percolation path formation that is responsible for failure. However, failure distributions can be strongly influenced by the variability in dielectric thickness due to manufacturing variations, which can lead to significant deviations from the Weibull statistics in experimentally obtained distributions, as shown in Figure 7.14. Care must be taken to account for these deviations to obtain accurate reliability estimates [93].

Reflecting the uncertainty in the mechanism of failure, there is active debate concerning models to describe failure times. One important class of models assumes that breakdown occurs as a result of field-induced bond breakage, leading to trap generation and breakdown [94,95]. The failure time is expressed as

$$t_f = C\,\exp(\lambda E_{ox}) \tag{7.5}$$

where C is a constant, E_{ox} is the field in the ILD, and λ is the field acceleration factor and is frequently called the "E" model. A second important class of models postulates that failure occurs as a result of current passage through the ILD, which catalyzes injection of Cu into the dielectric, leading to breakdown [96,97]. Since the current through the ILD is usually determined by a Frenkel-Poole mechanism, this leads to an expression for the failure time of the form

$$t_f = D\,\exp\left(\eta\sqrt{E_{ox}}\right) \tag{7.6}$$

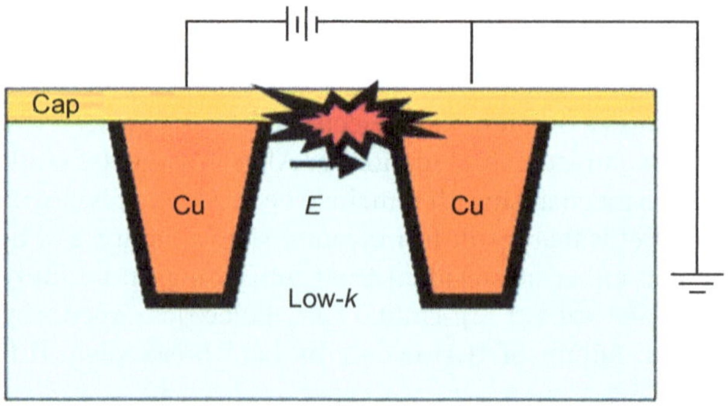

Figure 7.13 Schematic representation of the location of dielectric breakdown for a Cu/low-*k* dielectric. The breakdown occurs preferentially at the interface between the CMP polished low-*k* layer and the dielectric capping layer.

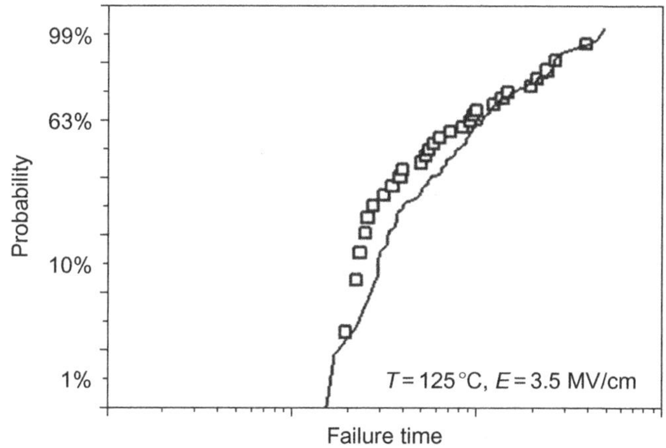

Figure 7.14 An example of a failure distribution for damascene Cu/low-*k* dielectric trench structures. There is a significant deviation from the expected Weibull distribution because of the effects of manufacturing-induced variability in the geometry of the conductor.

where D is a constant and η is a field acceleration factor. This model is colloquially known as the "square root E" model. This latter model gives more aggressive estimates of the failure time. The choice of failure time model is somewhat subjective, and it has a substantial impact on failure times extrapolated from accelerated test conditions of voltage and temperature. For conservative estimates of reliability, the E model is frequently used, while the more optimistic square root E model has found wide application for the most advanced Si technologies due to its apparently plausible physical description of the failure process.

To meet the RC performance requirements for advanced Si technologies, the capacitance of the dielectric must be reduced by lowering the dielectric constant, k. This reduction of k has been achieved by using porous SiO_2-based materials. However, these porous dielectrics are particularly vulnerable to breakdown due to their relatively low electrical breakdown strength. The pores are equivalent to shorting defects and so reduce the length of the percolation path required for breakdown. The lower limit of porous low-k reliability due to this reliability constraint is $k \sim 2.3$ [98,99]. To attain lower effective k, structures such as air gaps are being actively studied [100,101].

7.3.3 SER in Si circuits

7.3.3.1 Mechanisms and technology trends

Single event upsets (SEUs) in circuits encompass a range of phenomena associated with the interaction of energetic particles (α particles, cosmic ray neutrons and muons, energetic ions, and X-rays and γ rays) with the Si substrate. Charge generation in the active areas of transistors causes changes in internal voltages, which can lead to corruption of stored or transmitted data. Electrically, the most important consequences of SEU include temporary data loss in memories and flip-flops, transistor latchup, and corruption of logic waveforms. The errors associated with SEU are commonly referred to as SER since they usually do not cause system destruction and they can be recovered by rewriting the data.

SER were first discovered in the late 1970s in packaged DRAM [102] and were associated with alpha particles that are emitted from trace amounts of ^{238}U impurities in package materials and solder bumps. Subsequently, high-energy neutrons originating in the cosmic ray background were identified as an additional source of errors [103]. The passage of an α particle leads to the generation of electron-hole pairs. Neutrons interact via nuclear collisions to produce energetic charged species in the Si that then produce electron-hole pairs. Additional SER upset mechanisms include the capture of slow thermal neutrons by B^{10} isotopes, a reaction that produces an α particle and an energetic Li^{7+} ion [104]. More recently, memory SEUs have been identified in the most advanced Si technologies due to the interaction of cosmic ray muons with Si [105]. This latter mechanism represents a concern since muons are the most abundant particle in the cosmic ray shower. The impact of α particle SER on circuits is mitigated by control of purity of packaging materials. Neutron and muon SER requires specialized design techniques to eliminate or correct errors. Memory protection is achieved using error detection and correction techniques (ECC) for single bit upsets; bit interleaving, where bits in the same logical word are physically separated, is required to minimize multibit errors in the same word [106]. Protection of logic storage (flip-flop) elements requires design techniques such as triple modular redundancy [107], DICE design [108], or layout modification to reduce charge collection [109].

Scaling trends for SER are shown in Figure 7.15 covering the last 10 years of technology progression. These trends can be understood by considering that SER is determined by

$$\mathrm{SER} = F_{\mathrm{p}} A_{\mathrm{SA}} f(Q_{\mathrm{Q}}/Q_{\mathrm{C}}) \tag{7.7}$$

where F_{p} is the particle flux, A_{SA} is the area over which the charge collection occurs, and $f(Q_{\mathrm{Q}}/Q_{\mathrm{C}})$ is a function that describes the probability that a bit flip will occur as a result of the collection of charge Q_{C} at a sensitive node with critical charge Q_{Q}. The

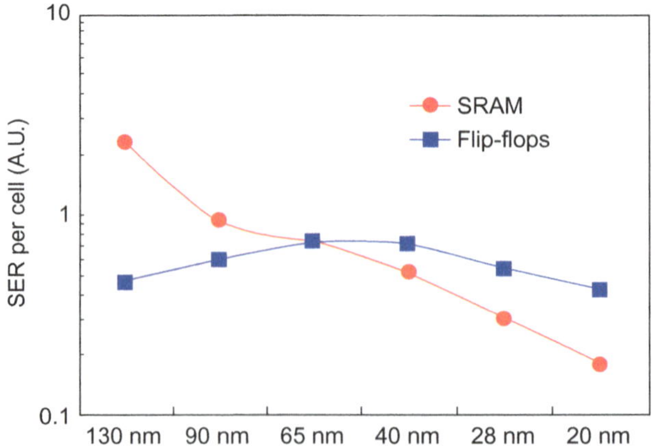

Figure 7.15 The effect of technology progression on the SER-related failure rates of SRAM and flip-flops (logic storage elements). For the most advanced technology nodes, SER decreases with scaling as a result of the decrease in SRAM and flip-flop cell size.

function $f(Q_Q/Q_C)$ depends on the energy spectrum of the incident particles and their energy transfer characteristics. SRAM shows a decreasing trend with scaling as area reduction has outpaced reduction in the critical charge. Flip-flop SER has remained approximately constant over the same technology range. The susceptibility of SRAM and flip-flops to thermal neutron SER decreases rapidly with technology progression and is not a significant concern below the 28 nm node. Limited data are available for muon SER; while it is clear that susceptibility increases for advanced nodes [110], at the present time, it is not clear how significant it will be compared to fast neutron and package alpha SER. The transition to the FinFET transistor architecture appears to be positive for SER trends [111]. Due to the restricted geometry of the fin, charge collection is suppressed compared to planar transistors, and error rates are significantly reduced, particularly for alpha particles [112]. SER of random logic is also of increasing interest as failure rates can approach that of flip-flops for GHz frequencies, and with the availability of ECC for memories, and hardened designs for flip-flops, it may become a dominant concern for logic circuit SER [113].

7.3.3.2 Simulation of circuit SER: virtual qualification

In digital circuits, a SEU may occur in any part of a circuit, that is, memory, flip-flop and latches, and random logic. To obtain an estimate of the SER of a circuit, firstly, it is necessary to estimate the SER for each part of the circuit separately. As discussed above, SER for memories and flip-flops can be obtained by performing accelerated SER tests. However, this is time-consuming and costly because of the need to design test circuits and perform accelerated testing and data analysis. Moreover, it is difficult if not impossible to experimentally determine the SER characteristics of the large number of different types of logic cell designs that are used in a large circuit. Therefore, to estimate SER of circuits, it is desirable to use simulation techniques. A fruitful approach is to first simulate the behavior of individual SRAM and flip-flop cells to SER and to then assemble these cells into the circuit and then simulate the overall response of the circuit [114]. Typically, simulations of SER behavior of SRAM and logic cells utilize TCAD together with a database of nuclear reaction data to determine the charge collection response to an energetic particle strike. The output of the charge collection simulation is then linked to a SPICE circuit simulator to predict the electrical consequences for the circuit [115]. By simulating all the various types of logic cells contained in a large circuit, it is possible to estimate the circuit SER. Figure 7.16 shows examples of the high level of accuracy that is attainable through the use of simulators to predict SER in SRAM and flip-flops.

7.4 Summary and conclusions

The rapid pace of reliability improvement achieved over the past 50 or so years underpins the rapid pace of technology progression of Si circuits as represented by Moore's law. This progress in reliability has been achieved through long-term,

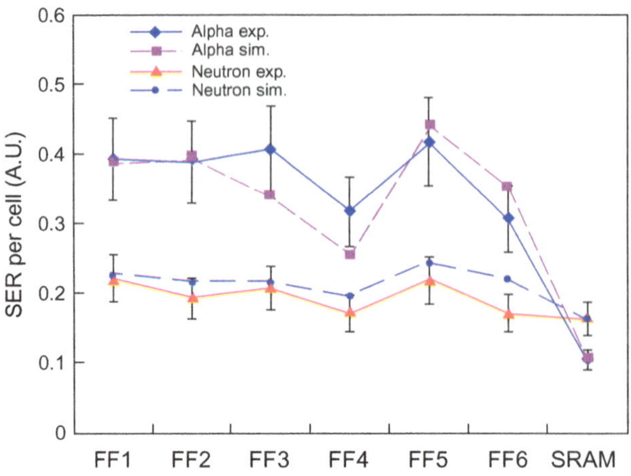

Figure 7.16 A comparison of experimental data for SER-related failure rates of several different designs of flip-flop and SRAM with simulations performed using the TFIT tool. It is now possible to accurately simulate SER of these basic circuit building blocks and to then synthesize these blocks into large circuits and simulate SER impact.

cross-disciplinary efforts that have led to a well-founded understanding of physical and electrical mechanisms of failure. Reliability is now considered from the very earliest stages of technology development and is a primary driver of decisions taken in the design of a transistor process technology. It is now widely recognized that the materials used in the manufacture of Si circuits are being pushed close to the limits imposed by reliability considerations of individual elements, such as single transistors or Cu conductors. As a consequence, it has become imperative to understand clearly how the degradation of individual elements impacts the reliability of a large circuit, which may contain billions of elements. Reliability is, therefore, evolving to be an inclusive consideration, not only for technology development and manufacturing but also for circuit design. Looking ahead, we can see that studies of reliability will continue to play an important role in the development of the Si industry: the imminent introduction of new materials and transistor structures, together their aggressive performance and power requirements, will need to be navigated while maintaining the high levels of circuit reliability we have become accustomed to.

The reader will notice from the reference list below that reliability topics are covered in a large variety of publications, in both journal and conference proceeding form. Dedicated publications dealing with Si circuit reliability, as well as microelectronic components in general, are available from IEEE, specifically Transactions on Device and Materials Reliability and the International Reliability Physics Symposium (IRPS), and Elsevier's Microelectronics Reliability. Microelectronic system reliability, and reliability theory, is covered extensively by IEEE Transactions on Reliability. Topics related to transistor reliability frequently appear in IEEE Transactions on Electron Devices, IEEE Electron Device Letters, and the International Electron Devices Meeting (IEDM). Research focused on the physics of materials used in the fabrication of transistors and interconnects appears in Journal of Applied Physics, Applied Physics Letters, and Journal of the Electrochemical Society.

Acknowledgments

The author wishes to thank Y. C. Sun for his support of this work and the Technology Reliability Physics Department of TSMC for their help in the preparation of this manuscript.

References

[1] F. Jensen, N.E. Petersen, Burn-In: An Engineering Approach to the Design and Analysis of Burn-In Procedures, John Wiley, New York, 1982.
[2] W. Riordan, R. Miller, J. Sherman, J. Hicks, Microprocessor reliability performance as a function of die location for a 0.25u, five layer metal CMOS logic process, in: IRPS, 1999, pp. 1–11.
[3] E. Snow, A.S. Grove, B.E. Deal, C.T. Sah, Ion transport phenomena in insulating films, J. Appl. Phys. 36 (1965) 1664–1673.
[4] C.R. Barrett, R.C. Smith, Failure Modes and Reliability of Dynamic RAMs, IEDM, Washington, DC, 1976.
[5] T.H. Ning, P.W. Cook, R.H. Dennard, C.M. Osburn, S.E. Schuster, H. Yu, 1 μm MOSFET VLSI technology, part IV—hot-electron design considerations, IEEE Trans. Electron Devices ED-26 (1979) 346–353.
[6] K.O. Jeppson, C.M. Svensson, Negative bias stress of MOS devices at high electric fields and degradation of MNOS devices, J. Appl. Phys. 48 (1977) 2004–2014.
[7] S.P. Fang, J.P. McVittie, Thin-oxide damage from gate charging during plasma processing, IEEE Electron Device Lett. 13 (1992) 288.
[8] S. Zafar, A. Callegari, E. Gusev, M.V. Fischetti, Charge trapping in high k gate dielectric stacks, in: IEDM Technical Digest, 2002, p. 517.
[9] R.P. Volertsen, Pragmatic procedure for the estimation of thin dielectric reliability, Qual. Reliab. Eng. Int. 8 (1992) 557–564.
[10] K. Okada, Extended time dependent dielectric breakdown model based on anomalous gate area dependence of lifetime in ultra thin silicon dioxide, Jpn. J. Appl. Phys. 36 (1-3B) (1997) 1443–1447.
[11] S.-H. Lee, H.-J. Cho, J.-C. Kim, S. H. Choi, Quasi-breakdown in ultrathin gate oxides under high field stress, in: IEDM Technical Digest, 1994, pp. 605–608.
[12] K. Okada, S. Kawasaki, Y. Hirofuji, New experimental findings on stress induced leakage current of ultrathin silicon dioxides, in: International Conference on Solid State Devices and Materials, 1994, pp. 565–567.
[13] M. Depas, T. Nigam, M.H. Heynes, Soft breakdown in ultrathin gate oxide layers, IEEE Trans. Electron Devices 43 (1996) 1499–1507.
[14] M. Depas, R. Degraeve, G. Groeseneken, M. Heynes, Reliability of ultrathin gate oxide below 3 nm in the direct tunneling regime, Jpn. J. Appl. Phys. 36 (38) (1997) 1602–1608.
[15] D.J. DiMaria, J.H. Stathis, Non-Arrhenius temperature dependence of reliability in ultrathin silicon dioxide films, Appl. Phys. Lett. 74 (12) (1999) 1752–1754.
[16] J.W. McPherson, Time-dependent dielectric breakdown physics—models revisited, Microelectron. Reliab. 52 (2012) 1753–1760.
[17] J.W. McPherson, D.A. Baglee, Acceleration factors for thin gate oxide stressing, in: Proceedings of the International Reliability Physics Symposium, March 1985, pp. 1–5.

[18] J.W. McPherson, D.A. Baglee, Acceleration factors for thin gate oxide breakdown, J. Electrochem. Soc. 132 (8) (1985) 1903–1908.
[19] P.E. Nicollian, W.R. Hunter, J. Hu, Experimental evidence for voltage driven breakdown models in ultra thin gate oxides, in: Proceedings of IRPS, 2000, pp. 7–15.
[20] J. Stathis, D. Dimaria, Reliability projection for ultra-thin oxides at low voltage, in: Technical Digest of Papers International Electron Devices Meeting, 1998, pp. 167–170.
[21] D. DiMaria, E. Cartier, D. Arnold, Impact ionization, trap creation, degradation, and breakdown in silicon dioxide films on silicon, J. Appl. Phys. 73 (1993) 3367–3384.
[22] E.Y. Wu, J. Sune, On voltage acceleration models of time to breakdown—part I: experimental and analysis methodologies, IEEE Trans. Electron Devices 56 (7) (2009) 1433–1441.
[23] D.R. Wolters, J.F. Verwey, Breakdown and wear-out phenomena in SiO_2 films, in: B. Barbottin, A. Vapaille (Eds.), Instabilities in Silicon Devices, Elsevier, Amsterdam, The Netherlands, 1986 (Chapter 6).
[24] B. Kaczer, R. Degraeve, G. Groeseneken, M. Rasras, S. Kubicek, E. Vandamme, G. Badenes, Impact of MOSFET oxide breakdown on digital circuit operation and reliability, in: IEDM, 2000, p. 22.7.1.
[25] R. Rodríguez, J.H. Stathis, B.P. Linder, S. Kowalczyk, C.T. Chuang, R.V. Joshi, G. Northrop, K. Bernstein, A.J. Bhavnagarwala, S. Lombardo, The impact of gate-oxide breakdown on SRAM stability, IEEE Electron Device Lett. 23 (9) (2002) 559.
[26] E. Xiao, J.S. Yuan, H. Yang, CMOS RF and DC reliability subject to hot carrier stress and oxide soft breakdown, IEEE Trans. Device Mater. Reliab. 4 (2004) 92.
[27] Y. Miura, Y. Matukura, Investigation of silicon-silicon dioxide interface using MOS structure, Jpn. J. Appl. Phys. 5 (1966) 180.
[28] A. Goetzberger, H.E. Nigh, Surface charge after annealing of Al-SiO_2-Si structures under bias, Proc. IEEE 54 (1966) 1454.
[29] A. Goetzberger, A.D. Lopez, R.J. Strain, On the formation of surface states during stress aging of thermal Si-SiO_2 interfaces, J. Electrochem. Soc. 120 (1973) 90–96.
[30] M.A. Alam, A critical examination of the mechanics of dynamic NBTI for PMOSFETs, in: IEDM, 2003, p. 14.4.1.
[31] M. Ershov, S. Saxena, H. Karbasi, S. Winters, S. Minehane, J. Babcock, R. Lindley, P. Clifton, M. Redford, A. Shibkov, Dynamic recovery of negative bias temperature instability in p-type metal-oxide-semiconductor field-effect transistors, Appl. Phys. Lett. 83 (2003) 1647.
[32] D.J. Breed, A new model for the negative voltage instability in MOS devices, Appl. Phys. Lett. 26 (1975) 116.
[33] V. Huard, F. Monsieur, G. Ribes, S. Bruyere, Evidence for hydrogen-related defects during NBTI stress in p-MOSFETs, in: IRPS, 2003, p. 178.
[34] M. Ershov, R. Lindley, S. Saxena, A. Shibkov, S. Minehane, J. Babcock, Transient effects and characterization methodology of negative bias temperature instability in p-MOS transistors, in: IRPS, 2003, p. 606.
[35] S.S. Tan, T.P. Chen, J.M. Soon, K.P. Loh, C.H. Ang, W.Y. Teo, Neighboring effect in nitrogen-enhanced negative bias temperature instability, in: SSDM, 2003, p. 70.
[36] T. Grasser, W. Gos, V. Sverdlov, B. Kaczer, The universality of NBTI relaxation and its implications for modeling and characterization, in: IRPS, 2007, p. 268.
[37] H. Küflüoglu, M.A. Alam, Theory of interface-trap-induced NBTI degradation for reduced cross section MOSFETs, IEEE Trans. Electron Devices 53 (2006) 1120.
[38] A.E. Islam, H. Kufluoglu, D. Varghese, S. Mahapatra, M.A. Alam, Recent issues in negative-bias temperature instability: initial degradation, field dependence of interface

trap generation, hole trapping effects, and relaxation, IEEE Trans. Electron Devices 54 (2007) 2143.

[39] J.H. Stathis, S. Zafar, The negative bias temperature instability in MOS devices: a review, Microelectron. Reliab. 46 (2006) 270.

[40] T. Grasser, B. Kaczer, W. Goes, T. Aichinger, P. Hehenberger, M. Nelhiebel, A two-stage model for negative bias temperature instability, in: IRPS, 2009, p. 33.

[41] N. Kimizuka, K. Yamaguchi, K. Imai, T. Iizuka, C.T. Liu, R.C. Keller, T. Horiuchi, NBTI enhancement by nitrogen incorporation into ultrathin gate oxide for 0.10-μm gate CMOS generation, in: VLSI, 2000, p. 92.

[42] E. Cartier, B.P. Linder, V. Narayanan, V.K. Paruchuri, Fundamental understanding and optimization of PBTI in nFETs with SiO_2/HfO_2 gate stack, in: IEDM, 2006, p. 1.

[43] K. Torii, H. Kitajima, T. Arikado, K. Shiraishi, S. Miyazaki, K. Yamabe, M. Boero, T. Chikyow, K. Yamada, Physical model of BTI, TDDB and SILC in HfO_2-based high-k gate dielectrics, in: IEDM, 2004, p. 129.

[44] R. Degraeve, M. Aoulaiche, B. Kaczer, P. Roussel, T. Kauerauf, S.A. Sahhaf, G. Groeseneken, Review of reliability issues in high-k/metal gate stacks, in: IPFA, 2008, p. 1.

[45] A. Kerber, S.A. Krishnan, E.A. Cartier, Voltage ramp stress for bias temperature instability testing of metal-gate/high-stacks, IEEE Electron Device Lett. 30 (2009) 1347.

[46] T. Grasser, B. Kaczer, P. Hehenberger, W. Gos, R. O'Connor, H. Reisinger, W. Gustin, C. Schunder, Simultaneous extraction of recoverable and permanent components contributing to bias-temperature instability, in: IEDM, 2007, p. 801.

[47] S. Pae, J. Maiz, C. Prasad, B. Woolery, Effect of BTI degradation on transistor variability in advanced semiconductor technologies, IEEE Trans. Device Mater. Reliab. 8 (2008) 519.

[48] J.C. Lin, A.S. Oates, H.C. Tseng, Y.P. Liao, T.H. Chung, K.C. Huang, P.Y. Tong, S. H. Yau, Y.F. Wang, Prediction and control of NBTI-induced SRAM Vccmin drift, in: IEDM, 2006, p. 1.

[49] J.C. Lin, A.S. Oates, C.H. Yu, Time dependent Vccmin degradation of SRAM fabricated with high-k gate dielectrics, in: International Reliability Physics Symposium, 2007, p. 439.

[50] B. Vaidyanathan, S. Bai, A.S. Oates, The relationship between transistor-based and circuit-based reliability assessment for digital circuits, in: International Reliability Physics Symposium, 2011, p. CR.4.1.

[51] Y. Chen, J. Zhou, S. Tedja, F. Hui, A.S. Oates, Stress-induced MOSFET mismatch for analog circuits, International Reliability Workshop Final Report, 2001, p. 41.

[52] S. Sze, Physics of Semiconductor Devices, John Wiley and Sons, New York, 1981.

[53] E. Takeda, C.Y. Yang, A. Miura-Hamada, Hot-Carrier Effects in MOS Devices, Academic Press, San Diego, 1995.

[54] A. Duncan, U. Ravaioli, J. Jakumeit, Full-band Monte-Carlo investigation of hot-carrier trends in the scaling of metal-oxide-semiconductor field-effect transistors, IEEE Trans. Electron Devices 45 (4) (1998) 867.

[55] C. Papadas, N. Revil, G. Ghibaudo, E. Vincent, On the threshold voltage shift after hot carrier injection in deep submicron N-channel MOSFETs—a quasi uniform approach, Microelectron. Eng. 28 (1-4) (1995) 361.

[56] R. Woltjer, G. Paulzen, H. Lifka, P. Woerlee, Positive oxide-charge generation during 0.25 μm PMOSFET hot-carrier degradation, IEEE Electron Device Lett. 15 (10) (1994) 427.

[57] W. Weber, I. Borchert, Hot-hole and electron effects in dynamically stressed n-MOSFETs, in: European Solid State Device Research Conference, Berlin, Springer-Verlag, New York, 1989, p. 719.

[58] C. Hu, S. Tam, F. Hsu, P. Ko, T. Chan, K. Terrill, Hot-electron-induced MOSFET degradation-model, monitor, and improvement, IEEE Trans. Electron Devices ED-32 (2) (1985) 375.

[59] S. Tam, P. Ko, C. Hu, Lucky-electron model of channel hot-electron injection in MOSFET's, IEEE Trans. Electron Devices ED-31 (9) (1984) 1116.
[60] D. Baglee, C. Duvvury, M. Smayling, M. Duane, Light doped drain transistors for advanced VLSI circuits, IEEE Trans. Electron Devices 32 (1985) 896–902.
[61] E. Philofsky, Purple plague revisited, in: Proceedings of the International Reliability Physics Symposium, 1970, p. 177.
[62] I.A. Blech, Rome Air Development Center, Contr. No. AF30(602)-3776, December 1965.
[63] G.S. Prokop, R.R. Joseph, Electromigration failure at aluminum-silicon contacts, J. Appl. Phys. 43 (1972) 2595.
[64] J. Klema, R. Pyle, E. Domangue, Reliability implications of nitrogen contamination during deposition of sputtered aluminum/silicon metal films, in: IEEE International Reliability Physics Symposium, 1984, p. 1.
[65] A.L.S. Loke, J.T. Wetzel, R. Changsup, W.J. Lee, S.S. Wong, Copper drift in low-K polymer dielectrics for ULSI metallization, in: IEEE VLSI Technology Symposium, 1998, p. 26.
[66] I.A. Blech in Rome Air Develop. Center, Contract No. AF30(602)-3776, December 1965.
[67] R. Rosenberg, Inhibition of electromigration damage in thin films, J. Vac. Sci. Technol. 9 (1972) 263.
[68] I. Ames, F.M. D'Heurle, R.E. Horstmann, Reduction of electromigration in aluminum films by copper doping, IBM J. Res. Dev. 14 (1970) 461.
[69] S. Vaidya, A. Sinha, Effect of texture and grain structure on electromigration in Al-0.5% Cu thin films, Thin Solid Films 75 (3) (1981) 253.
[70] S.C. Lee, A.S. Oates, Identification and analysis of dominant electromigration failure modes in copper/low-k dual damascene interconnects, in: Proceedings of the International Reliability Physics Symposium (IRPS), 2006, p. 107.
[71] J.R. Black, Electromigration—a brief survey and some recent results, IEEE Trans. Electron Devices 16 (1969) 338.
[72] J.J. Clement, Vacancy supersaturation model for electromigration failure under dc and pulsed dc stress, J. Appl. Phys. 71 (1992) 4264.
[73] A.S. Oates, Electromigration transport mechanisms in Al thin-film conductors, Appl. Phys. Lett. 66 (1995) 1475.
[74] I.A. Blech, Electromigration in thin aluminum films on titanium nitride, J. Appl. Phys. 47 (1976) 1203.
[75] A.S. Oates, Electromigration failure distribution of contacts and vias as a function of stress conditions in submicron IC metallizations, in: International Reliability Physics Symposium, 1996, p. 164.
[76] A.S. Oates, M.H. Lin, The scaling of electromigration lifetimes, in: Proceedings of the International Reliability Physics Symposium (IRPS), 2012, p. 6B.2.1.
[77] A.S. Oates, M.H. Lin, The impact of trench width and barrier thickness on scaling of the electromigration short-Length effect in Cu/low-k interconnects, in: Proceedings of the International Reliability Physics, 2013, p. 3F.1.1.
[78] C.-K. Hu, et al., Effects of overlayers on electromigration reliability improvement for Cu/low K interconnects, in: Proceedings of the International Reliability Physics Symposium (IRPS), 2004, p. 222.
[79] K.L. Lee, C.K. Hu, K.N. Tu, In situ scanning electron microscope comparison studies on electromigration of Cu and Cu (Sn) alloys for advanced chip interconnects, J. Appl. Phys. 78 (1995) 4428.
[80] S. Yokogawa, H. Tsuchiya, Y. Kakuhara, K. Kikuta, Analysis of Al doping effects on resistivity and electromigration of copper interconnects, Trans. Device Mater. Reliab. 8 (2008) 216.

[81] Y. Hayashi, N. Matsunaga, M. Wada, S. Nakao, K. Watanabe, A. Sakata, H. Shibata, Low resistive and highly reliable copper interconnects in combination of silicide-cap with Ti-barrier for 32 nm-node and beyond, in: Proceedings of the International Interconnect Technology Conference (IITC), 2009, p. 252.
[82] T. Watanabe, H. Nasu, T. Sui, G. Minmihaba, N. Kurashima, A. Gawase, M. Shimada, Y. Yoshimizu, Y. Uozumi, H. Shibata, Self-formed barrier technology using CuMn alloy seed for copper dual-damascene interconnect with porous-SiOC/porous-PAr hybrid dielectric, in: Proceedings of the International Interconnect Technology Conference (IITC), 2007, p. 7.
[83] C. Christiansen, B. Li, M. Angyal, T. Kane, V. McGahay, Y.Y. Wang, S. Yao, Electromigration-resistance enhancement with CoWP or CuMn for advanced Cu interconnects, in: Proceedings of the International Reliability Physics Symposium (IRPS), 2011, p. 3E.3.1.
[84] M.H. Lin, A.S. Oates, The effects of Al doping and metallic-cap layers on electromigration transport mechanisms in copper nanowires, Trans. Device Mater. Reliab. 11 (2011) 540.
[85] T. Usami, et al., Highly reliable interface of self-aligned CuSiN process with low-k Sic barrier dielectric (k3.5) for 65 nm node and beyond, in: Proceedings of the International Interconnect Technology Conference (IITC), 2006, p. 125.
[86] O. Aubel, et al., Comprehensive reliability analysis of CoWP Metal Cap unit processes for high volume production in sub-μm dimensions, in: Proceedings of the International Reliability Physics Symposium (IRPS), 2008, p. 675.
[87] J.W. McPherson, C.F. Dunn, A model for stress-induced metal notching and voiding in very large-scale-integrated Al–Si (1%) metallization, J. Vac. Sci. Technol. B 5 (1987) 1321.
[88] E.T. Ogawa, J.W. McPherson, J.A. Rosal, K.J. Dickerson, T.-C. Chiu, L.Y. Tsung, M. K. Jain, T.D. Bonifield, J.C. Ondrusek, W.R. McKee, Leakage, breakdown, and TDDB characteristics of porous low-k silica-based interconnect dielectrics, in: IEEE International Reliability Physics Symposium, 2003, p. 312.
[89] C.C. Lee, A.S. Oates, A new stress migration failure mode in highly scaled Cu/low-interconnects, IEEE Trans. Device Mater. Reliab. 12 (2012) 529.
[90] A.S. Oates, S. Lytle, The effect of stress-induced voiding on electromigration, J. Appl. Phys. 71 (1992) 174.
[91] A.S. Oates, Electromigration in stress-voided Al alloy conductors, in: International Reliability Physics Symposium, 1993, p. 297.
[92] J. Noguchi, T. Saito, N. Ohashi, H. Ashihara, H. Maruyama, M. Kubo, H. Yamaguchi, D. Ryuzaki, K.-I. Takeda, K. Hinode, Impact of low-K dielectrics and barrier metals on TDDB lifetime of Cu interconnects, in: Proceedings of the International Reliability Physics Symposium, 2001, pp. 355–359.
[93] S.C. Lee, A.S. Oates, Geometric variability of nanoscale interconnects and its impact on the time-dependent breakdown of Cu/low-k dielectrics, IEEE Trans. Device Mater. Reliab. 10 (2010) 307.
[94] J.W. McPherson, H.C. Mogul, Underlying physics of the thermal chemical E model in describing low-field time-dependent dielectric breakdown in SiO_2 thin film, J. Appl. Phys. 84 (3) (1998) 1513–1523.
[95] J. Kim, E.T. Ogawa, J.W. McPherson, Time dependent dielectric breakdown characteristics of low-k dielectric (SiOC) over a wide range of test areas and electric fields, in: Proceedings of the International Reliability Physics Symposium, 2007, pp. 399–404.
[96] F. Chen, O. Bravo, K. Chanda, P. Mclaughlin, T. Sullivan, J. Gill, J. Lloyd, R. Kontra, J. Aitken, A comprehensive study of low-k SiCOH TDDB phenomena and its reliability

lifetime model development, in: Proceedings of the International Reliability Physics Symposium, 2006, pp. 46–53.
[97] N. Suzumura, S. Yamamoto, D. Kodama, K. Makabe, J. Komori, E. Murakami, S. Maegawa, K. Kubota, A new TDDB degradation model based on Cu ion drift in Cu interconnect dielectrics, in: Proceedings of the International Reliability Physics Symposium, 2006, pp. 484–489.
[98] S.C. Lee, A.S. Oates, Reliability limitations to the scaling of porous low-k dielectrics, in: Proceedings of the International Reliability Physics Symposium, 2011, p. 2F.5.1.
[99] S.-C. Lee, A.S. Oates, K.-M. Chang, Fundamental understanding of porous low-k dielectric breakdown, in: Proceedings of the International Reliability Physics Symposium, 2009, pp. 481–485.
[100] J. Nogochi, T. Oshima, T. Matsumoto, S. Uno, K. Sato, Multilevel interconnect with air-gap structure for next-generation interconnections, IEEE Trans. Electron Devices 56 (11) (2009) 2675–2682.
[101] M. Pantouvaki, F. Sebaai, K. Kellens, D. Goossens, B. Vereecke, J. Versluijs, R. Caluwaerts, K. Marrant, H. Bender, A. Moussa, H. Struyf, G.P. Beyer, Dielectric reliability of 70 nm pitch air-gap interconnect structures, Microelectron. Eng. 88 (7) (2011) 1618–1622.
[102] T. May, M. Woods, Alpha-particle-induced soft errors in dynamic memories, IEEE Trans. Electron Devices ED-26 (1979) 2–9.
[103] J.F. Ziegler, W.A. Lanford, Effect of cosmic rays on computer memories, Science 206 (4420) (1979) 776–788.
[104] R.C. Baumann, E.B. Smith, Neutron-induced boron fission as a major source of soft errors in deep submicron SRAM devices, in: Proceedings of the IEEE 38th Annual International Reliability Physics Symposium, 2000, pp. 152–157.
[105] B.D. Sierawski, M.H. Mendenhall, R.A. Reed, M.A. Clemens, R.A. Weller, R.D. Schrimpf, E.W. Blackmore, M. Trinczek, B. Hitti, J.A. Pellish, R.C. Baumann, S.-J. Wen, R. Wong, N. Tam, Muon-induced single event upsets in deep-submicron technology, IEEE Trans. Nucl. Sci. 57 (6) (2010) 3273–3278.
[106] B. Gold, M. Ferdman, B. Falsafi, K. Mai, Mitigating multi-bit soft errors in L1 caches using last store prediction, in: Proceedings of the Workshop on Architectural Support for Gigascale Integration, June 2007.
[107] J.F. Wakerly, Transient failures in triple modular redundancy systems with sequential modules, IEEE Trans. Comput. 34 (6) (1976) 889–895.
[108] T. Calin, M. Nicolaidis, R. Velazco, Upset hardened memory design for submicron CMOS technology, IEEE Trans. Nucl. Sci. 43 (6) (1996) 2874–2878.
[109] H.-H.K. Lee, K. Lilja, M. Bounasser, P. Relangi, I.R. Linscott, U.S. Inan, S. Mitra, LEAP: layout design through error-aware transistor positioning for soft-error resilient sequential cell design, in: Proceedings of the IEEE International Reliability Physics Symposium (IRPS), 2010, pp. 203–212.
[110] B.D. Sierawski, R.A. Reed, M.H. Mendenhall, R.A. Weller, R.D. Schrimpf, S.-J. Wen, R. Wong, N. Tam, R.C. Baumann, Effects of scaling on muon-induced soft errors, in: Proceedings of the IEEE International Reliability Physics Symposium, 2011, pp. 247–252.
[111] Y.-P. Fang, A.S. Oates, Thermal neutron-induced soft errors in advanced memory and logic devices, IEEE Trans. Device Mater. Reliab. 14 (1) (2014) 583–586.
[112] N. Seifert, B. Gill, S. Jahinuzzaman, J. Basile, V. Ambrose, Q. Shi, R. Allmon, A. Bramnik, Soft error susceptibilities of 22 nm trigate devices, IEEE Trans. Nucl. Sci. 59 (6) (2012) 2666–2673.

[113] N.N. Mahatme, et al., Comparison of combinational and sequential error rates for a deep submicron process, IEEE Trans. Nucl. Sci. 58 (2011) 2719–2725.

[114] H. Chapman, E. Landman, A. Margalit-Ilovich, Y.-P. Fang, A.S. Oates, D. Alexandrescu, O. Lauzeral, A multi-partner soft error rate analysis of an InfiniBand Host Channel Adapter, in: Proceedings of 2010 IEEE Workshop on Silicon Errors in Logic-System Effects (SELSE), 2010.

[115] Y.-P. Fang, A.S. Oates, K. Belhaddad, V. Correas, I. Nofal, O. Lauzeral, Neutron and alpha soft error rate simulations for memory and logic devices at advanced technologies, in: Proceedings of 2013 IEEE Workshop on Silicon Errors in Logic-System Effects (SELSE), 2013.

Reliability of emerging nanodevices

8

N. Raghavan, K.L. Pey
Singapore University of Technology and Design (SUTD), Singapore

8.1 Introduction to emerging nanodevices

Any modern-day electronic circuit comprises of two key components—the "logic unit" that serves as the computational brain of the circuit and the "memory unit" that, as the name suggests, stores the processed data in a digital binary format representing "1" and "0." Both these units have to communicate with each other, and a reliable interface is needed to realize the robust design of an integrated circuit. The fundamental device that assists in logic function is the transistor, illustrated in Figure 8.1 (metal-oxide-semiconductor field-effect transistor (MOSFET)), which is of two types—the *p-type* and the *n-type*, both of which complement each other in their polarity dependence for driving current [2,3]. The *p-type* transistor (rich in holes as carriers of current) has a silicon substrate doped with acceptor impurities (typically boron, B), and the *n-type* transistor (rich in electrons as carriers of current) has a silicon substrate doped with donor impurities (typically phosphorus (P) or arsenic (As)) [3]. All transistors, which are four-point terminal devices, have a threshold voltage (V_{TH}) (Figure 8.1(f)) [1], which is the minimum turn-on voltage to be applied to the gate (denoted as "G" $\rightarrow$ ($V_G > V_{TH}$)) so as to allow current to flow through from the drain (D) to the source (S) terminal, while the substrate/bulk (B) terminal is usually grounded. The value of V_{TH} is positive for NMOS (*n*-type channel) and negative for PMOS (*p*-type channel). When the PMOS and NMOS are connected as shown in Figure 8.1(g) with a certain power supply voltage (V_{DD}), the simple circuit comprising of two complementary transistors functions as an inverter, i.e., a low voltage at the input (representative of binary "0") is converted to a high voltage at the output (representative of binary "0") and vice versa. This "inverter" combination is referred to as the complementary metal-oxide semiconductor (CMOS) technology, and this is the simplest unit that can be combined with many such units in different combinations to realize all types of logic functionalities.

Analogously, the unit that is fundamental to the "memory" of the circuit is either a transistor or a capacitor (Figure 8.2) [4–6]. The memory unit could be volatile (such as static random access memory and dynamic random access memory, which require constant refresh to store binary data) or nonvolatile (where different technologies such as flash memory (Figure 8.2(b)) [4], resistive RAM (RRAM) (Figure 8.2(c) and (d)) [5,6], and magnetic RAM exist to store data by different storage mechanisms (using

Reliability Characterisation of Electrical and Electronic Systems. http://dx.doi.org/10.1016/B978-1-78242-221-1.00008-3

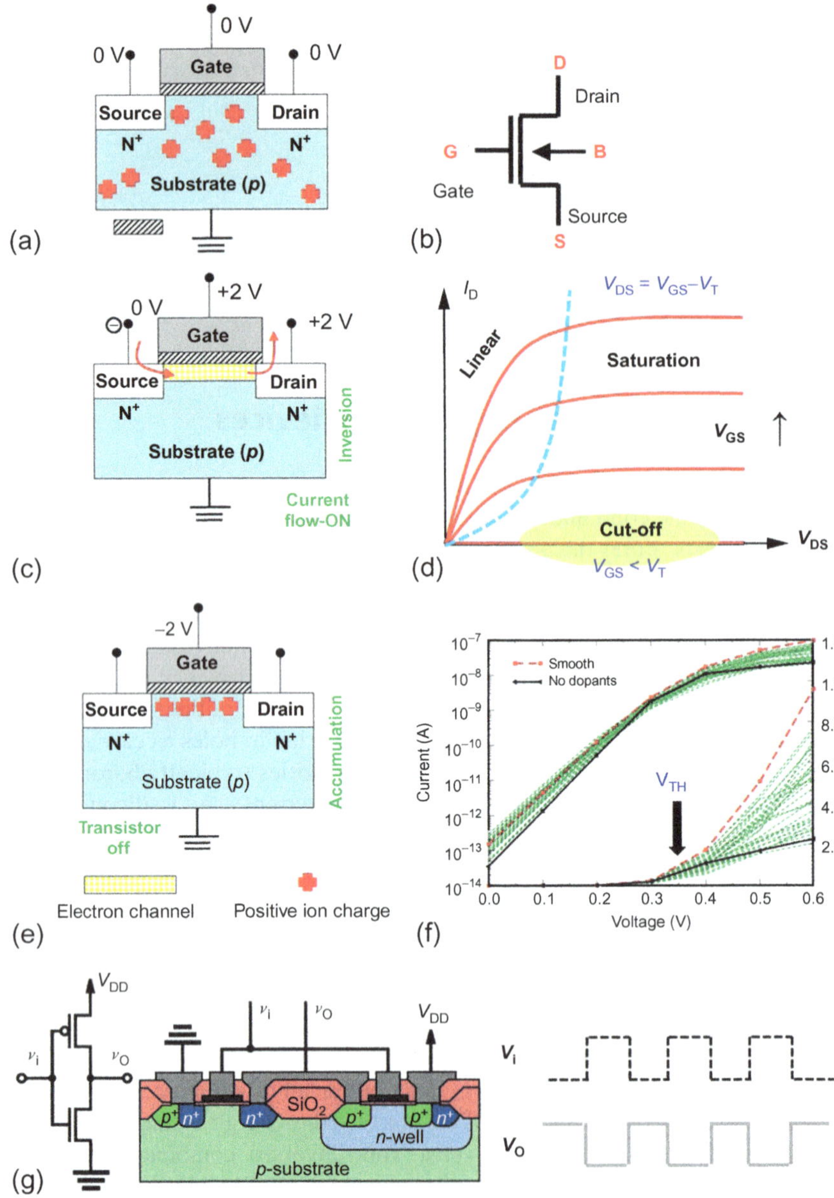

Figure 8.1 (a) Schematic and (b) symbol of a transistor with the four terminals (drain (D), source (S), gate (G), and body (B)). (c) Carrier distribution in the transistor in the inversion regime. (d) The drain current-drain voltage ($I_D - V_D$) characteristic of the device with three modes of operation depending on the relative values of ($V_G - V_{TH}$) and V_D. In general, the transistor operates in the linear region as a switch and the saturation region as an amplifier. (e) Carrier distribution in the transistor when biased in the accumulation mode. (f) The transfer characteristic ($I_D - V_G$) of the device with the *x*-intercept of the slope (tangent) in the midvoltage range representing the V_{TH}. (g) Simplest circuit consisting of a serially connected PMOS and NMOS that functions as a voltage inverter. The Si substrate is *p*-type in this case and an *n*-well is fabricated by doping so that the PMOS device can be embedded into it. Note that a PMOS has an *n*-type substrate and NMOS has a *p*-type substrate.
(f) From Ref. [1]—reprinted with permission from Springer®.

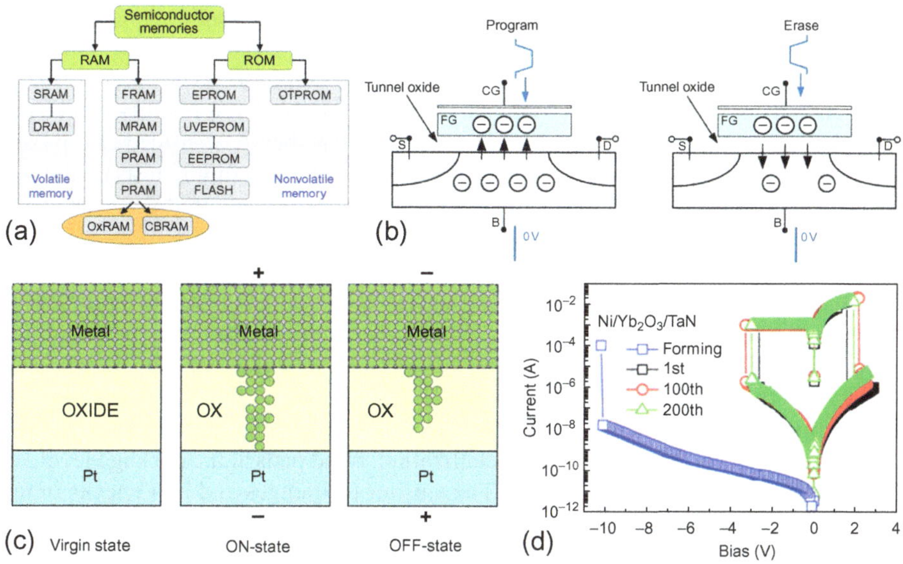

Figure 8.2 (a) Different classifications of memory devices including both volatile and nonvolatile types. Every memory technology has a different mechanism for storing the binary data bits. (b) Illustrating the functioning of a flash memory (which looks very similar to the standard transistor) where the program and erase (P/E) pulses inject and eject charge from the floating gate (FG), which is a charge storing layer sandwiched between two insulators on both sides. The P/E voltage is applied to the control gate (CG) [4]. (c) Illustration of the mechanism of data storage in RRAM (which resembles a simple metal-insulator-metal capacitor) with the virgin device having ideally zero defects in the dielectric and the ON and OFF states having connected and disconnected cluster of defects (oxygen vacancies) induced by bipolar voltage stress [5]. (d) Experimental I–V switching between low- and high-resistance states (LRS and HRS) in Yb_2O_3-based transition metal oxide (TMO) RRAM.
Reprinted with permission from Ref. [6]. Copyright 2012, The Electrochemical Society.

charge carriers, oxygen vacancy defects, metal filaments, electron spin state, etc., with prolonged retention) [7].

Whether it is the logic or the memory unit, there are a few key design metrics to be considered when integrated circuits are fabricated and used to realize all the modern-day electronic gadgets we have such as the laptop, mobile phone, and camera. These metrics include performance, reliability, stability, low-power operation, high integration density (for realizing portable gadgets), low cost, and safety. Most design and fabrication activities are usually aimed at enhancing the performance and reducing the power and device footprint. However, reliability often takes a backstage and is tested only after all the other design metrics are found to be within certain threshold criteria. Although reliability looks like a back-end activity, its relevance and criticality cannot be undermined, as it has a big impact on the sales of the company's products, its warranty policy, and customer base and satisfaction. A high-performance product will not sell well if its reliability is poor and vice versa. Both performance and reliability have to go hand in hand. But it turns out in many cases that boosting one downgrades the other, and these trade-offs need to be accounted for while optimizing the design metrics of any new technology. In fact, the philosophy of reliability has undergone drastic changes in recent times, and there is increased emphasis on design for reliability at an early stage of product design and development so that reliability can be proactively built-in in the design stage (rather than being confined to passive postproduction investigations with feedback from accelerated failure tests on the final product) with improved choice of material and device architecture by means of various modeling and simulation tools.

While various chapters in this book have dealt with these aspects of reliability design and reliability methods, our focus here is to examine the various failure issues that exist in current nanoscale logic and memory technology and understand their root causes of failure. The material, process, and architectural aspects of advanced semiconductor technology nodes will be discussed. We will consider specific case studies that look at different failure mechanisms in logic at the front-end device level, metal interconnects at the back-end metallization, bond pads at the package level, and failure issues in memory technology. The various challenges and bottlenecks in reliability assessment of nanodevices will also be addressed so that key challenges for future microelectronic reliability studies are identified. We shall first begin with the evolution of nanodevices and the various reliability concerns that exist.

8.2 Material and architectural evolution of nanodevices

The development of semiconductor technology over the past 3-4 decades has been guided by an empirical relationship known as "Moore's law" [8], which predicts a trend that the number of transistors in a chip consistently double up in every 18 months. This empirical study turned out to be a guideline for the industry to benchmark its development and advancement of microelectronics. Until about a few years back, the adherence to Moore's law that required aggressive downscaling of the transistor dimensions was smooth and successful based on the guidelines of the International Technology Roadmap for Semiconductors (ITRS) [9]. However, as the length of the transistor channel reduced to 32 nm and the thickness of the gate insulator (dielectric) in the MOSFET had to be brought down to <1 nm (corresponding to just 2-3 monolayers of atoms) [10], the limits of downscaling with planar devices were being sensed as the ultrathin dielectrics gave rise to high tunneling leakage currents during circuit operation. Furthermore, the mobility of the electrons and holes that give rise to drive current in the transistor was also being limited by the interface scattering of the silicon with the dielectric. The roadblocks to further downscaling and performance enhancement had to be tackled by two approaches—(a) exploration of alternative materials and (b) exploration of different device architectures.

From a material point of view, the substrate that has remained as silicon for many decades is now being replaced by silicon germanium (SiGe) (or strained Si) [11] and materials from the III-V groups of the periodic table such as GaAs [12]. The doping of Si with Ge induces additional mechanical strain in the channel that enhances carrier mobility, and similarly, III-V materials tend to have a much higher intrinsic mobility that helps in the realization of circuits with faster speed (higher-frequency operation)

[13]. Another option considered was silicon-on-insulator technology [14] where the silicon substrate had an oxide layer underneath so as to reduce bulk leakage currents significantly. The dielectric in the transistor that used to be grown SiO_2 has been replaced by high-permittivity deposited materials such as HfO_2 (hafnium oxide) so that the effective oxide thickness can be lowered while maintaining a higher physical thickness of the insulator, thereby reducing leakage current [15]. The gate electrode that used to be highly doped polysilicon is now replaced by metal electrodes such as TiN and TaN for improved work function tuning and better control over the V_{TH} value [16]. With this drastic change in the materials used for fabricating the transistor and with downscaling, the failure mechanisms and their severity could be very different from what it was a few years back. Every material has a different intrinsic defect generation rate induced by voltage and temperature stress, and the interfaces of different materials induce additional sources of instability that could cause unanticipated early failures, while the process conditions are continually optimized and fine-tuned so as to try and reduce the as-processed (time zero) defect density of the fabricated gate stacks. The reliability study of the MOSFET with these new material combinations requires in-depth analysis, and research is still in its incipient stages to understand these new material failure physics in greater detail. Figure 8.3 summarizes the various material innovations and alterations to the conventional transistor.

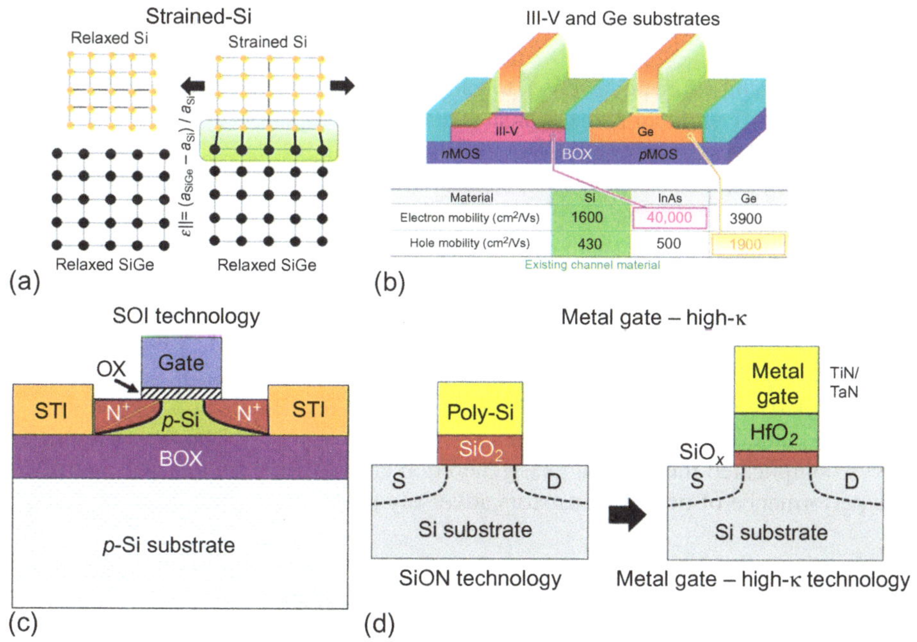

Figure 8.3 Summary of the material innovations and alterations to the conventional transistor ranging from (a) strained-Si (using SiGe substrate), (b) III-V (GaAs/InAs) and Ge-based high-mobility substrate [13], (c) silicon-on-insulator (SOI) technology to minimize substrate leakage losses, and (d) replacement of polysilicon with metal gate and $SiO_2$with high-κ dielectric for better equivalent oxide thickness (EOT) scaling and suppression of gate leakage tunneling currents.

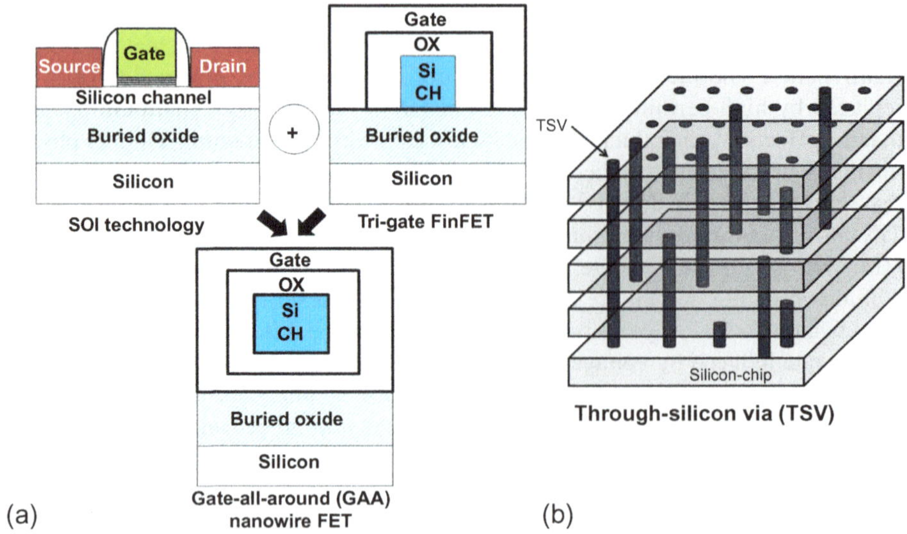

Figure 8.4 Architectural evolution of the integrated circuit stack including (a) shift to FinFET and realization of GAA nanowire FET (using SOI architecture) so as to achieve better channel control of the gate and (b) introduction of through-silicon via (TSV) for interdie/interwafer connectivity to enable 3-D heterogeneous technology integration [19].

From an architecture perspective, to gain better control of the gate over the channel carriers, the conventional planar transistors are being replaced by FinFETs and gate-all-around (GAA) devices [17,18], as illustrated in Figure 8.4(a). With these unique architectures, the electric field and potential distributions are no longer the same and therefore defect generation and failure kinetics will be non-random and more localized around sharp corners (with concentrated field lines). The impact of these architectural modifications on the device reliability also needs to be investigated. Another example of architectural change is the possible inclusion of through-silicon via (TSV, also shown in Figure 8.4(b)) [19], typically made of copper (Cu), in order to achieve 3-D integration of different technologies such as logic, memory, MEMS, and optical on a single platform whereby these TSVs serve as connecting pathways for communication between different functional units stacked vertically [20]. The presence of these TSVs complicated the thermal heat generation and sink mechanism and also affects the performance of nearby transistors adversely [21].

8.3 Failure mechanisms in nanodevices

Having addressed the rapid material and architectural changes that the integrated circuit has been undergoing over the past few years, let us now look into the most common failure modes and mechanisms in integrated circuits. Since integrated circuits consist of the transistor devices (referred to as "front end"), which are connected to

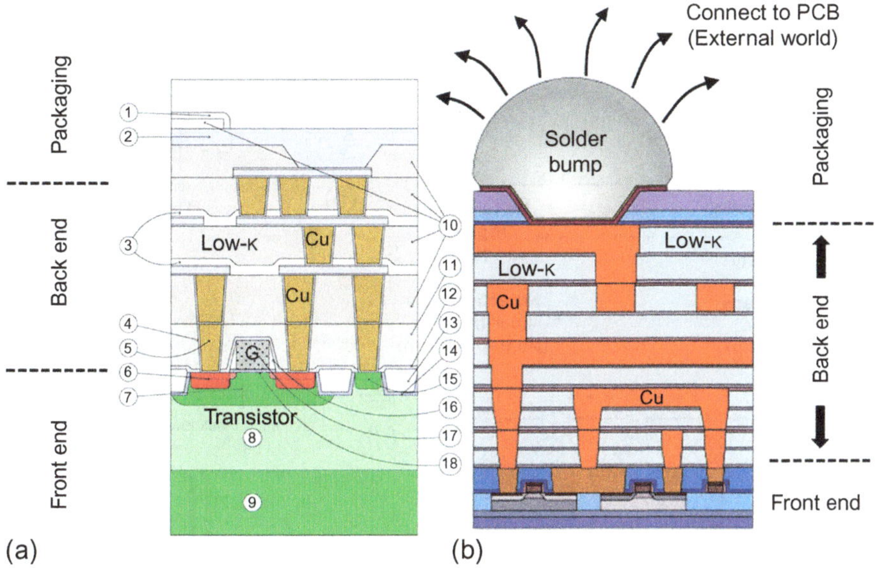

Figure 8.5 Cross section of the integrated circuit (die/chip) showing the (a) front-end transistors connected through multiple layers of Cu metallization at the back end [23] and (b) connection of the chip to the external world (printed circuit board) using the solder bump technology at the package level [22].

each other by many levels of copper interconnect metallization (referred to as "back end") and then to the external world through solder bond pads (known as "packaging"), we will investigate the failure mechanisms separately for the front-end, back-end, and packaging domains. The complete cross section of the integrated circuit (die) is shown in Figure 8.5 [22,23] with the front-end, back-end, and package-level domains indicated. Moreover, the failure issues in nonvolatile memory (NVM) technology will also be addressed briefly, considering the *flash memory*, *phase-change memory (PCM)*, and *RRAM* technologies.

8.3.1 Front-end failure mechanisms

The front-end failure mechanisms mostly involve the ultrathin dielectric in the transistor. When the transistor is subjected to voltage stress during operation, the dielectric that functions as an insulator starts to slowly degrade due to field-induced and temperature-assisted bond breaking of the Hf—O (in HfO_2) or Si—O (in SiO_2) bonds [24]. When a defect is generated by bond breaking, the oxygen ion (O^{2-}) is liberated from the dielectric, leaving behind an oxygen vacancy (denoted as V_0) [25]. These defects serve as "stepping stones" that enable more electrons to tunnel through the dielectric, thereby increasing the leakage current and enhancing the power dissipation of the circuit (refer to Figure 8.6(a)). The carriers can also get trapped in these defect sites resulting in *random telegraph noise (RTN)* fluctuations of the gate and drain

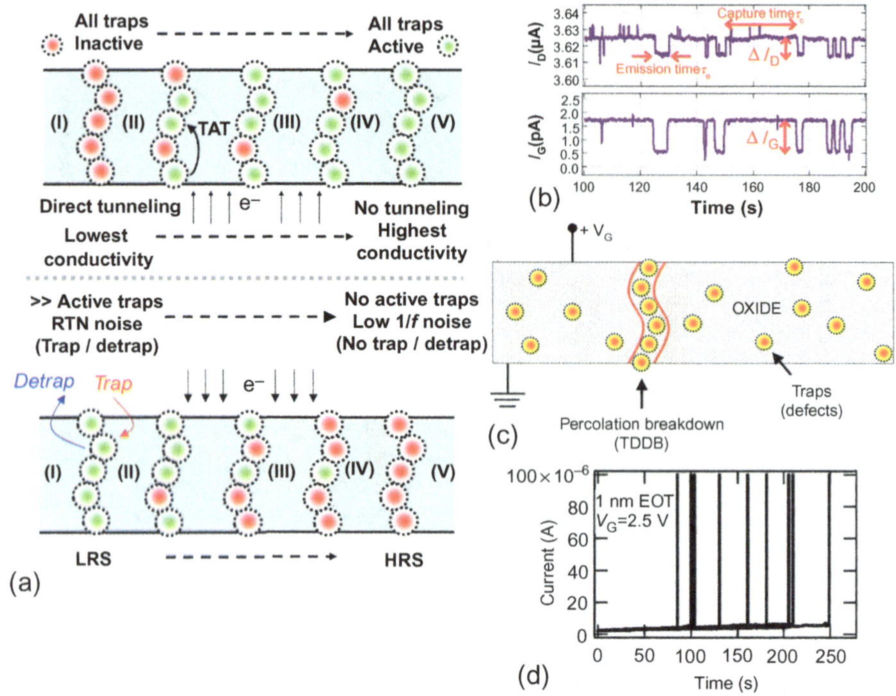

Figure 8.6 (a) Generation of defects in the dielectric that can align to form a percolation path. These defects cannot assist in conduction (inactive) when they have an electron trapped (carrier capture) into them and, conversely, they assist in trap-assisted tunneling (TAT) (active) when there is no charge trapped or when detrapping (carrier emission) has occurred. If all the traps are active, the dielectric is fully percolated and allows significant current to leak through, hampering the performance of the transistor. (b) Random telegraph noise (RTN) signals in the gate and drain current due to stochastic carrier capture and emission phenomenon. (c) Illustration of the random defect generation in the dielectric. After a prolonged duration of stress, a certain cluster of defects is aligned such that there is full connectivity between the gate and the substrate. This is referred to as the "percolation path." (d) Typical current–time stress patterns during TDDB stress [26]. The times to failure follow a Weibull distribution. (b) From Ref. [27]—reprinted with permission from Elsevier®.

current [28] as evident in Figure 8.6(b) [27]. The defects that are generated randomly (according to the Poisson distribution) across the stressed dielectric continue to be generated until a point is reached when a connecting path of defects linking the gate and substrate is formed. This is called the "percolation path" [29] (Figure 8.6(c)), and once this percolation event occurs, a sudden jump in the leakage current is observed and this whole process is called *time-dependent dielectric breakdown (TDDB)* [30]. Figure 8.6(d) shows typical trends of measured gate leakage resulting in the TDDB event [26].

If the current jump is arrested by a low compliance setting, then the oxide is minimally degraded and this scenario is referred to as *soft breakdown (SBD)*. However, if

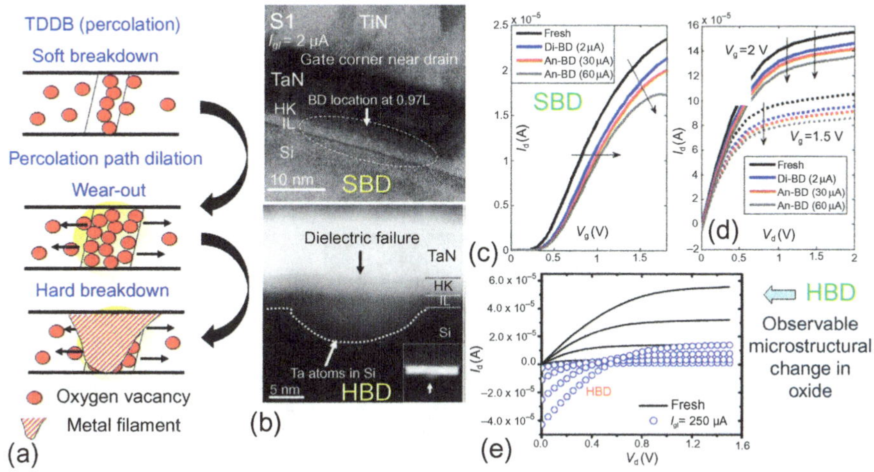

Figure 8.7 (a) Schematic of the different stages of defect generation in the dielectric ranging from soft breakdown, to wear-out (both due to oxygen vacancies), to finally hard breakdown when metal atoms/ions start to migrate into the dielectric leading to ohmic behavior. (b) TEM micrograph of two devices capped to very low (SBD) and very high compliance (HBD) during the TDDB stress. Clearly, there are no morphological change for SBD, but evidence of metal migration in the HBD phase. (c, d) The $I_D - V_G$ and $I_D - V_D$ trends still show good transistor behavior in the SBD mode, (e) while the trends completely disappear in the HBD mode.

the transient stage of percolation is uncontrollable, then the dielectric suffers substantial damage that is called *hard breakdown (HBD)* [31]. For SBD, the transistor functionality is still maintained, while for HBD, the transistor is completely malfunctional [32] and behaves like a resistor (Figure 8.7) [33]. Typically, SBD events are more prevalent in ultrathin dielectrics, while HBD occurs for thick oxides. This is because in thick dielectrics, even prior to forming a percolation path, there are way too many defects created in the dielectric that cause the percolated region to dilate and "explode" during TDDB. Transistors where the dielectric has suffered only an SBD event can continue to be operational. However, the only downside is the increased variability in the performance trends due to enhanced RTN from the connected defect sites. Further stressing of the SBD region can cause the percolation path to slowly wear out (referred to as progressive breakdown (PBD) or post-BD) and dilate with time, and finally, the transistor again suffers a HBD event (Figure 8.7). The TDDB and PBD mechanisms are dependent on the activation energy of the chemical bonds in the dielectric as well as their permittivity and dipole moment [34].

The physical defect signature for TDDB has been observed for various gate stacks using a high-resolution transmission electron microscope [35] to probe the defect site (shown by schematic in Figure 8.8(a)) [36]. As shown in Figure 8.7(b), the initial BD is very soft without any obvious physical defect detectable. However, an elemental oxygen scan using electron energy loss spectroscopy (EELS) reveals a drop in the

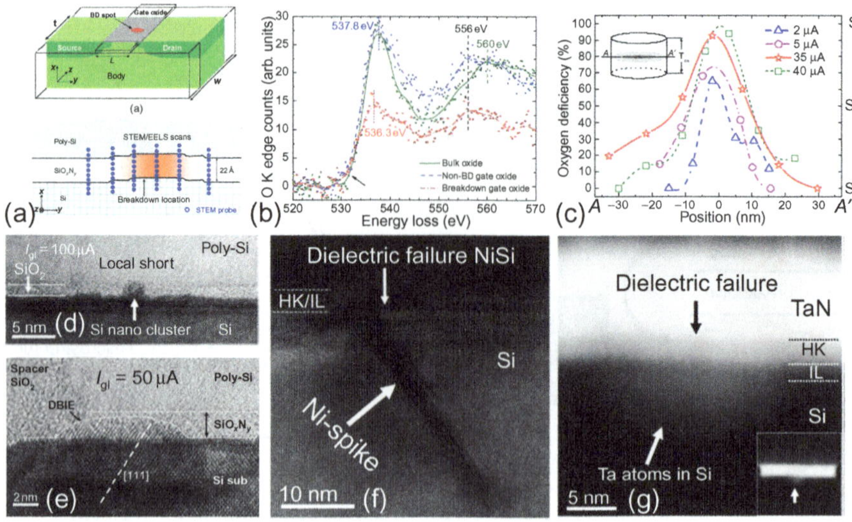

Figure 8.8 (a) Methodology of TEM analysis for a failed sample where a reference region is always used for comparison where no degradation is believed to have taken place during the stress [36]. The breakdown location is roughly determined by the weighted value of the source and drain leakage currents. (b) EELS scan of oxygen elemental mapping shows that the degraded oxide has lower O count, suggesting that the defects are nothing but oxygen vacancies that arise due to bond breaking of the dielectric bonds by the thermochemical model.
(c) Gaussian distribution of the oxygen deficiency at the percolation path as a function of a different compliance current. (d, e) Formation of a silicon nanowire and silicon epitaxial protrusion into the dielectric leading to effective oxide thinning in polysilicon-gated stacks. Migration of (f) Ni and (g) Ta into the oxide for metal gate stacks after high-compliance breakdown test.
(b) Reprinted with permission from Ref. [25]. Copyright 2008, American Institute of Physics®.
(c) Reprinted with permission from Ref. [37]. Copyright 2009, American Institute of Physics®.
(d, e) Reprinted with permission from Ref. [24]. Copyright 2013, American Institute of Physics®. (f, g) Reprinted with permission from Refs. [33,38]. Copyright 2010, American Institute of Physics®.

oxygen count [25] (indicating that the defects are just oxygen vacancies) (Figure 8.8 (b)) [39]. The oxygen count distribution drops further and expands laterally for increased oxide degradation (Figure 8.8(c)) [37]. In the event of a catastrophic HBD, a physical morphology of the defect is observable (Figures 8.7 and 8.8). This could be due to either a complete oxygen depletion in the percolation core leading to silicon (for SiO_2) / hafnium (for HfO_2) nanowire formation in the oxide (Figure 8.8(d)) or a silicon epitaxial protrusion into the dielectric (Figure 8.8(e)) [35] (due to joule heating and thermomigration [40]) or spiking of the metal from the gate into the oxide (for nickel [38]- and tantalum [33]-based gate electrodes), resulting in a metal filament in the dielectric (Figure 8.8(f) and (g)) making the transistive action "non-existent."

Another mechanism that affects the V_{TH} value (and hence the drain (drive) current and transconductance as well) as a function of time is called *negative bias temperature instability* (NBTI) [41], which occurs most frequently in PMOS devices when biased in the inversion regime of conventional operation. This arises from the bond breaking of Si—H bonds creating "interface traps" at the interface of the silicon with the oxide. In addition to this, "trapping of holes" (inversion carriers) in the pre-existing bulk defects in the dielectric also contributes to the V_{TH} shift [42], as illustrated in Figure 8.9(a)–(c) [43,44]. The overall degradation of V_{TH} follows a logarithmic or power law function of time [42]. While the first mechanism is deemed to be permanent, the second one on hole trapping is considered to be recoverable [41]. So, the overall NBTI process contains a permanent and recoverable component that makes this mechanism a complex one to study and model. In current 22 nm-based technology, the BTI mechanism is the most serious and dominant one affecting reliability at

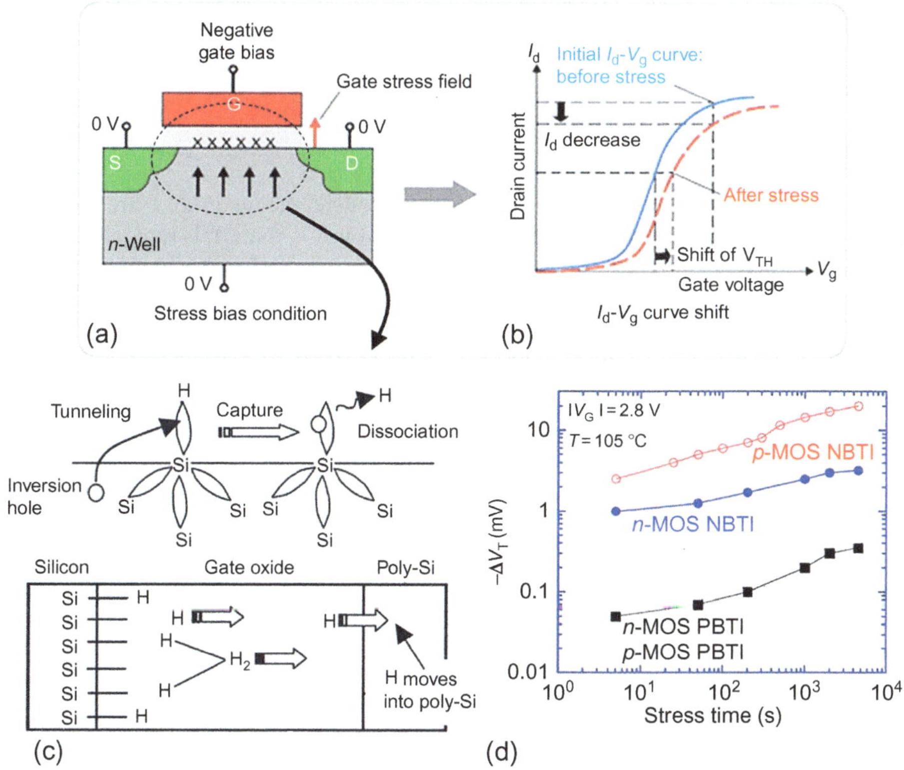

Figure 8.9 (a) Generation of defects at the interface of silicon and the dielectric due to Si—H bond breaking during stress for NBTI in PMOS devices. (b) These generated defects and trapping of holes cause an increase in the V_{TH} and decrease in drain current, I_D, of the transistor (leading to reduced circuit speed) [43]. (c) The mechanism of bond breaking and H diffusion through the dielectric is illustrated here. The model is popularly known as the reaction-diffusion model. (d) Experimental power law trend in degradation of V_{TH} with time for both NBTI and PBTI mechanisms.

(c) From Ref. [44]—reprinted with permission from Elsevier®. (d) From Ref. [45]—reprinted with permission from Elsevier®.

the front end. Similar to the NBTI in PMOS, there is an analogous failure mechanism in NMOS devices called *positive bias temperature instability* (PBTI) (Figure 8.9(d)) [45] where the V_{TH} is believed to increase predominantly due to trapped electrons at defect sites [46], which reside in the bulk of the high-κ dielectric. The trapping and detrapping of electrons result in a partially repeatable degradation and recovery phenomenon.

The last critical mechanism is known as "hot carrier injection (HCI)" [47] whereby the electrons that are accelerated by the lateral electric field between the source and the drain acquire sufficient kinetic energy when they reach the drain end. These energetic carriers are able to overcome the Si/SiO_2 barrier and trigger defect generation at the interface by breaking Si—H bonds and also get trapped in pre-existing defects, which results in a degradation of the V_{TH}, subthreshold slope and transconductance leading to lower drive current, and operating frequency of the integrated circuit. The mechanism of HCI is illustrated in Figure 8.10. We shall now change gear to focus on the failure mechanisms at the back end.

8.3.2 Back-end failure mechanisms

The back end includes two key structural components in the integrated circuit—metal line interconnects and low-κ dielectrics. The use of low-κ materials replacing SiO_2 in the past helps reduce the RC time constant delay leading to faster signal propagation across the complex multilevel interconnect network. One of the most critical and extensively studied mechanism for interconnect failure is *electromigration (EM)* [48]. It refers to the momentum transfer of multiple electrons to atoms under high current density that causes the atoms to migrate from one end (cathode side) to the other (anode side) leading to metal extrusion at the anode and voiding at the cathode side (Figure 8.11(a)). These voids nucleate and then grow further, causing resistance to gradually increase to a point where there is a complete open circuit with catastrophic jump in resistance, leading to circuit failure. EM is a relevant mechanism in both aluminum (Al) and copper (Cu) interconnects, where Cu replaced Al due to its lower resistivity and higher melting point. While grain boundaries in these polycrystalline thin films were the major path of atomic flux for Al with the lowest activation energy, the Cu transport was dominant at its interface with the high-resistivity (TaN/TiN) metal liner and dielectric [48]. There are two key mechanisms of failure in EM with different locations of failure—one at the via end and one at the line end (Figure 8.11(b)

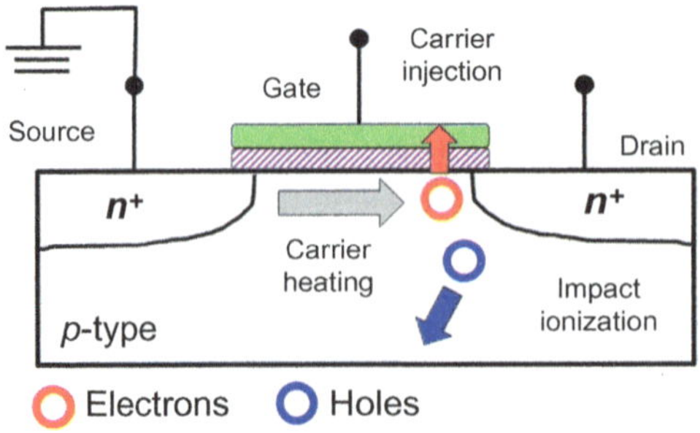

Figure 8.10 Underlying mechanism for hot carrier degradation that involves accelerated transport of carriers in the transistor channel with high energy (carrier heating) and generation of electron-hole pairs by impact ionization close to the drain end. The generated electrons then get injected into the dielectric causing additional defect creation at the interface or trapping in the bulk.

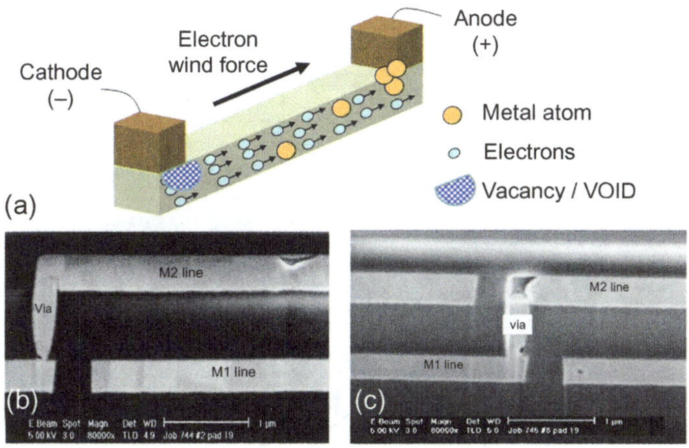

Figure 8.11 (a) Illustration showing the driving force for electromigration that involves the transport of metal atoms from the cathode to the anode by momentum transfer of electrons (electron wind force). The depletion of metal atoms at the cathode end causes voiding. (b, c) SEM micrograph showing voiding in the line and via corner, corresponding to the bimodal EM failure distribution observed.
From Ref. [48]—reprinted with permission from Elsevier®.

and (c)) [48]. The presence of these two failure mechanisms in different proportions often results in a bimodal failure distribution for EM [49]. The EM resilience of interconnects depends on various factors including the alloy composition, thin film microstructure, dimensions (length of interconnect and via tapering), vacancy reservoir effect due to extended metal line beyond the via pillars [50], redundancy in design, direction of current flow (DC/AC), and interface with adjacent materials (liners and dielectric). It is well known that EM is driven by two forces that counteract each other—one being the electron wind force that causes voiding and the other being back-stress force due to mechanical stress gradients. For interconnects lower than a critical length of L_{eff} [51], these two forces perfectly balance each other (the back-stress force is higher in shorter lines due to steeper stress gradients), and hence, the metal line is said to be immune to EM effects. The most popular failure time estimation model for EM is Black's equation, proposed way back in the 1960s [52]. This equation includes a power law dependence on current density and Arrhenius dependence on the temperature.

Another failure mechanism involving the same voiding process, but caused by the fabrication process and not the current density during operation, is called *stress migration (SM)* [53,54] (Figure 8.12). This results from the high-temperature processing of interconnect stacks wherein due to the different coefficients of thermal expansion of the metal line, liner, and dielectric materials, the fabrication process induces a hydrostatic stress gradient (Cu lines end up having a large tensile stress). The overall structure undergoes gradual stress relaxation over a period of time by means of directional diffusion of vacancies, which can result in open-circuit failure [55]. This mechanism is

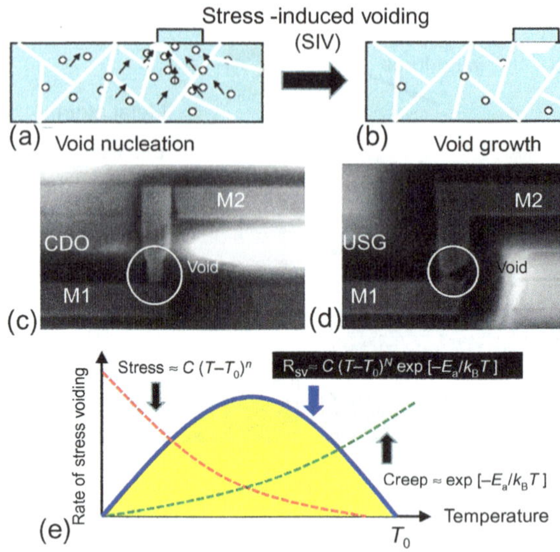

Figure 8.12 (a, b) Schematic showing the gradual movement of vacancies in the interconnect to accommodate for stress relaxation of the hydrostatic stress induced by the fabrication process. The vacancies agglomerate to form bigger voids close to the via edge. (c, d) SEM micrographs showing the different locations and directions of voiding near the via. (e) Illustration showing the dominance of diffusion and hydrostatic stress in different temperature regimes with rate of SIV growth being most intense at midrange temperature values.
(c, d) From Ref. [55]—reprinted with permission from Elsevier®.

also sometimes referred to as *stress-induced voiding (SIV)*, as shown in Figure 8.12(c) and (d) where the void can grow either vertically at the via edge or horizontally under the via (the second process being more detrimental) [55]. SIV is a two-stage process that involves accumulation of vacancies (referred to as void nucleation) followed by void growth (Figure 8.12(a) and (b)). Diffusion (creep) and stress are the two key driving forces needed for SIV, and since stress is high at low temperature, while diffusion is significant only at high temperature, SIV effect is detrimental in the circuit at intermediate temperature ranges with high rate of void growth, as illustrated in Figure 8.12 (e) [56].

Similar to the dielectric breakdown that occurs in ultrathin high-κ dielectrics ($t_{ox} \sim$ 1-4 nm) in the front end, it is also possible to have a *TDDB breakdown of the low-κ dielectric* in the interconnect stack when the metal lines are subject to voltage stress (Figure 8.13). While the physics of breakdown in the back end is not as well understood as it is in the front end, there are a few plausible explanations proposed that include copper lateral migration (Figure 8.13(b)) [57] across the defective dielectric interfaces [58], field-induced defect creation [59], and carrier energy-related impact damage [60]. There are phenomenological models that have been proposed for each of these, but all of them converge only at accelerated stress conditions and their extrapolated lifetimes can vary by many orders of magnitude at field operation voltage levels [61].

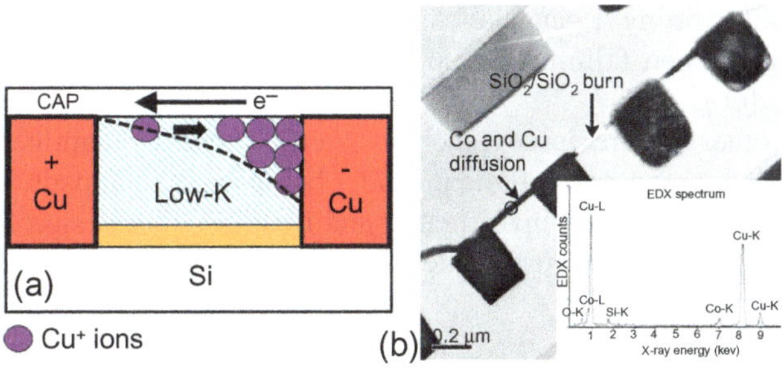

Figure 8.13 (a) One possible mechanism for low-κ breakdown that involves ionic migration of Cu^+ ions along the imperfect interface between the dielectric and the metal cap (barrier) layers. (b) SEM image showing the migration of copper resulting in a short of the two lateral electrodes during the TDDB.
Reprinted with permission from Ref. [57]. Copyright 2008, American Institute of Physics®.

8.3.3 Package-level failure mechanisms

At the package level, the solder bumps serve as the primary medium of contact between the Cu lines on the die and the Cu lines on the printed circuit board. These solder bumps are now made of Pb-free materials such as Sn-Ag and Sn-Cu (specifically Cu_6Sn_5) intermetallics [62], considering that Pb is not environment friendly. While the Cu and Al interconnects required current densities of the order of 1-10 MA/cm^2, solders are prone to EM for current density as low as 10^4 A/cm^2 [63]. *Solder electromigration* is a major failure mechanism that has been intensely studied [64]. It is more complex due to the presence of intermetallics and also the unique pattern of current crowding (Figure 8.14(a) and (b)) [65] that occurs at one corner of the bump in the cathode side. The voids begin to nucleate at one end of the bump and

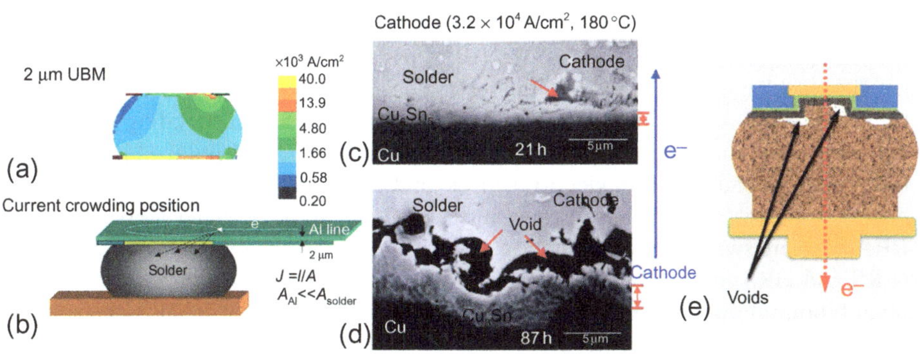

Figure 8.14 (a, b) Current density profile in a solder bump with the current crowding taking place at one corner of the solder-metal line contact area [65]. (c, d) Nucleation and growth of voids in the solder at the cathode side after prolonged duration of stressing [65]. (e) Schematic showing the potential locations of voiding [66].

laterally expand forming a pancake-shaped defect (Figure 8.14(c) and (d)) [65,67], finally leading to open failures. Some potential locations of voiding are shown in Figure 8.14(e) [66].

There are other failure issues such as thermal stress delamination, moisture-induced corrosion, creep, and fatigue in solder bumps and bond pads. However, since most of these have been well investigated and the reliability design for these mechanisms is already in place, further detailed insight into these mechanisms is not warranted here.

8.3.4 Failure mechanisms in memory technology

Our focus thus far has been limited to only logic devices. However, it is well known that memory makes up about 70% of the circuit in modern-day electronic gadgets. Therefore, the unique failure mechanisms pertaining to NVM also deserve attention. There are three main reliability criteria for any NVM device that stores the binary data "0" or "1" by different physical principles. The first one refers to *endurance*, which is the number of times the device can be cycled between 0 and 1 given a particular pulse width and frequency for the voltage input signal. The second is *retention*, which refers to the duration for which given data are stored without being disturbed due to unexpected events and/or gradual device degradation such as charge leakage or vacancy (ionic) diffusion. The last one is *read disturb immunity*, indicative of the probability that the state of the memory would be disturbed and altered when the device is probed by voltage for reading the data already stored. All these three criteria have to be tested and optimized for any new memory technology. This requires extensive tests and material/process engineering. All these three reliability metrics are generally triggered by a single mechanism that causes loss of memory state or memory window.

For flash memory, the mechanism involves charge trapping in the dielectric and creation of additional trap sites in the tunnel dielectric sandwiched between the floating gate (FG) and the substrate. This charge trapping phenomenon directly affects the V_{TH} value and hence the memory window (Figure 8.15(a) and (b)) [4,69]. The performance of the flash memory is also affected by parasitic leakage of charge by tunneling or other mechanisms. For PCM, a unique material GST ($Ge_2Sb_2Te_5$) is used that can transit between the crystalline and the amorphous phases corresponding to different conductivity values upon electrical and thermal stress. The EM-induced ionic motion of Ge and Sb causes the device to degrade and get stuck in the low-resistance state during write/erase cycling [70]. The latest NVM technology under consideration is RRAM where switching between the low-resistance state and the high-resistance state (LRS and HRS) occurs by repeated nucleation and partial dissolution of a filament upon bipolar/unipolar stressing where the filament could comprise of either oxygen vacancies or metal filaments depending on the compliance used for capping the current as well as the material stack (Ni-based stacks result in metallic filament due to Ni migration into the dielectric and TiN-based stacks involve reversible transport of oxygen ions) (Figure 8.15(c)) [33,71]. In the case of an oxygen vacancy filament, degradation of the RRAM occurs due to partial oxidation of the metal gate

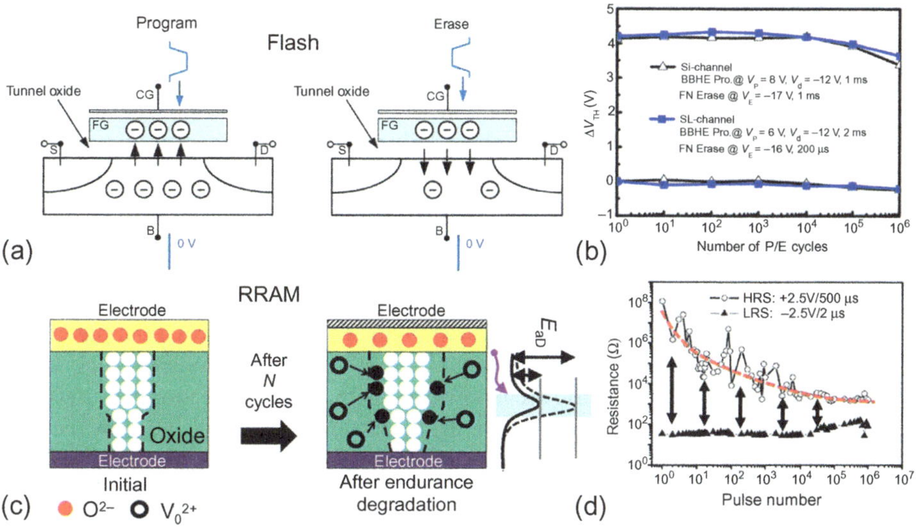

Figure 8.15 (a, b) Operation of flash memory by repeated program/erase pulses [4] whereby the degradation occurs (resulting in reduced V_{TH} window) due to charge trapping and additional defect creation. (c, d) Operation of RRAM involving oxygen vacancies where, upon cycling many times, there is an increasing imbalance between available oxygen ions and vacancies as some ions cause oxidation of the metal electrode and become immobile. This leads to reduced memory window as the experimental endurance trend shows.
(a, b) From Ref. [69]—reprinted with permission from Elsevier®. (c, d) From Ref. [68]—reprinted with permission from Elsevier®.

(which is ideally a very good oxygen reservoir) and enhanced barrier to oxygen ion diffusion from the electrode to the filament (percolation path). As a result, the number of mobile O^{2-} ions gradually decreases, while the number of vacancies ($V_O{}^{2+}$) increases (due to repeated stress), causing an imbalance in the ion-vacancy count that degrades the LRS-HRS memory window (Figure 8.15(d)) [68]. For a metallic filament, the incomplete rupture results in metallic nanoparticles or nanocrystals being embedded in the dielectric as the number of switching cycles increases. These nanocrystals that behave as trapping centers cause the HRS resistance level to gradually fall down, causing an eventual collapse of the memory window.

In this section, we have provided a complete overview of the various failure mechanisms inherent in logic and memory devices both at the device, interconnect, and package levels. The different driving forces for failure have been identified, and suitable measures to enhance the robustness of the device/circuit to these mechanisms can be initiated using design-for-reliability approach considering various material and process engineering options. The approaches used to enhance the reliability are not covered here as it is in itself another big topic that requires a dedicated chapter. In Section 8.4, we will present the challenges and issues involved in the reliability study of nanoelectronic devices that will provide some insight on the direction toward which future reliability research should spearhead.

8.4 Reliability challenges: opportunities and issues

With the constant and drastic changes in the process flow and materials used in the development of every successive downscaled semiconductor technology node, there is always the issue of whether any observed failure mode or mechanism in a new node during the reliability test phase is an *intrinsic* one or a *process-induced extrinsic* issue. Testing a circuit's reliability for an extrinsic failure mode is not a fair assessment because extrinsic issues are to be eliminated as far as possible before any new technology is qualified for commercial production. Very often, it is difficult to identify whether an observed failure is an intrinsic or extrinsic one. This complicates the true assessment of the system's reliability. If we plot the time to failure of a sample of units on a statistical plot (pertaining to the particular failure mechanism—such as the Weibull distribution for dielectric breakdown and lognormal for electromigration), then the presence of extrinsic issues will make the distribution "bimodal" whereby the early failure component should correspond to the extrinsic case. When a few samples from the extrinsic distribution are subjected to in-depth failure analysis, the root cause of these process-induced failures can be identified and hopefully eliminated by alternative design of material and process flow conditions. It is worth noting that extrinsic failures are very common in studies on dielectric breakdown [72] and electromigration [73].

Most of the statistical models consider the defect generation to be uniform so as to simplify the analysis and make the reliability assessment more straightforward. However, with the shift toward 3-D architecture in both the front end (replacement of planar FET with FinFETs) and the back end (use of through-silicon vias to bond and integrate wafers/dies of different technologies to realize system-on-chip application), we have induced *non-random degradation phenomena* due to concentrated field/flux lines, at sharp corners and high-aspect-ratio structures. Additionally, in the case of the MOSFET, the polycrystalline microstructure of the HfO_2 high-κ dielectric (while SiO_2 that was previously used remained amorphous in all cases) makes the situation more complicated as grain boundaries serve as a segregating corner for pre-existing oxygen vacancy defects that serve as localized weak-link paths for faster TDDB compared to the bulk grain region [74].

It is very common to have *more than one failure mechanism* being triggered in a device for the same set of test conditions (with a common driving force). In such cases, the reliability model has to account for the coexistence of these two mechanisms and also incorporate their dependency on each other wherein one failure mechanism may exacerbate the failure rate of the second one and vice versa. Two good examples are the interdependency of TDDB and NBTI in the front end [75,76] and SM-EM [56] in the back end. These dependencies are quite hard to model accurately and reliability assessment becomes all the more challenging due to such issues.

All reliability assessment procedures conclude with an *extrapolation* of the accelerated failure data set by means of an empirical or phenomenological model. Although most extrapolation models fit the accelerated stress data very well, their estimation at field operating conditions tends to deviate by many orders of magnitude [77,78]. As a result, even a carefully planned and rigorously executed accelerated stress test can become useless if the extrapolation model is chosen wrongly. The best way to choose an extrapolation model is to use the physics of failure as the basis. Models that are based on the observed physics of failure can be considered to be more reliable as opposed to simple empirical power law or exponential/inverse-exponential trend models. However, the compromise involved is that these physical models are more complex and require sophisticated optimization techniques for estimation of the different parameter values.

The next cause for concern is *whether the mechanism of failure at accelerated and field-stress conditions is the same*. Although we generally (blindly) assume this to be true, it may not always be the case. For example, in ultrathin dielectrics where carrier injection into the dielectric and carrier fluence have a direct impact on the defect generation efficiency, the current density can be very different at low and high voltages due to the dominance of direct tunneling (through the entire trapezoidal dielectric barrier) and Fowler-Nordheim tunneling (through the triangular barrier (Figure 8.16))

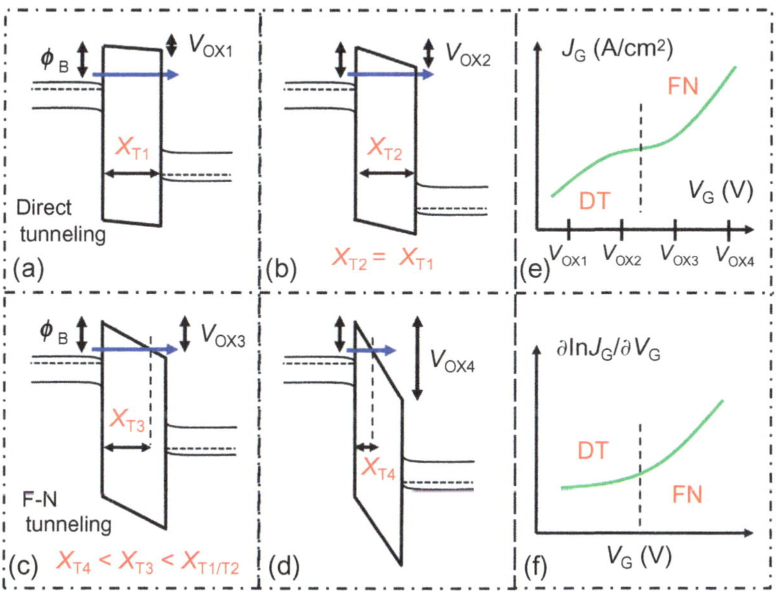

Figure 8.16 (a, b) Energy band diagram of the dielectric for low stress levels, $V_{OX2} > V_{OX1}$, wherein the tunnel barrier is trapezoidal in shape implying that for both V_{OX1} and V_{OX2} ($V_{OX2} > V_{OX1}$), the electron has to tunnel through the same barrier making the current highly insensitive to voltage stress. However, with increasing stress levels of V_{OX3} and V_{OX4} in (c, d), the oxide bend is sharper and the conduction mechanism shifts to Fowler-Nordheim tunneling through a triangular barrier (shorter and varying tunnel distance), which determines the leakage current. In this regime, small changes in voltage stress cause a big change in the current. (e, f) This drastic difference in sensitivity of current at low and high voltage implies that the mechanism and severity of defect generation kinetics at operating field voltage and accelerated stress levels cannot be described by a simple empirical extrapolation model. This complicates the reliability study and assessment of ultrathin dielectrics.

respectively [79]. As a result, a direct translation of the accelerated stress reliability function to field-level lifetime distribution cannot be done. The reliability model has to be modified to account for these changes in the fundamental conduction mechanism as a function of the voltage, temperature, and dielectric thickness.

Physical analysis of the failed device sounds straightforward to discuss than it actually is in real practice. Given that the defects in CMOS are generally in the nanoscale dimension, sophisticated techniques such as TEM, EELS, and energy dispersive X-ray are required so as to unveil the chemistry inherent in the thermodynamics and kinetics of the failure phenomenon. For FETs, the application of TEM has to be done after a focused-ion beam milling to narrow down and isolate the failure site for in-depth analysis. Each of these failure analysis (FA) investigations takes a long time to execute, and the success rate of such studies is also relatively low because pinpointing the exact failure site is an arduous task. In spite of the challenges associated with FA, it is an important activity that provides us with a wealth of information and enables us to develop better physically representative models. To enhance the success rate of FA, it is key to design and fabricate innovative test structures that enable easier localization of the defect site and force the failure to occur at a particular location by purposely making it more prone to degradation.

With downscaling of transistor dimensions toward 20 nm and dielectric thickness ~1-2 nm, *variability* is also a key concern. A high reliability, which generally comes at the expense of higher variability, may not necessarily be a good option. It is necessary to consider trade-offs in the performance, reliability, and variability all at once while designing a robust device and circuit. Of late, reliability is also being interpreted as time-dependent variability [80]. In other words, degradation of a device adds in additional variability on top of the time-zero process-induced variability that is inherent in the fabricated stack. It is important to identify design factors and process parameters that will help enhance reliability and reduce variability at the same time.

Lastly, a new school of thought on reliability assessment known as *prognostics and system heath management (PHM)* for electronic products and systems [81] has been under development over the last few years. This field talks about the dynamic estimation of reliability of a product or system in real time by operating it at normal field use conditions and tracking its degradation trend using physics of failure models or data-intensive models (such as neural networks and genetic algorithms) or a combination of both [81]. The parameters of the reliability model that enable estimation of the remaining useful life distribution are constantly updated in real time as and when new data are available from the sensors that track various degradation indexes of the product/system. Prognostic principles have been applied successfully to reliability assessment of high-power transistors [82], lithium ion batteries [83], and solder bumps [84]. Their application to nanoscale transistors is nevertheless very challenging, but our recent work [85] shows that the gate leakage current noise spectrum and the power law frequency exponent (slope (α) in the logarithmic scale of the power spectral density plot where $\alpha \rightarrow 1$ and $\alpha \rightarrow 2$ represent $1/f$ noise and RTN, respectively) could serve as a good qualitative indicator of the state of degradation of the transistor. For more details on PHM, readers can refer to Chapter 3 in this book by Hendricks et al. and Prof Michael Pecht, who is the founder and director of the CALCE research center at the University of Maryland.

8.5 Summary and conclusions

In this chapter, we have looked at the rapid evolution ongoing in the development of logic and memory device technology in terms of the new materials and device architectures that have been considered for improvement of device performance and *I-V* characteristics. The challenges that arise in reliability studies of nanoelectronic devices and systems as a result of these material, process, and architectural modifications were highlighted. We then went in-depth into looking at the various failure mechanisms at the front end, back end, and package level of CMOS logic technology, giving insight into the physics of failure and observed electrical trends and/or physical morphology of the defect. The reliability bottlenecks in memory technology were also briefly touched upon, and in the last section, we identified the various challenges and bottlenecks that currently exist for a robust reliability study in the field of nanoelectronics. These challenges will hopefully serve as a motivating factor for further research in this area so that future technologies can be tested more efficiently and effectively and their remaining useful life predicted with greater accuracy.

References

[1] A. Asenov, A.R. Brown, G. Roy, B. Cheng, C. Alexander, C. Riddet, U. Kovac, A. Martinez, N. Seoane, S. Roy, Simulation of statistical variability in nano-CMOS transistors using drift-diffusion, Monte Carlo and non-equilibrium Green's function techniques, J. Comput. Electron. 8 (3–4) (2009) 349–373, http://dx.doi.org/10.1007/s10825-009-0292-0.

[2] A. Agarwal, J. Lang, Foundations of Analog and Digital Electronic Circuits, Morgan Kaufmann, San Francisco, CA, USA, 2005.

[3] S.M. Sze, Physics of Semiconductor Devices, second ed., John Wiley & Sons, New York, 1981

[4] P. Hulbert, Keithley Instruments Inc., Alternative NVM Technologies Require New Test Approaches, EE Times, 2012. http://www.eetimes.com/document.asp?doc_id=1280120.

[5] F. Zhuge, B. Fu, H. Cao, Advances in resistive switching memories based on graphene oxide, in: J.R. Gong (Ed.), New Progress on Graphene Research, InTech, Croatia, http://dx.doi.org/10.5772/51260, InTech, Publishers Inc.

[6] S. Mondal, J.L. Her, F.H. Ko, T.M. Pan, The effect of Al and Ni top electrodes in resistive switching behaviors of Yb_2O_3-based memory cells, ECS Solid State Lett. 1 (2) (2012) P22–P25, http://dx.doi.org/10.1149/2.005202ssl.

[7] Y. Fujisaki, Review of emerging new solid-state non-volatile memories, Jpn. J. Appl. Phys. 52 (2013) 040001, http://dx.doi.org/10.7567/JJAP.52.040001.

[8] J.R. Powell, The quantum limit to Moore's Law, Proc. IEEE 96 (8) (2008) 1247–1248, http://dx.doi.org/10.1109/JPROC.2008.925411.

[9] International Technology Roadmap for Semiconductors (ITRS), 2012.

[10] K.L. Pey, N. Raghavan, X. Li, W.H. Liu, K. Shubhakar, X. Wu, M. Bosman, New insight into the TDDB and post breakdown reliability of novel high-k gate dielectric stacks, in: IEEE International Reliability Physics Symposium (IRPS), 2010, pp. 354–363, http://dx.doi.org/10.1109/IRPS.2010.5488805.

[11] J.L. Hoyt, H.M. Nayfeh, S. Eguchi, I. Aberg, G. Xia, T. Drake, E.A. Fitzgerald, D. A. Antoniadis, Strained silicon MOSFET technology, in: IEEE International Electron

Device Meeting (IEDM), 2002, pp. 23–26, http://dx.doi.org/10.1109/IEDM.2002.1175770.
[12] K. Rajagopalan, J. Abrokwah, R. Droopad, M. Passlack, Enhancement-mode GaAs n-channel MOSFET, IEEE Electron Device Lett. 27 (12) (2006) 959–962, http://dx.doi.org/10.1109/LED.2006.886319.
[13] T. Tezuka, Research and development of high-mobility and non-planar channel CMOS technology, Green Nanoelectronics Center (GNC). http://www.yokoyama-gnc.jp/english/research/cmos.html, 2010.
[14] J.P. Colinge, Silicon-on-Insulator Technology: Materials to VLSI, Kluwer Academic Publishers, Boston, 1997.
[15] M. Houssa, High-κ Gate Dielectrics, Taylor and Francis, New York, 2003.
[16] A. Yagishita, T. Saito, K. Nakajima, S. Inumiya, K. Matsuo, T. Shibata, Y. Tsunashima, K. Suguro, T. Arikado, Improvement of threshold voltage deviation in damascene metal gate transistors, IEEE Trans. Electron Devices 48 (8) (2001) 1604–1611, http://dx.doi.org/10.1109/16.936569.
[17] J. Appenzeller, Carbon nanotubes for high-performance electronics—progress and prospect, Proc. IEEE 96 (2) (2008) 201–211, http://dx.doi.org/10.1109/JPROC.2007.911051.
[18] N. Singh, A. Agarwal, L.K. Bera, T.-Y. Liow, R. Yang, S.C. Rustagi, C.H. Tung, R. Kumar, G.Q. Lo, N. Balasubramanian, D.L. Kwong, High-performance fully depleted silicon nanowire (diameter $\leq$ 5 nm) gate-all-around CMOS devices, IEEE Electron Device Lett. 27 (5) (2006) 383–386, http://dx.doi.org/10.1109/LED.2006.873381.
[19] J. Kawakita, T. Chikyow, Ultra-fast forming organic conductive material for low cost 3D micro wiring in advanced electronics, NIMS, Japan, 2011. http://www.nims.go.jp/mana/research/results/2011/09211/.
[20] M. Motoyoshi, Through-silicon via (TSV), Proc. IEEE 97 (1) (2009) 43–48, http://dx.doi.org/10.1109/JPROC.2008.2007462.
[21] A. Mercha, G. Van der Plas, V. Moroz, I. De Wolf, P. Asimakopoulos, N. Minas, S. Domae, D. Perry, M. Choi, A. Redolfi, C. Okoro, Y. Yang, J. Van Olmen, S. Thangaraju, D. Sabuncuoglu Tezcan, P. Soussan, J.H. Cho, A. Yakovlev, P. Marchal, Y. Travaly, E. Beyne, S. Biesemans, B. Swinnen, Comprehensive analysis of the impact of single and arrays of through silicon vias induced stress on high-κ/metal gate CMOS performance, in: IEEE International Electron Devices Meeting (IEDM), 2010, pp. 2.2.1–2.2.4, http://dx.doi.org/10.1109/IEDM.2010.5703278.
[22] Cepheiden, Schematic structure of a CMOS chip, 2006.
[23] V. Ottesen, A cross section of a simple integrated circuit, showing a single MOSFET transistor, 2011.
[24] N. Raghavan, A. Padovani, X. Li, X. Wu, V.L. Lo, M. Bosman, L. Larcher, K.L. Pey, Resilience of ultra-thin oxynitride films to percolative wear-out and reliability implications for high-κ stacks at low voltage stress, J. Appl. Phys. 114 (2013) 094504. http://dx.doi.org/10.1063/1.4819445.
[25] X. Li, C.H. Tung, K.L. Pey, The nature of dielectric breakdown, Appl. Phys. Lett. 93 (7) (2008) 072903. http://dx.doi.org/10.1063/1.2974792.
[26] R. Ranjan, Failure mechanisms in HfO_2 high-κ gate stack MOSFETs, PhD Thesis, School of Electrical and Electronic Engineering (EEE), Nanyang Technological University (NTU), 2007.
[27] B. Kaczer, M. Toledano-Luque, W. Goes, T. Grasser, G. Groeseneken, Gate current random telegraph noise and single defect conduction, Microelectron. Eng. 109 (2013) 123–125, http://dx.doi.org/10.1016/j.bbr.2011.03.031.

[28] W.H. Liu, A. Padovani, L. Larcher, N. Raghavan, K.L. Pey, Analysis of correlated gate and drain random telegraph noise in post soft-breakdown TiN/HfLaO/SiO_x nMOSFETs, IEEE Electron Device Lett. 35 (2) (2014) 157–159, http://dx.doi.org/10.1109/LED.2013.2295923.
[29] J.H. Stathis, Percolation models for gate oxide breakdown, J. Appl. Phys. 86 (1999) 5757–5766, http://dx.doi.org/10.1063/1.371590.
[30] S. Lombardo, J.H. Stathis, B.P. Linder, K.L. Pey, F. Palumbo, C.H. Tung, Dielectric breakdown mechanisms in gate oxides, J. Appl. Phys. 98 (2005) 121301, http://dx.doi.org/10.1063/1.2147714.
[31] S. Sahhaf, R. Degraeve, P.J. Roussel, T. Kauerauf, B. Kaczer, G. Groeseneken, TDDB reliability prediction based on the statistical analysis of hard breakdown including multiple soft breakdown and wear-out, in: IEEE International Electron Devices Meeting (IEDM), 2007, pp. 501–504, http://dx.doi.org/10.1109/IEDM.2007.4418984.
[32] K.L. Pey, V.L. Lo, C.H. Tung, W. Chandra, L.J. Tang, D.S. Ang, R. Ranjan, New insight into gate dielectric breakdown induced MOSFET degradation by novel percolation path resistance measurements, in: IEEE International Electron Devices Meeting, 2004, pp. 717–720, http://dx.doi.org/10.1109/IEDM.2004.1419271.
[33] X. Li, K.L. Pey, M. Bosman, W.H. Liu, T. Kauerauf, Direct visualization and in-depth physical study of metal filament formation in percolated high-κ dielectrics, Appl. Phys. Lett. 96 (2) (2010) 022903, http://dx.doi.org/10.1063/1.3291621.
[34] L. Vandelli, A. Padovani, L. Larcher, G. Bersuker, Y. Jung, P. Pavan, A physics-based model of the dielectric breakdown in HfO_2 for statistical reliability prediction, in: IEEE International Reliability Physics Symposium (IRPS), 2011, pp. GD.5.1–GD.5.4, http://dx.doi.org/10.1109/IRPS.2011.5784582.
[35] N. Raghavan, A. Padovani, X. Wu, K. Shubhakar, M. Bosman, L. Larcher, K.L. Pey, The 'buffering' role of high-κ in post breakdown degradation immunity of advanced dual layer dielectric gate stacks, in: IEEE International Reliability Physics Symposium (IRPS), Monterey, California, 2013. http://dx.doi.org/10.1109/IRPS.2013.6532020, pp. 5A.3.1–5A.3.8.
[36] X. Li, Nanoscale characterization of advanced gate stacks using transmission electron microscopy and electron energy loss spectroscopy, PhD Thesis, in: School of Electrical and Electronic Engineering (EEE), Nanyang Technological University (NTU), 2009.
[37] X. Li, C.H. Tung, K.L. Pey, V.L. Lo, The physical origin of random telegraph noise after dielectric breakdown, Appl. Phys. Lett. 94 (2009) 132904. http://dx.doi.org/10.1063/1.3114410.
[38] X. Li, W.H. Liu, N. Raghavan, M. Bosman, K.L. Pey, Resistive switching in NiSi gate metal-oxide-semiconductor transistors, Appl. Phys. Lett. 97 (2010) 202904. http://dx.doi.org/10.1063/1.3516466.
[39] X. Li, C.H. Tung, K.L. Pey, V.L. Lo, The chemistry of gate dielectric breakdown, in: IEEE International Electron Devices Meeting (IEDM), 2008, pp. 1–4, http://dx.doi.org/10.1109/IEDM.2008.4796813.
[40] T.A.L. Selvarajoo, R. Ranjan, K.L. Pey, L.J. Tang, C.H. Tung, W.H. Lin, Dielectric-breakdown-induced epitaxy: a universal breakdown defect in ultrathin gate dielectrics, IEEE Trans. Device Mater. Reliab. 5 (2) (2005) 190–197, http://dx.doi.org/10.1109/TDMR.2005.846674.
[41] M.A. Alam, S. Mahapatra, A comprehensive model of PMOS NBTI degradation, Microelectron, Reliability 45 (1) (2005) 71–81. http://dx.doi.org/10.1016/j.microrel.2004.03.019.
[42] S. Mahapatra, N. Goel, S. Desai, S. Gupta, B. Jose, S. Mukhopadhyay, K. Joshi, A. Jain, A. E. Islam, M.A. Alam, A comparative study of different physics-based NBTI models,

IEEE Trans. Electron Devices 60 (3) (2013) 901–916, http://dx.doi.org/10.1109/TED.2013.2238237.

[43] T. Nishimura, Solve MOSFET characteristic variation and reliability degradation issues, Electronic Design News (EDN), 2013. http://www.edn.com/design/test-and-measurement/4419732/Solve-MOSFET-characteristic-variation-and-reliability-degradation-issues.

[44] S. Mahapatra, M.A. Alam, P.B. Kumar, T.R. Dalei, D. Varghese, D. Saha, Negative bias temperature instability in CMOS devices, Microelectron. Eng. 80 (2005) 114–121, http://dx.doi.org/10.1016/j.mee.2005.04.053.

[45] D.K. Schroder, Negative bias temperature instability: what do we understand?, Microelectron. Reliab. 47 (6) (2007) 841–852, http://dx.doi.org/10.1016/j.microrel.2006.10.006.

[46] D.P. Ioannou, S. Mittl, G. La Rosa, Positive bias temperature instability effects in nMOSFETs With HfO_2/TiN gate stacks, IEEE Trans. Device Mater. Reliab. 9 (2) (2009) 128–134, http://dx.doi.org/10.1109/TDMR.2009.2020432.

[47] G. Groeseneken, R. Degraeve, T. Nigam, G. Van den bosch, H.E. Maes, Hot carrier degradation and time-dependent dielectric breakdown in oxides, Microelectron. Eng. 49 (1999) 27–40, http://dx.doi.org/10.1016/S0167-9317(99)00427-X.

[48] C.M. Tan, A. Roy, Electromigration in ULSI interconnects, Mater. Sci. Eng. R Rep. 58 (2007) 1–75, http://dx.doi.org/10.1016/j.mser.2007.04.002.

[49] R.G. Filippi, P.-C. Wang, A. Brendler, P.S. McLaughlin, J. Poulin, B. Redder, J.R. Lloyd, J.J. Demarest, The effect of a threshold failure time and bimodal behavior on the electromigration lifetime of copper interconnects, in: IEEE International Reliability Physics Symposium (IRPS), 2011, pp. 1–3, http://dx.doi.org/10.1109/IRPS.2009.5173295.

[50] P. Lamontagne, D. Ney, Y. Wouters, Effect of reservoir on electromigration of short interconnects, in: IEEE International Integrated Reliability Workshop (IIRW), 2010, pp. 46–50, http://dx.doi.org/10.1109/IIRW.2010.5706484.

[51] V.M. Dwyer, Analysis of critical-length data from electromigration failure studies, Microelectron, Reliability 51 (2011) 1568–1572. http://dx.doi.org/10.1016/j.microrel.2011.06.036.

[52] W. Li, C.M. Tan, Black's equation for today's ULSI interconnect electromigration reliability—a revisit, in: International Conference of Electron Devices and Solid-State Circuits (EDSSC), 2011, pp. 1–2, http://dx.doi.org/10.1109/EDSSC.2011.6117717.

[53] Y.K. Lim, R. Arijit, K.L. Pey, C.M. Tan, C.S. Seet, T.J. Lee, D. Vigar, Stress migration reliability of wide Cu interconnects with gouging vias, in: Reliability Physics Symposium, 2005. Proceedings. 43rd Annual. 2005 IEEE International, 2005, pp. 203–208, http://dx.doi.org/10.1109/RELPHY.2005.1493085.

[54] A. Sekiguchi, J. Koike, K. Maruyama, Microstructural influences on stress migration in electroplated Cu metallization, Appl. Phys. Lett. 83 (10) (2003) 1962–1964, http://dx.doi.org/10.1063/1.1609238.

[55] Z. Gan, W. Shao, S.G. Mhaisalkar, Z. Chen, H. Li, The influence of temperature and dielectric materials on stress induced voiding in Cu dual damascene interconnects, Thin Solid Films 504 (2006) 161–165, http://dx.doi.org/10.1016/j.tsf.2005.09.162.

[56] A. Heryanto, K.L. Pey, Y.K. Lim, W. Liu, N. Raghavan, J. Wei, C.L. Gan, M.K. Lim, J.B. Tan, The effect of stress migration on electromigration in dual damascene copper interconnects, J. Appl. Phys. 109 (1) (2011) 013716, http://dx.doi.org/10.1063/1.3531393.

[57] T.L. Tan, C.L. Gan, A.Y. Du, C.K. Cheng, J.P. Gambino, Dielectric degradation mechanism for copper interconnects capped with CoWP, Appl. Phys. Lett. 92 (20) (2008) 201916, http://dx.doi.org/10.1063/1.2929388.

[58] F. Chen, O. Bravo, K. Chanda, P. McLaughlin, T. Sullivan, J. Gill, J. Lloyd, R. Kontra, J. Aitken, A Comprehensive study of low-κ SiCOH TDDB phenomena and its reliability lifetime model development, in: IEEE International Reliability Physics Symposium, 2006, pp. 46–53, http://dx.doi.org/10.1109/RELPHY.2006.251190.

[59] J. Kim, E.T. Ogawa, J.W. McPherson, Time dependent dielectric breakdown characteristics of low-k dielectric (SiOC) over a wide range of test areas and electric fields, in: IEEE International Reliability Physics Symposium, 2007, pp. 399–404.

[60] J.R. Lloyd, On the physical interpretation of the impact damage model in TDDB of low-κ dielectrics, in: IEEE International Reliability Physics Symposium (IRPS), 2010, pp. 943–946.

[61] M. Baklanov, P.S. Ho, E. Zschech (Eds.), Advanced Interconnects for ULSI Technology, John Wiley & Sons, United Kingdom, 2012.

[62] D.R. Frear, J.W. Jang, J.K. Lin, C. Zhang, Pb-free solders for flip-chip interconnects, JOM 53 (6) (2001) 28–33, http://dx.doi.org/10.1007/s11837-001-0099-3.

[63] K.N. Tu, Recent advances on electromigration in very-large-scale-integration of interconnects, J. Appl. Phys. 94 (9) (2003) 5451–5473, http://dx.doi.org/10.1063/1.1611263.

[64] C. Chen, H.M. Tong, K.N. Tu, Electromigration and thermomigration in Pb-free flip-chip solder joints, Annu. Rev. Mater. Res. 40 (2010) 531–555, http://dx.doi.org/10.1146/annurev.matsci.38.060407.130253.

[65] K.N. Tu, Electronic Thin Film Reliability, vol. 396, Cambridge University Press, New York, 2011.

[66] H. Ceric, IμE, TU Wien, Annual review report, 2013.

[67] L. Zhang, S. Ou, J. Huang, K.N. Tu, S. Gee, L. Nguyen, Effect of current crowding on void propagation at the interface between intermetallic compound and solder in flip chip solder joints, Appl. Phys. Lett. 88 (1) (2006) 012106, http://dx.doi.org/10.1063/1.2158702.

[68] Q. Zhou, J. Zhain, HfO_x bipolar resistive memory with robust endurance using ZrN_x as buttom electrode, Appl. Surf. Sci. 284 (2013) 644–650.

[69] L.J. Liu, K.S. Chang-Liao, Y.C. Jian, T.K. Wang, M.J. Tsai, Performance enhancement in p-channel charge-trapping flash memory devices with Si/Ge super-lattice channel and band-to-band tunneling induced hot-electron injection, Thin Solid Films 533 (2013) 1–4, http://dx.doi.org/10.1016/j.tsf.2012.11.115.

[70] A. Debunne, K. Virwani, A. Padilla, G.W. Burr, A.J. Kellock, V.R. Deline, B. Jackson, Evidence of crystallization–induced segregation in the phase change material Te-rich GST, J. Electrochem. Soc. 158 (10) (2011) H965–H972, http://dx.doi.org/10.1149/1.3614508.

[71] N. Raghavan, R. Degraeve, A. Fantini, L. Goux, S. Strangio, B. Govoreanu, G. Groeseneken, M. Jurczak, Microscopic origin of random telegraph noise fluctuations in aggressively scaled RRAM and its impact on read disturb variability, in: IEEE International Reliability Physics Symposium (IRPS), 2013. http://dx.doi.org/10.1109/IRPS.2013.6532042, pp. 5E.3.1–5E.3.7.

[72] R. Degraeve, J.L. Ogier, R. Bellens, P.J. Roussel, G. Groeseneken, H.E. Maes, A new model for the field dependence of intrinsic and extrinsic time-dependent dielectric breakdown, IEEE Trans. Electron Devices 45 (2) (1998) 472–481, http://dx.doi.org/10.1109/16.658683.

[73] M. Gall, C. Capasso, D. Jawarani, R. Hernandez, H. Kawasaki, P.S. Ho, Statistical analysis of early failures in electromigration, J. Appl. Phys. 90 (2) (2001) 732–740, http://dx.doi.org/10.1063/1.1377304.

[74] K. Shubhakar, K.L. Pey, S.S. Kushvaha, S.J. O'Shea, N. Raghavan, M. Bosman, M. Kouda, K. Kakushima, H. Iwai, Grain boundary assisted degradation and breakdown

study in cerium oxide gate dielectric using scanning tunneling microscopy, Appl. Phys. Lett. 98 (7) (2011) 072902, http://dx.doi.org/10.1063/1.3553190.
[75] H. Luo, Y. Wang, J. Velamala, Y. Cao, Y. Xie, H. Yang, The impact of correlation between NBTI and TDDB on the performance of digital circuits, in: IEEE 54th International Midwest Symposium on Circuits and Systems (MWSCAS), 2011, 2011, pp. 1–4, http://dx.doi.org/10.1109/MWSCAS.2011.6026408.
[76] N. Rahim, D. Misra, Correlation of negative bias temperature instability and breakdown in HfO_2/TiN gate stacks, ECS Trans. 28 (2) (2010) 323–330, http://dx.doi.org/10.1149/1.3372586.
[77] E.Y. Wu, J. Suñé, On voltage acceleration models of time to breakdown—part I: experimental and analysis methodologies, IEEE Trans. Electron Devices 56 (7) (2009) 1433–1441, http://dx.doi.org/10.1109/TED.2009.2021721.
[78] E.Y. Wu, J. Suñé, On voltage acceleration models of time to breakdown—part ii: experimental results and voltage dependence of weibull slope in the FN regime, IEEE Trans. Electron Devices 56 (7) (2009) 1442–1450, http://dx.doi.org/10.1109/TED.2009.2021725.
[79] P.E. Nicollian, Errors in projecting gate dielectric reliability from Fowler–Nordheim stress to direct-tunneling operation. IEEE Electron Device Lett. 30 (11) (2009) 1185–1187, http://dx.doi.org/10.1109/LED.2009.2030698.
[80] M. Toledano-Luque, B. Kaczer, J. Franco, P.J. Roussel, T. Grasser, T.Y. Hoffmann, G. Groeseneken, From mean values to distributions of BTI lifetime of deeply scaled FETs through atomistic understanding of the degradation, in: IEEE Symposium on VLSI Technology (VLSIT), 2011, pp. 152–153.
[81] M. Pecht, Prognostics and Health Management of Electronics, John Wiley & Sons Ltd., USA, 2008.
[82] N. Patil, D. Das, K. Goebel, M. Pecht, Identification of failure precursor parameters for insulated gate bipolar transistors (IGBTs), in: IEEE International Conference on Prognostics and Health Management, 2008, pp. 1–5, http://dx.doi.org/10.1109/PHM.2008.4711417.
[83] Q. Miao, L. Xie, H. Cui, W. Liang, M. Pecht, Remaining useful life prediction of lithium-ion battery with unscented particle filter technique, Microelectron. Reliab. 53 (6) (2013) 805–810, http://dx.doi.org/10.1016/j.microrel.2012.12.004.
[84] P. Lall, M. Hande, C. Bhat, N. Islam, J. Suhling, J. Lee, Feature extraction and damage-precursors for prognostication of lead-free electronics, Microelectron. Reliability 47 (12) (2007) 1907–1920, http://dx.doi.org/10.1109/ECTC.2006.1645736.
[85] N. Raghavan, K.L. Pey, D.D. Frey, Noise-based prognostic design for real-time degradation analysis of nanodevice dielectric breakdown, in: IEEE Annual Reliability and Maintainability Symposium (RAMS), 2013, pp. 1–7, http://dx.doi.org/10.1109/RAMS.2013.6517656.

Design considerations for reliable embedded systems

9

R.A. Shafik, A. Das, S. Yang, G. Merrett, B.M. Al-Hashimi
University of Southampton, Southampton, United Kingdom

9.1 Introduction

Technology scaling has enabled the fabrication of efficient and low-power electronic devices for current and future generations of embedded systems. Examples of such systems include IBM's 22-nm [1] chip and Intel's emerging 16-nm [2] processor with promises to provide with unprecedented integration capacity and performance. However, with these technological advances, embedded systems design is increasingly being confronted with emerging challenges. A major challenge for such systems design is the drastically increased number of hardware faults manifested in the form of logic upsets or timing violations, exacerbated further by manufacturing imperfections and harsh operating environments [3]. These faults affect the stored values and timing of signal transfers at circuit level, leading to incorrect execution or even failure at higher levels in embedded systems. Operating reliably in the presence of these hardware faults is challenging, particularly for high-availability, safety-critical systems [4–6].

To improve reliability of embedded electronic systems, designers need to ensure continued functionality of the system in the presence of various hardware faults. This is carried out through detection of faults, followed by their correction or tolerance mechanisms. Fault detection deals with identifying the presence of one or more faults in the system, which are then suitably corrected by reinstating the original logic or tolerated by mitigating the impact of the detected fault. Over the years, researchers in academia and industry have proposed various design approaches to enable fault detection and tolerance. While industries have preferred pragmatic approaches through the traditional approach, such as radiation hardening [7], hardware redundancy [8,9], or information redundancy-based error detection and correction (EDAC) coding [10] methods, academic research efforts have mostly been based on cost-effective approaches to achieve overall system reliability at circuit and system level for current and future generations of embedded systems.

This chapter gives a comprehensive account of the design considerations for reliable embedded systems design. Section 9.2 outlines the sources and trends of hardware faults, and Section 9.3 gives the reliable system design principles and practices. Section 9.4 highlights the design trade-offs involved in low-cost reliable design practices and reviews some of the existing academic and industrial research

Reliability Characterisation of Electrical and Electronic Systems. http://dx.doi.org/10.1016/B978-1-78242-221-1.00009-5

efforts in low-cost reliable system design. Section 9.5 presents design challenges of current and future generations of embedded systems to motivate further focused research, and Section 9.6 concludes the chapter.

9.2 Hardware faults

Figure 9.1 depicts a typical embedded system showing hardware faults in different architectural components: processors, interconnects, and memories. Faults in the processors generally lead to erroneous computation or even failures, while faults in interconnects can lead to loss of communication and incoherence between the processors' executions. Faults in memory components are considered serious as they can last for indefinite time and cause series of computations to be carried out erroneously or even cause system failure. These faults manifest in two forms: logic and timing faults. In the following, these faults are further detailed, followed by the trends of different hardware faults and their occurrences.

9.2.1 Logic faults

Logic faults are perturbations in the circuit that may lead to erroneous logic. Depending on their nature and duration of occurrence, logic faults can be classified into the following two categories: transient logic faults and permanent logic faults. Transient logic faults, also known as soft errors, do not represent a physical defect in the circuit. These faults manifest themselves during the operation of a circuit due to various operating conditions. High-energy radiation particle events are major contributors

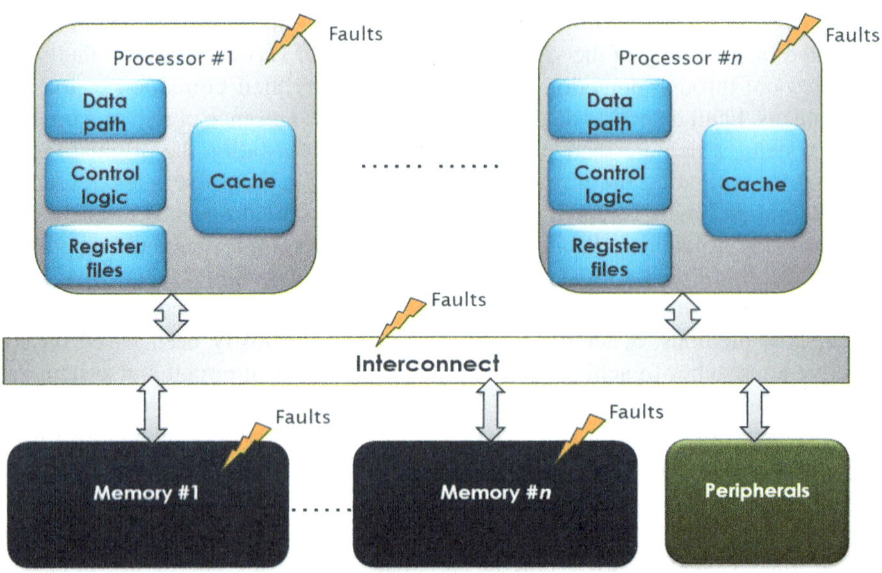

Figure 9.1 Embedded system architecture in the presence of hardware faults.

of transient logic faults, among others. Such events produce a burst of hole-electron pairs in a transistor junction that is large enough to overcome the critical charge required to cause the circuit to change logic state. Since technology scaling and low-power design techniques drastically reduce this critical charge further, modern devices are vulnerable to transient faults [11]. Single-event upset (SEU) is the most popular transient fault model, which implies that a logic upset is caused by a single radiation event [12,13]. However, recent studies [14] suggest that due to high packing density enabled by modern technology nodes, multiple circuits can change state at once by a single radiation event, leading to multibit upset (MBU). Increased logic upsets in the form of SEUs or MBUs in modern technology nodes are a major reliability concern for embedded systems applications as these mainly take place in circuit storage elements, such as memories and registers [12].

Permanent logic faults are generally attributed to device-level imperfections, introduced by either improper lithographic process, systematic aberrations during manufacturing, or even postmanufacturing burn-in process [11]. Modern semiconductor manufacturers use advanced postmanufacturing test and diagnosis techniques at predesigned integrated circuit interfaces to ensure that permanent faults or defects caused by manufacturing processes are identified and faulty chips are discarded/repaired before the actual product shipment. Test and diagnosis techniques are not within the scope of this chapter; hence, interested readers are referred to [15,16] for further details on manufacturing test and diagnosis techniques. An emerging logic fault concern in modern electronic devices is permanent faults due to accelerated device wear-outs. Such wear-outs are caused by increased high power densities resulting from high device activity and stress [11]. Due to increased power densities, these devices are subjected to elevated operating temperatures, leading to transistor-level degradations. Major degradation mechanisms include electromigration (EM) in conductors as ions experience momentum transfer between conducting electrons and diffusing metal atoms, time-dependent dielectric breakdown (TDDB) due to tunneling current, and negative bias temperature instability (NBTI) as threshold voltage increases in PMOS devices. These degradations eventually manifest themselves in the form of permanent logic faults and hence reduce the lifetime of the electronic device [17].

9.2.2 Timing faults

In a synchronous digital circuit, when the delays of the combinational logic paths exceed the operating clock period, the circuit fails to generate expected outcomes by the clock event. This leads to timing errors, which are caused by one or more of the following three phenomena [18]. First, static process variations due to manufacturing aberrations can result in circuit-level mismatch between the desired circuit timing characteristics and the actual circuit timing behavior in the critical or near-critical paths. Second, switching between operating conditions (voltage, frequency, and temperature) in low-power designs can lead to dynamic variations, which can affect the circuit timing behavior temporarily. Finally, wear-out mechanisms, such as NBTI, can gradually increase the circuit delay over its lifetime. Since timing

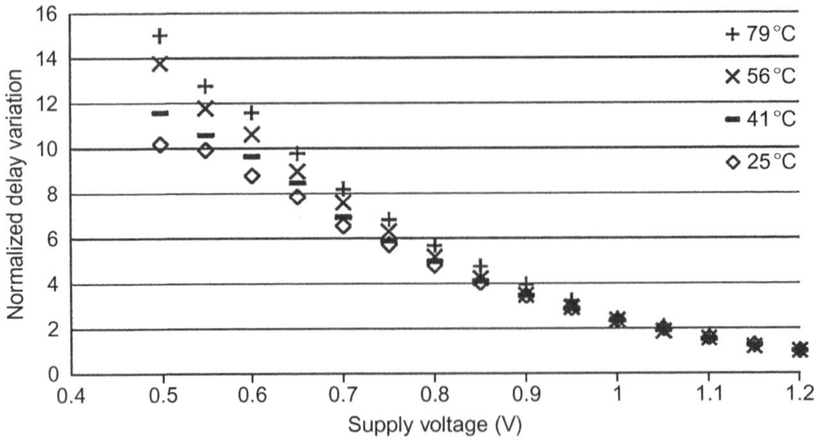

Figure 9.2 Measured results of intra-die PVT variation on delay of the test chip [19].

dictates the event-based signal transfer and flow in synchronous circuits, timing faults are a challenging reliable design concern.

To get an insight into the cause of timing faults, Figure 9.2 shows normalized delay variations of a ring oscillator circuit implemented with NAND gates [19]. The delay variations are measured for a range of supply voltages with different operating temperatures. Normalization is carried out with respect to the measured oscillator delays operating with a 1.2 V supply voltage at 25 °C. It can be seen that normalized delay variation is the lowest at nominal supply (1.2 V) and room temperature (25 °C). However, it increases by up to 10 × at the same temperature when supply voltage is reduced from 1.2 to 0.5 V. When temperature is also increased to 79 °C, the delay increases further (up to 15 ×). With such increased delays, the circuit cannot operate at higher operating frequencies at the lower supply voltages, since the probability of timing faults increases significantly with increased delay variation.

9.2.3 Trends of hardware faults

Figure 9.3(a) shows the trends of device failure rates over its lifetime, highlighting the impact of technology scaling [20]. As can be seen, due to manufacturing silicon or metal defects, devices experience high failure rates during early burn-in process. These defects are rigorously tested for in postmanufacturing testing phase and the failing devices are discarded from the shipment line. The good circuits that are delivered to customers also experience failures during operational lifetime, which are dominated by the environmentally induced transient faults (Section 9.2.1). As the device continues to experience stress during the operating lifetime, device-level mechanisms, such as EM, NBTI, and TDDB, start to accelerate the wear-out, eventually leading to permanent faults (Section 9.2.1). Note that the normalized device failure rates increase significantly throughout the entire device lifetime due to technology scaling. The number of manufacturing defects increases due to aggravated process variations,

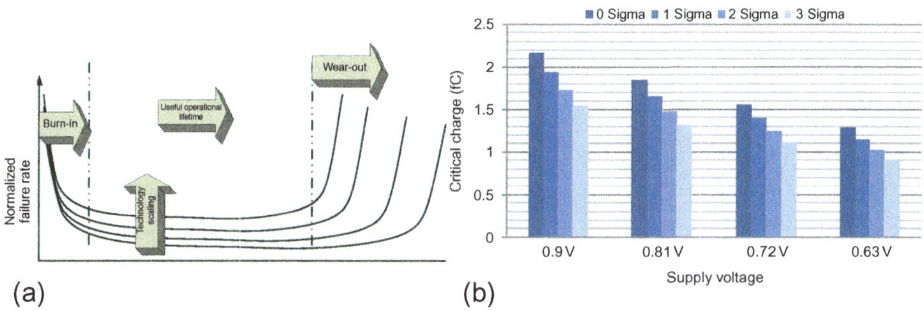

Figure 9.3 (a) Trends of failure rates with technology scaling [20] and (b) impact of process variation and voltage scaling on device critical charge.
Reprinted from Ref. [19].

exacerbated by the limitations of the existing lithographic technologies. The transient failure rates during operational lifetime also increase with technology scaling due to drastic reduction of the device critical charge (i.e., the minimum charge that is needed to change the circuit logic levels). Moreover, useful operating lifetime is also incrementally being shortened due to accelerated device wear-outs caused by increased stress and device power densities. Such increase in the hardware failures rates poses serious challenges to systems designers to ensure reliability of embedded systems with modern technology nodes.

For a given technology node, voltage scaling and manufacturing process variations affect the critical charge and eventually the vulnerability of a device. To illustrate the effect of supply voltage scaling, Figure 9.3(b) shows measured critical charges of a flip-flop in the presence of process variations, using 45-nm CMOS technology with 0.9 V nominal supply voltage [19]. As can be seen, the critical charge of the flip-flop reduces with the reduction in supply voltage. The figure also shows the reduction in the worst-case critical charge of the flip-flop at 1σ, 2σ, and 3σ variations at different supply voltages. These results demonstrate that the critical charge of a flip-flop is negatively affected not only because of reduction in the supply voltage but also by process variation. Such reduction in critical charge significantly increases the vulnerability in the presence of radiation events.

9.3 Reliable design principles

Reliability in an electronic system is traditionally achieved through redundancy or error hardened hardware design. In the following, a brief account of key reliable design principles is presented highlighting some classical design methods.

9.3.1 Hardware redundancy

Traditional reliable design methods employ hardware redundancy to achieve desired reliability of a system. In such methods, one or more additional hardware resources are incorporated along with the original hardware resource to facilitate error detection,

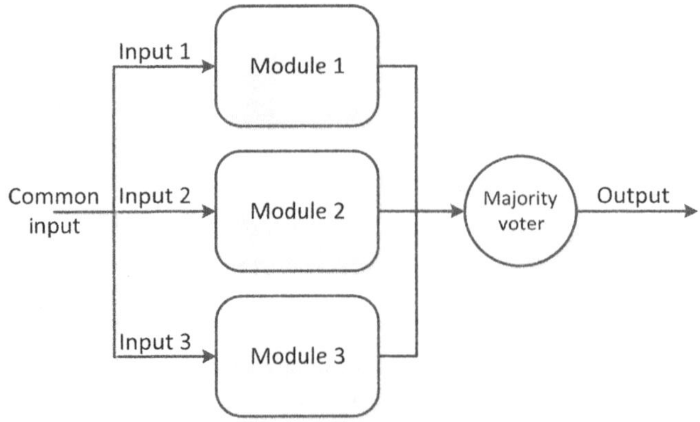

Figure 9.4 Triple modular redundancy (TMR) method.

followed by correction or tolerance. A classical example is the triple-modular redundancy (TMR) method [21], as shown in Figure 9.4. In TMR method, three instances of the same hardware module are used with the same input sequences. In fault-free circumstances, all three are expected to generate the same output. However, when one of the modules is faulty, the other two modules are expected to agree in terms of their outputs for the given input and continue system operation without failure. To facilitate such agreement between the outputs of the modules, a majority voting system is incorporated in TMR method with the ability of selecting any two conforming modules out of the three in the event of fault in one of the modules (Figure 9.4).

TMR method ensures continued functionality in the presence of transient faults. This makes it highly suitable for mission-critical or failure-sensitive applications. An example of TMR method in such applications is Maxwell's SCS750 computing system used in space [9]. The system uses three IBM PowerPC processors together with majority voting system to ensure computing without failure in the presence of faults.

TMR method has two main limitations. Firstly, it has high (200%) area overhead. Moreover, since all modules operate at the same time, the overall energy increases significantly [22]. Secondly, when one of the modules goes permanently faulty, the effectiveness of the method can no longer be validated. This is because, with a permanent failure in one of the modules, the system cannot ensure fault detection or tolerance when one of the other modules is affected by a fault. Similarly, when the majority voting circuit fails due to one or more faults, the system can no longer ensure correct selection of the outputs. Due to these limitations, TMR method is usually mission time (the time required for completing a given mission)-limited, defined by classical reliability evaluation techniques [18].

To increase effective fault-free mission time, many other hardware redundancy methods have been proposed over the years, which include higher-order *N*-modular redundancy (NMR) and standby-sparing, etc. [21]. In NMR methods, an arbitrary

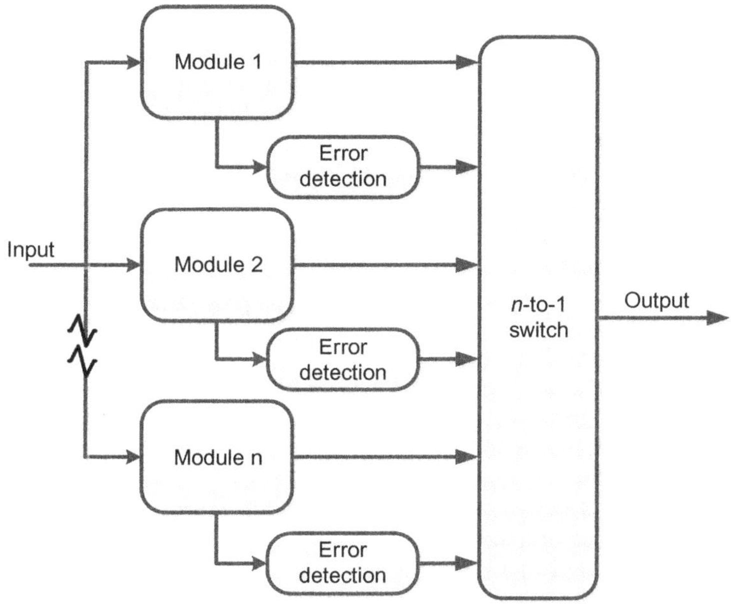

Figure 9.5 Standby-sparing method for fault detection and tolerance.

N number of modules are redundantly placed alongside the original module to achieve increased mission time at the cost of high system costs. In standby-sparing methods, each module incorporates a fault detection mechanism as shown in Figure 9.5. When a module out of *N* similar redundant modules detects a fault, its outputs are deactivated and a spare modules outputs are activated by the *n*-to-1 switch. As the number of spare modules increases, the fault-free mission time of the system also increases. Both NMR and standby-sparing methods, however, suffer from high silicon area and energy consumption overheads [21].

9.3.2 Error hardening

Since transient faults usually dominate the hardware faults in many space applications, designers have traditionally also resorted to error hardening methods. Using these methods, the transient faults can be mitigated at either physical or device level. In physical-level hardening method, the system-on-chip is packaged and shielded against radioactivity [23]. The shielding is carried out using depleted atoms, such as borons, in the external glass passivation layers, which can drastically intercept and decay energy of radiation (neutron and alpha) particles. In device-level error hardening methods, the target circuit is deliberately designed with devices with higher radiation noise margins. This can be achieved by carefully designing the devices with higher capacitances (and hence critical charges) or selecting the devices with higher radiation tolerance. For example, bipolar integrated circuits generally have few orders of magnitude higher radiation tolerance than circuits designed with

CMOS devices [24]. Another effective device-level error hardening method is to design CMOS-based integrated circuits with nontraditional substrate materials with higher noise margin and radiation tolerance, such as silicon carbide and gallium nitride. The use of these substrate materials ensures that the devices have higher critical charge requirement to inflict logic state change, which drastically reduces the occurrence of transient logic faults even in harsh radiation environments [7].

9.3.3 EDAC codes

Using EDAC codes is a popular method for reliable design of electronic circuits. Due to high vulnerability in the presence of faults and storage capability of logic circuitry, EDAC coding method is particularly suitable for memory devices [25]. In this method, extra information codes are incorporated along with the original circuit logic to identify the presence of one or more transient or permanent faults and correct them [26]. The length of these codes varies with the coding technique used and the target fault detection and correction capability of the circuit. To achieve correction capability of k errors per codeword, the coding technique must ensure that any two codewords are at least different by $2k+1$ bits, otherwise known as the Hamming distance of the coding technique [27]. Example coding techniques include Hamming codes [27], single-error correction and double-error detection (SECDED) codes [28], etc.

In memory circuits, EDAC codes can be integrated at byte or word level. Figure 9.6 shows a typical memory architecture with such support for integrating EDAC codes. When a memory byte or word is being written into a memory location, the CODEC unit encodes the necessary extra information bits and stores them as EDAC bits

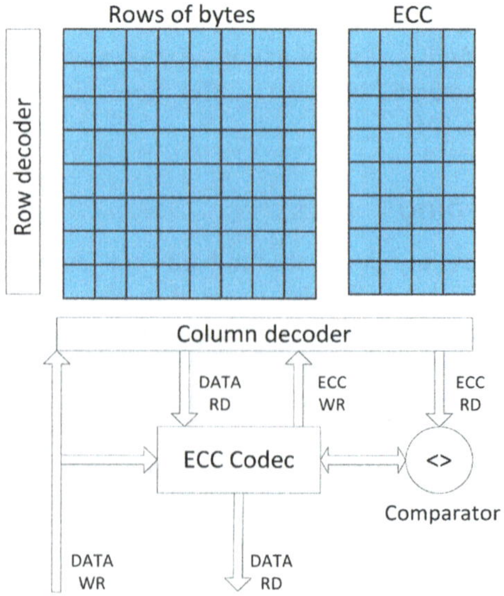

Figure 9.6 Memory architecture with EDAC support.

alongside the original byte or word. When the word is being read out, the CODEC unit detects fault and performs necessary corrections, if there is any. Intel dual-core Xeon-7100 system has EDAC features to incorporate fault detection and correction [29].

9.3.4 Re-execution and application checkpointing

Re-execution is another popular transient fault tolerance method. This method requires incorporating fault detection capabilities in different parts of the system. Upon detection of faults, re-execution is carried out from the last execution point to mitigate the impact of detected faults, as shown in Figure 9.7(a). Such repetition of execution incurs overheads in terms of execution time, which can be minimized by carrying out re-execution during idle times. However, availability of such idle times can depend on the type of computation being carried out. Many modern processors incorporate re-execution features to facilitate transient fault tolerance, for example, academic prototypes [30,31], ARM's Cortex™-R [32], and Intel's Xeon [33].

Application checkpointing is another effective method similar to re-execution. Using this method, application traces are saved at regular checkpoint intervals, as shown in Figure 9.7(b). In the event of faults, application execution is repeated (or rolled back) from the last known good checkpoint. However, application checkpointing requires additional hardware support to enable interrupts. Moreover, to ensure that the stored application checkpoints are fault-free, reliable storage components need to be incorporated. The determination of the checkpointing interval poses another challenge to system designers. This is because frequent checkpointing incurs high execution time overheads as application execution is frequently interrupted. On the other hand, longer checkpointing intervals can cause large re-execution times in the event of a fault occurrence. Due to such trade-offs between execution time and fault tolerance achieved, system designers have to decide on the interval based on the fault tolerance needs with a performance constraint [34]. Similar to the re-execution method, application checkpointing has been featured in reliable computing systems, such as checkpointing-based RAID systems [35].

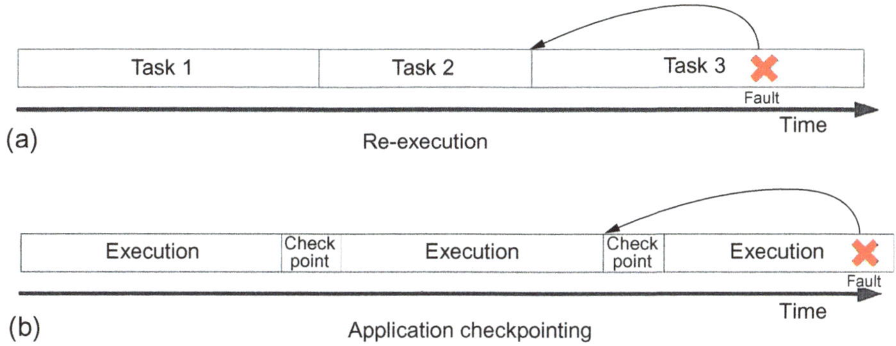

Figure 9.7 (a) Re-execution and (b) application checkpointing.

9.3.5 Industrial practices

To meet the reliability demands of various applications, industries have generally favored traditional and pragmatic methods. An example of such methods is ARM's Cortex-R4 [36] embedded processor, used in performance-oriented and reliable hard disk drive controllers, wireless baseband processors, and electronic control units for automotive systems. Figure 9.8 shows the block diagram of the Cortex-R4, showing the configuration and different protection measures in the event of faults. As can be seen, a redundant core is used for duplicate execution and lockstepping. The two cores execute the same program from level 1 cache instructions and data. Additional comparison and synchronization logic synchronizes and inspects every cycle of operation to detect a difference in the execution outcomes that would indicate that a transient or permanent fault has occurred. When a fault is detected, the processor goes into fail-safe mode and identifies the fault by looking into the last instruction in the level 1 cache line. The instruction that generated faulty execution is invalidated and refetched through write-through operation. Apart from core-level lockstepping and re-execution, the Cortex-R4 also features error correction coding (ECC)-enabled level 2 caches and tightly coupled memories (TCMs) to provide high integrity data storage and transfer in the system.

Another example of effective industrial methods used in reliable superscalar processor design is Oracle and Fujitsu's fifth generation SPARC64 [37]. The processor has been used by Fujitsu in their low-power computing systems with high-availability and high-reliability needs. To enable such high reliability, the processor incorporates fault detection mechanism in the data path, different levels of caches, and memory, as shown in Figure 9.9. As can be seen, different EDAC coding protection mechanisms

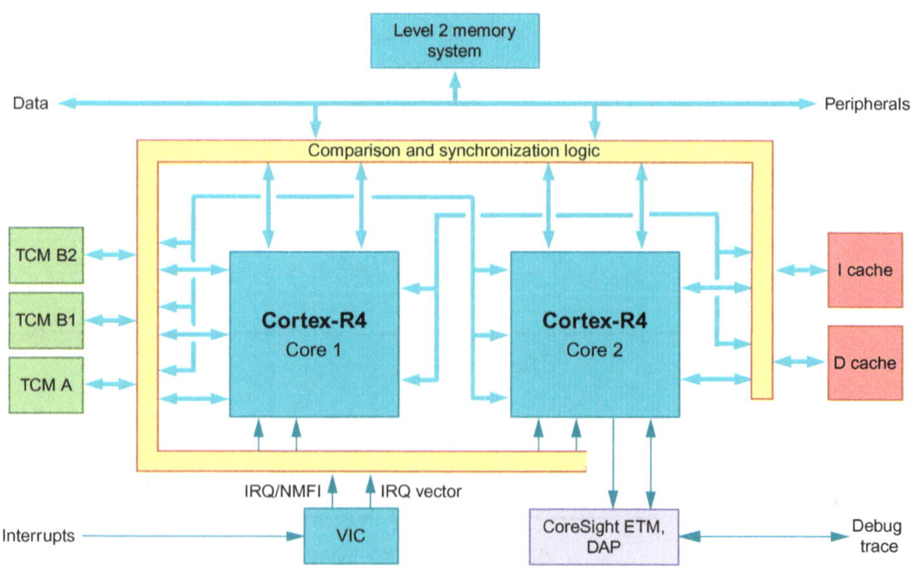

Figure 9.8 Duplication and lockstepping method in ARM's Cortex-R [36] processors.

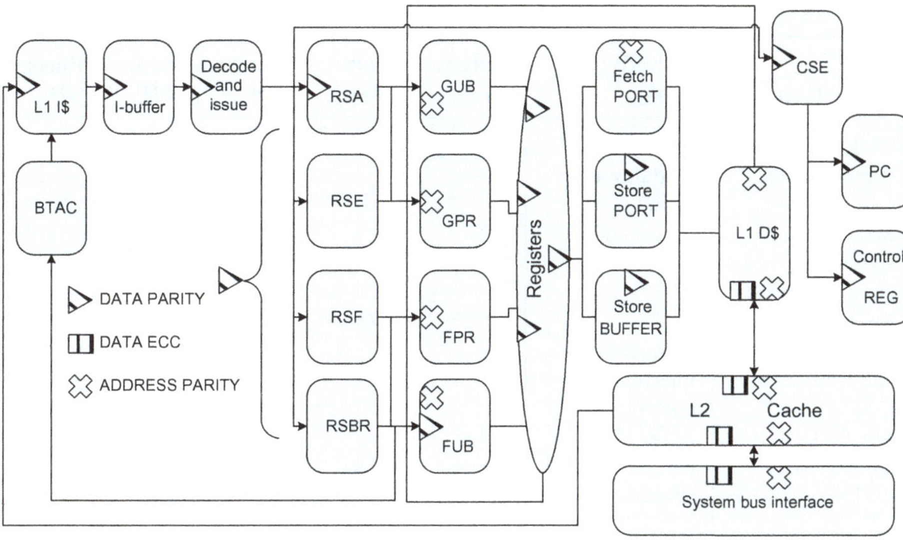

Figure 9.9 Multilevel reliability approach in Fujitsu SPARC4 processors (RSA, reservation station for address generation; GPR, general-purpose register; FPR, floating-point register; GUB, general-purpose update buffer; FUB, floating-point update buffer; CSE, commit stack entries; PC, program counter register).
Reprinted from Ref. [37].

(parity and data ECC) are used in different integer and floating-point registers and execution units and their update buffers in the processor core and its pipeline stages. When a fault is detected during any of the four pipeline stages, the instruction is invalidated, refetched, and retried. Due to such multilevel fault detection and tolerance mechanisms, the SPARC64 [37] is highly effective in high-availability computing that requires working without failures most of the time.

9.3.6 Design trade-offs

Table 9.1 summarizes some of the classical methods applied in various applications showing their efficiencies in terms of fault detection and tolerance coverages and the overheads incurred. Columns 1 and 2 show the method applied in an application and the hardware/software configurations associated with the method, while columns 3 and 4 show the reliability of the system in terms of detection coverage and fault tolerance evaluated with comprehensive fault injection campaigns. Columns 5-7 show the area, performance, and energy overheads incurred by the methods. As can be seen, the TMR method applied in a reliable computing [21] provides the best possible reliability with 99% detection coverage and high (93%) fault tolerance of all injected faults. However, such high reliability is achieved within a predefined mission time of the system (as described in Section 9.3.1) at the cost of high chip area (~200%)

Table 9.1 **Design trade-offs of traditional reliable design methods**

Method/Appl.	HW/SW config.	Detect. cov.	Fault toler.	Area OH	Perf. OH	Energy OH
TMR: reliable computing [21]	Two other modules and voter	99%	93%	~200%	~15%	214%
Standby-sparing: computing [38]	Extra module and comparator	92%	78%	~113%	~12%	~180%
Re-execution: RAID storage and computing [35]	HW redundancy, interrupt control	~99%	~85%	Up to 127%	~17%	Up to 150%
SECDED: memory only [39]	EDAC CODEC per 256 bytes (Figure 9.6)	Two faults per word	One fault per word	~18%	~28%	25%
OECNED: memory only [28]	EDAC CODEC per 256 bytes (Figure 9.6)	Nine faults per word	Eight faults per word	~93%	~55%	~220%

overhead. Since the redundant modules also operate at all times, higher energy overhead (214%) is also incurred in this method. The standby-sparing method used in computing units [38] and re-execution method used in RAID systems [35] both achieve similar high reliability at low performance overheads. However, both incur large area and energy overheads due to duplication and hardware redundancy for fault detection. The EDAC coding method, such as SECDED used in memory protection [39], gives limited fault detection (two faults per word) and correction (one fault per word) capabilities at low system overheads. However, when the EDAC coding length is increased in OECNED, higher fault detection (nine faults per word) and tolerance capabilities are obtained at high energy (~220%) and area (~93%) overheads (Table 9.1).

9.4 Low-cost reliable design

Traditional methods are characterized by high costs in terms of area and energy overheads (Section 9.3). To control costs, designers need to assess the target reliability and carefully address the costs incurred due to increased system overheads, redesign, and validation requirements [40]. However, controlling system costs using traditional methods is becoming incrementally difficult for current and future generations of embedded systems. This is because continued technology scaling has exponentially increased the device-level power densities and vulnerabilities in the presence of hardware faults and manufacturing process variations. As a result, achieving high reliability in terms of fault detection and correction capabilities for these systems has emerged as a crucial design challenge considering the cost implications. Over the years, researchers have proposed a number of approaches highlighting such design trade-offs at various design levels: microarchitectural, system, and software. In the following, some of these approaches are briefly reviewed.

9.4.1 Microarchitectural approaches

Introducing hardware or logic redundancy at microarchitectural level has been a popular research topic over the years. Such redundancy is generally incorporated in critical components in the system to ensure an overall reliability target is achieved at low cost. Figure 9.10 shows an example microarchitectural approach, which uses built-in soft error resilience (BISER) architecture for detecting and correcting transient faults in latches/flip-flops and combinational logic. The faults in latches/flip-flops are corrected using a C-element as shown in Figure 9.10(a). During normal operation, when the clock is enabled, the latch output is driven by the combinational logic. Hence, latches are not susceptible to transient faults during this time. However, when clock is not enabled, C-OUT will hold the correct value. Any fault in either latch during this time will be detected through the disagreement between A and B, but will not affect C-OUT value. The cost of adding this redundant latch is minimized by reusing scan cells (generally used for postmanufacturing tests) in the circuit. As shown in Figure 9.10(b), fault tolerance in combinational logic circuits in BISER is incorporated through duplication and re-execution. When duplicate combinational units disagree on an output due to a fault, both units are re-executed to mitigate the impact of the detected fault. Due to such built-in microarchitectural redundancies, BISER enables more than an order of magnitude reduction in chip-level fault rates with minimal area impact, 6-10% chip-level power overhead, and 1-5% performance overhead [41].

Another effective microarchitectural transient fault detection and mitigation approach has been proposed in Ref. [42], as shown in Figure 9.11. The approach consists of a state monitoring block and an error correction block (CRC or Hamming). The scan chain inputs to the circuit are controlled by a multiplexer with a "sel" signal, which allows the scan chains to be recirculated or conditionally corrected (inverted),

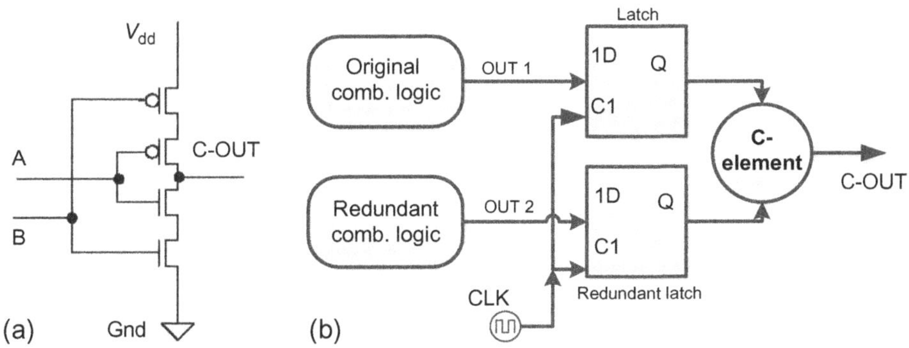

Figure 9.10 (a) Scan reuse for mitigating transient faults in latches using C-element and (b) duplication and re-execution for combinational circuits [41].

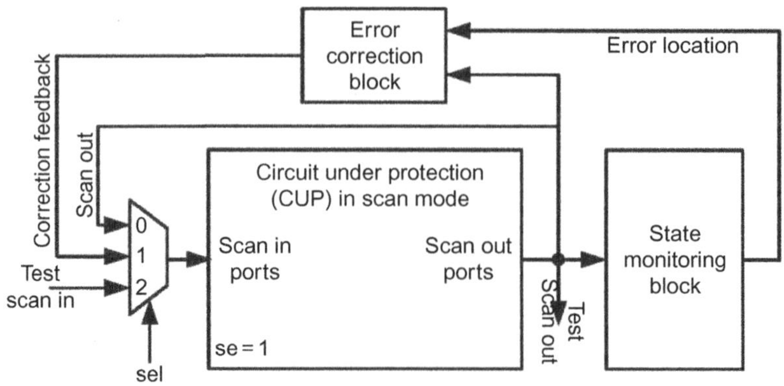

Figure 9.11 Microarchitectural approach for state monitoring and recovery [42].

or used as conventional manufacturing scan inputs. The state monitoring block checks the states of the circuit against the stored parity bits. When errors are detected, the state monitoring block sends the error locations to the error correction block, which corrects the corrupted states and feeds back to the circuit. This approach benefits from reusing the scan chains for system-state monitoring and recovery in the presence of logic faults. Similar microarchitectural approaches, such as [26,43,44], have also shown effective logic fault detection and tolerance approaches.

The detection and mitigation of timing faults has been an emerging research concern over the years due to timing variability in combinational logic circuits [30]. Such variability is particularly pronounced for dynamic voltage/frequency scaling (DVFS)-enabled systems (see Section 9.2.2). To ensure that DVFS can be effectively applied at lower supply voltages in the presence of significantly increased timing variations, a number of approaches have been proposed. RAZOR [45] is one such approach, which uses circuit- and architectural-level redundancies for *in situ* timing fault detection and mitigation. As shown in Figure 9.12, RAZOR introduces shadow latch and comparators alongside the main flip-flops in the critical path of the system. When the two

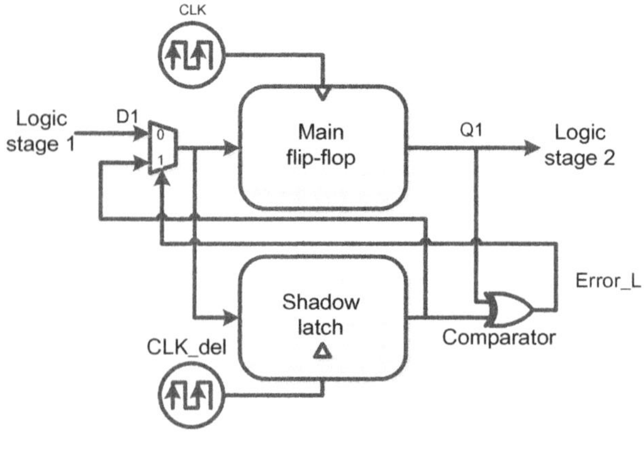

Figure 9.12 RAZOR flip-flop for detecting and tolerating timing faults [45].

latches experience different delays, their logic inputs are latched in different times at the input of the output comparator. This generates a disagreement at the comparator output, indicating detection of a timing fault. The fault is then mitigated in a pipelined processor system by rolling back the last instruction that experienced the timing fault. RAZOR has been reported to improve operating frequencies at lower operating voltages, facilitating robust low-power computing.

9.4.2 System-level approaches

System-level design approaches for reliability optimization work at a higher level of abstraction compared to microarchitectural approaches. These approaches effectively employ hardware/software codesign techniques to achieve target system-wide reliability at low cost. The hardware techniques generally enable fault detection or reliability monitoring in the system, while software techniques facilitate fault mitigation/correction or reliability improvement mechanisms. Key software techniques include architecture allocation, task mapping, and scheduling. Architecture allocation deals with identifying the suitable number of cores and their interconnect architecture. Mapping deals with the distribution of software tasks of a given application among the processing cores of a multiprocessor system, while scheduling considers ordering the tasks on a processor satisfying the data and control dependencies. System-level approaches proposed over the years have generally focused on unrestricted optimization problem, that is, the best effort to allocate resources to improve performance measured in terms of throughput or *makespan* [46,47]. Recently, studies are being conducted on reliability- and energy-driven optimization objectives under a given performance constraint. The mapping and scheduling problem (both unrestricted and restricted counterparts) are NP-hard, and as such, heuristics are adopted to simplify the complexity. As shown in Figure 9.13, reliability-aware system-level optimization approaches can be classified into two categories—runtime management (i.e., online) and design-time optimization (offline), as discussed below.

9.4.2.1 Runtime reliability management

Multiprocessor systems execute applications with different performance requirements. These applications exercise the hardware differently due to the computation being carried out. The runtime approaches adapt to these performance and workload variations using the power control levers or the scheduling decisions to perform reliability management based on the continuous feedback from the hardware performance monitoring unit at runtime. A slack borrowing approach is proposed in Ref. [48] to dynamically manage peak temperature for an MPEG-2 decoder. A reinforcement learning-based adaptive approach is proposed in Ref. [49] to optimize temperature by controlling task mapping based on the temperature of the current iteration. A neural network-based adaptive approach is proposed in Ref. [50] to reduce peak temperature. Both these approaches rely on the *HotSpot* tool for temperature prediction. The relationship between temperature and voltage and frequency of operation is formulated in Refs. [51,52]. Based on this, an online heuristic is proposed to determine the voltage

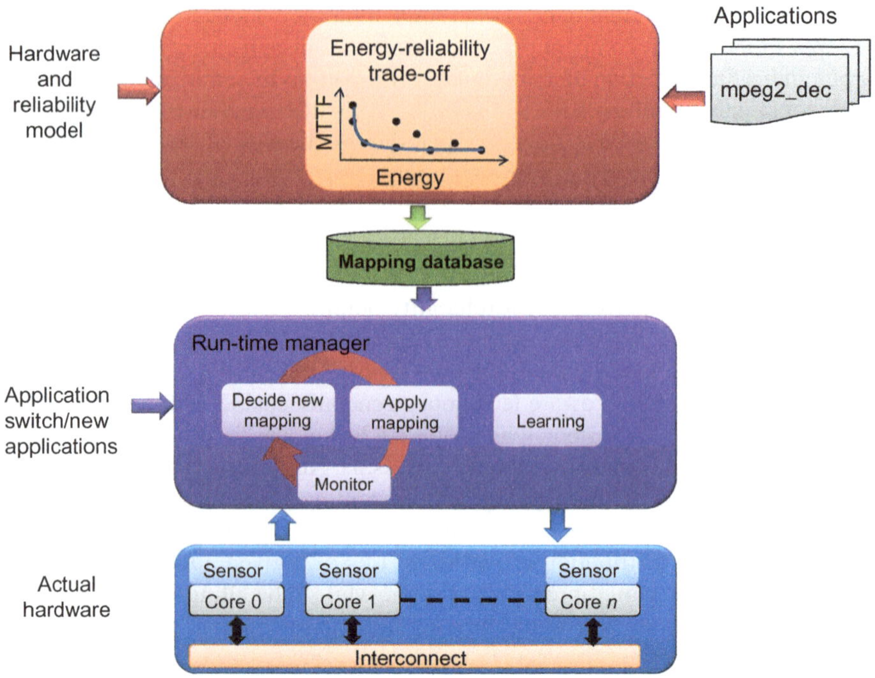

Figure 9.13 System-level design techniques.

and frequency of the cores to minimize the temperature. This approach does not perform learning, resulting in reperforming the same optimization for the same environment. A reinforcement learning algorithm is proposed in Ref. [53] to manage performance/thermal trade-offs by sampling temperature data from the onboard thermal sensors. A distributed learning agent is proposed in Ref. [54] to optimize peak temperature with a given power budget. The approach is implemented on FPGA with temperature measurement using an external thermal gun. A control-theoretical approach is presented in Ref. [55] to maximize the reliability of a system through the use of long-term and short-term controllers. A reinforcement learning-based approach is presented in Ref. [56] to jointly optimize peak temperature and thermal cycling in order to maximize the lifetime reliability.

9.4.2.2 Design-time reliability optimization

The design-time approaches exploit the full knowledge of the application to optimize reliability offline using worst-case considerations. These approaches are used to estimate the reliability budget early in the design cycle. A simulated annealing-based heuristic is proposed in Ref. [57] to determine task mapping that maximizes lifetime reliability. An ant colony optimization approach is proposed in Ref. [58] as an alternative. The approach in Ref. [59] uses Markov decision process to determine the availability of a multiprocessor system considering wear-out as the failure mechanism. The approach in Ref. [60] determines the resource utilization, which maximizes lifetime reliability and minimizes the occurrences of transient and intermittent faults. The approach in Ref. [61] optimizes reliability and energy jointly by selective slowdown of the tasks of a given application. There are also approaches to minimize temperature, which indirectly leads to lifetime improvement. An optimal phased steady-state mixed integer linear programming approach is proposed in Ref. [62] to minimize the temperature of a multiprocessor system by exploiting the slack from the task execution. A design-time and runtime hybrid approach is proposed in Ref. [63], where the temperature of a system is determined at design-time using linear programming. Based on this, strategies are developed for dynamic thermal optimization. A design optimization with task duplication and processor hardening for intermittent and transient faults is proposed in Ref. [64]. A checkpoint-based approach is proposed in Ref. [65] for transient fault tolerance of hard-soft real-time tasks on a static multiprocessor system. A genetic algorithm is proposed in Ref. [66] to optimize for cost, time, and dependability. Tasks are replicated to improve transient fault tolerance, and task graphs with replicated tasks are scheduled on a multiprocessor system. A multiobjective evolutionary algorithm is proposed in Ref. [67] to determine the mapping of tasks to processing elements considering permanent and transient faults.

A system-level synthesis flow is proposed in Ref. [68] to mitigate transient faults. Fault management requirements (e.g., fault detection/tolerance) of different tasks of an application are incorporated in the task graph representation, and a scheduling approach is proposed to map the modified application graph on a multiprocessor system (consisting of GPPs and/or dedicated hardware modules) satisfying the performance requirement. A multiobjective evolutionary algorithm is proposed in Ref. [69] to jointly optimize reliability, area, and latency while deciding on application mapping on GPPs and custom modules. The proposed approach optimizes for permanent faults on the processors only. A hardware-software codesign approach is presented in Ref. [70]. Every task of an application is specified by different implementation alternatives such as GPP and ASIC with each implementation differing in area, cost, and reliability figures.

In a recent study [71], fault tolerance of hardware-software hybrid tasks is proposed. Fault tolerance is incorporated as rollback and recovery for software tasks (running on GPP) and TMR for hardware tasks. This approach is built on the precomputed decisions of tasks that need to be run on GPPs and those requiring hardware implementation. The approach in Ref. [72] determines the application task mapping and scheduling on a reconfigurable multiprocessor system to determine the number of checkpoints for the software tasks, simultaneously maximizing the lifetime reliability.

9.4.3 Software approaches

Hardware redundancy is expensive in terms of area and energy overheads. Moreover, for many commercial off-the-shelf-based systems, hardware modification or redesign options may be limited. For such systems, an alternative is to use compiler-based optimizations or modification to introduce software-implemented error tolerance [73]. An

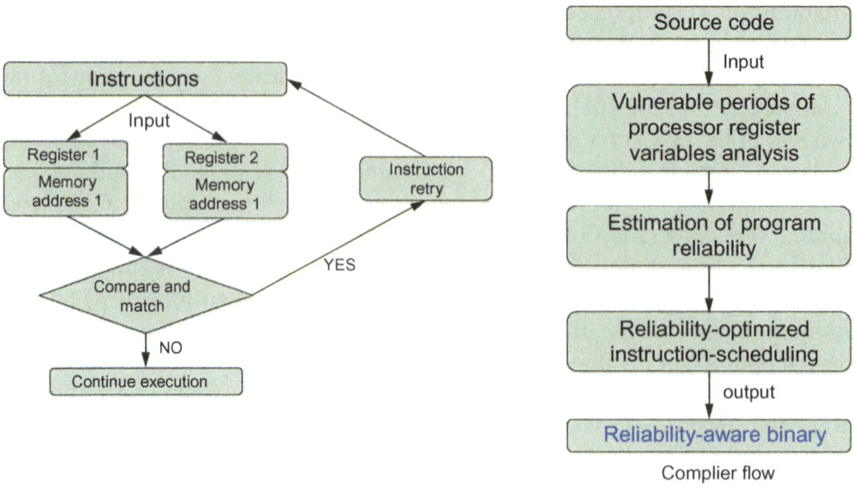

Figure 9.14 Software-based fault tolerance approaches: (a) instruction duplication and retry [74] and (b) reliability-aware instruction scheduling [75].

example of this is instruction duplication approach (called ED4I), proposed in Ref. [74], as shown in Figure 9.14(a). Such duplication is facilitated by compiler-generated usage of different registers and memory locations for each instruction. The outputs of these instructions are then compared to detect errors. Reis et al. further enhanced EDDI in their software-implemented fault tolerance (SWIFT) approach by taking out memory replications with an assumption of ECC-enabled memory unit [76]. As a result, SWIFT achieved lower overall system overheads with reasonable error detection capabilities.

Since duplication for each instruction can also be expensive in terms of performance and energy overheads [76], estimating the comparative vulnerability has been suggested by Rehman et al. [75]. As shown in Figure 9.14(b), using compiler-based approaches, each incoming instruction can be weighed in terms of its vulnerable periods. Depending on its estimated vulnerability, the instruction can then be scheduled with duplication, if necessary. This can effectively reduce the overall application failure probability by 35% at low performance and energy cost.

Recently, there is growing interest in evaluating the impact of transient faults at application level rather than architectural level to reduce system design cost through embedding lightweight fault tolerance mechanisms [77]. One such approach has been reported in Ref. [78], where it has been shown that faults at architectural level do not always lead to faults at application level. For example, using the peak signal-to-noise ratio (PSNR) as a fidelity metric in a MPEG-2 video decoder, it was shown that up to 73% of architectural faults injected in the processor's registers and memories do not affect the application or lead to gracefully degraded PSNR, while only 27% of faults result in execution failures. Because of such relaxed requirements of application-level correctness, acceptable application-level reliability was exploited for minimizing power consumption.

9.5 Future research directions

Frontier technology scaling is likely to continue. With such scaling, future generations of embedded systems are expected to feature many interconnected processing cores on a single chip. The emergence of such systems will potentially enable performance, scalability, and flexibility at unprecedented levels. However, increased device-level power densities in these system will render prohibitive operating temperatures, which will require some of the cores to be shut down as a compensation measure, originally envisaged as the "dark silicon age" [79]. Moreover, unreliability issues, such as increased process variability and device wear-outs, will drastically increase the failures in these systems. Hence, to advance the development of reliable many-core systems at low energy cost, major research efforts are required. In the following section, a few such research directions are discussed.

9.5.1 Cross-layer system adaptation

Many-core systems are expected to serve requirements of multiple applications, each with different performance and reliability targets, as shown in Figure 9.15. To achieve the specified requirements of these applications in the presence of logic and timing faults, cross-layer interactions between the layers (applications and system software and hardware) are critical for system-wide adaptation. In such a system, each application indicates its requirements to the respective runtime manager. The runtime manager will then carry out necessary hardware control tasks, such as hardware resource allocation, mapping/scheduling of the application tasks, and processor voltage/frequency scaling. The runtime manager will also interact with the hardware processor core to evaluate the impact of these control actions through performance counters and reliability monitors (i.e., detected faults and core temperature). These evaluations will guide the runtime to adapt and improve the system reliability at low cost through making better

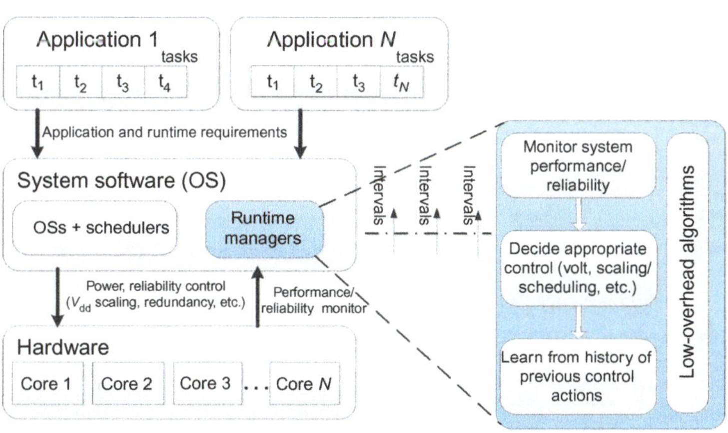

Figure 9.15 Cross-layer approach for runtime system adaptation.

decisions. To ensure runtime systems can simplify such decisions during runtime, previously recorded control actions can also be learned occasionally (Figure 9.15).

9.5.2 *Quality-of-experience-aware design*

The architectural reliability requirements of future applications are likely to be more challenging due to degraded device reliability with technology scaling. Hence, for many user-oriented applications, designers will need to address the user's quality of experience (QoE) as a reliability target. Such QoE-based reliability optimization has obvious advantages in terms of low cost as opposed to architectural reliability optimization approaches [78]. Figure 9.16 shows QoE examples of three different applications: video decoder, webpage rendering, and audio decoder. As can be seen, the video QoE is evaluated using its PSNR, webpage rendering application is measured in terms of delays, and finally, audio is evaluated in terms of sampling rates. Each application will be studied extensively by the application developer to suitably find the ranges of acceptable QoEs for the end users. These QoE ranges can then be exploited by the system designer toward providing acceptable reliability and QoE of an application at low power cost. For example, for a video decoder, lower PSNRs can be traded off for reduced power consumption through aggressive voltage scaling. Although such scaling can drastically increase the vulnerability of the decoder, it can still render acceptable QoE in terms of PSNR. Similarly, computing components can be shut down during runtime to improve reliability or to reduce energy consumption depending on the expected QoE in a webpage rendering or audio decoder application (Figure 9.16).

To reduce the architectural vulnerability significantly while achieving performance and energy advantages, currently, there is a lot of interest in approximate computing systems [80]. These systems rely on the fact that reduced precision or incorrect outputs due to hardware faults in many components can still provide good enough QoE at system level [81].

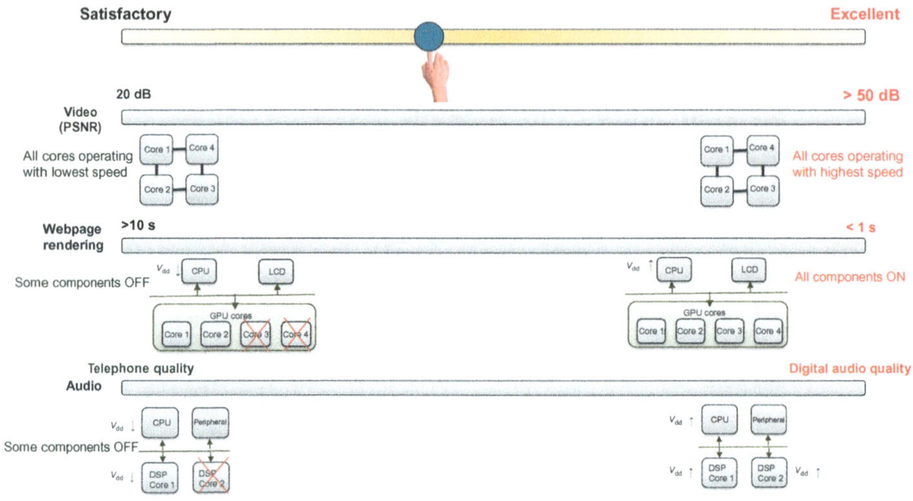

Figure 9.16 Quality-of-experience-aware reliable and power-efficient design.

9.5.3 *Programming models*

To address the performance and scalability challenges of future many-core computing systems, a paradigm shift from traditional *ad hoc* programming model to a more systematic parallel programming model is highly necessary. Such programming models will enable a holistic runtime control for a given application with an aim to achieve reliability optimization at low cost. A great deal of research is already underway for programming models, such as OpenMP[1] and OpenCL.[2]

9.5.4 *Reliable design automation*

To advance reliability evaluation of a design, recently, standardization initiatives, such as RIIF and IEC 61508 [82], have been taken. These initiatives have formulated systematic modeling approaches toward achieving target reliability, particularly for safety- and mission-critical applications. However, systematic design automation driven by such initiatives remains an area where significant research is further needed. Figure 9.17 shows an envisioned approach showing possible design automation steps. The design automation will start with a given reliability specification. The design vulnerability will then be assessed and evaluated against such specification, identifying

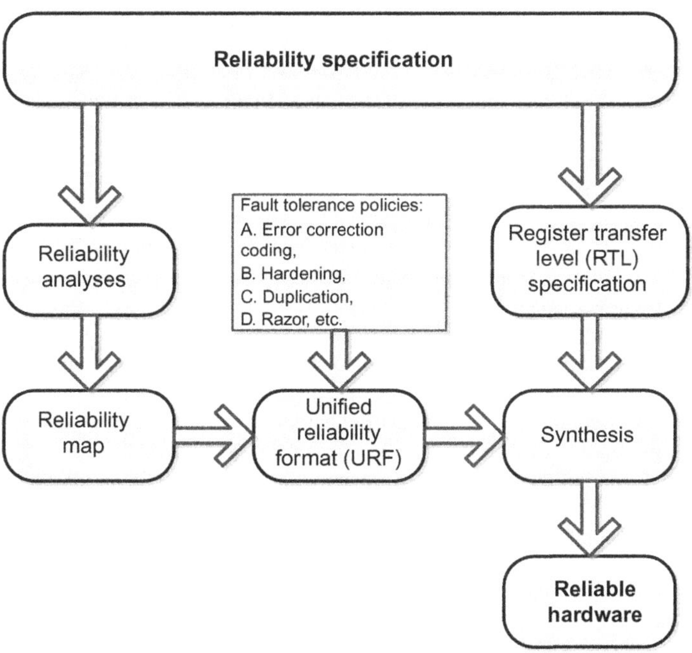

Figure 9.17 Reliable hardware design automation.

[1] http://openmp.org/wp/.

[2] https://www.khronos.org/opencl/.

the vulnerable components of the system for the given application. This will generate a reliability map with component vulnerabilities in the system, to be used together with a unified reliability format (URF) to guide the usage of predefined fault tolerance policies. For example, for highly vulnerable components, duplications may be needed at architectural level, while nonvulnerable components may need no redundancy at all. The URF and the system RTL will then be used as inputs to the synthesis process, which will generate application-tailored reliable hardware.

9.6 Conclusions

Reliability in the presence of hardware faults is an emerging design challenge for modern embedded systems. This is expected to be further aggravated for future many-core systems as device-level reliability is expected degrade further with technology scaling. This chapter has discussed traditional design practices and presented an overview of academic and industrial research efforts toward low-cost reliable design. Moreover, some important directions toward future focused research have also been presented.

References

[1] R. Kim, S. Holmes, S. Halle, V. Dai, J. Meiring, A. Dave, M.E. Colburn, H.J. Levinson, 22 nm technology node active layer patterning for planar transistor devices, J. Micro/Nanolithogr. MEMS MOEMS 9 (1) (2010) 013001–013004.

[2] S.R. Shenoy, A. Daniel, Intel architecture and silicon cadence: the catalyst for industry innovation, Technical Report, Intel Corp., April 2009.

[3] P. Hazucha, C. Svensson, Impact of CMOS technology scaling on the atmospheric neutron soft error rate, IEEE Trans. Nucl. Sci. 47 (6) (2000) 2586–2594.

[4] J. Gray, Why do computers stop and what can be done about it? in: Proc. of Symposium on Reliability in Distributed Software and Database Systems, January 2006, pp. 3–12.

[5] P. Kudva, J. Rivers, Balancing new reliability challenges and system performance at the architecture level, IEEE Des. Test Comput. (2012) 99.

[6] S. Mitra, P. Sanda, N. Seifert, Soft errors: technology trends, system effects, and protection techniques, in: IOLTS '07: Proceedings of the 13th IEEE International OnLine Testing Symposium, IEEE Computer Society, Washington, DC, 2007, p. 4.

[7] T. Calin, M. Nicolaidis, R. Velazco, Upset hardened memory design for submicron CMOS technology, IEEE Trans. Nucl. Sci. 43 (6) (1996) 2874–2878.

[8] X. She, P.K. Samudrala, Selective triple modular redundancy for single event upset (SEU) mitigation, in: Proc. of NASA/ESA Conf. on Adaptive Hardware and Systems (AHS), 2009, pp. 344–350.

[9] L. Longden, R. Thibodeau, C. Hillman, P. Layton, M. Dowd, Designing a single board computer for space using the most advanced processor and mitigation technologies, ESA Publ. 507 (1) (2002) 313–316.

[10] R.M. Goodman, M. Sayano, The reliability of semiconductor RAM memories with on-chip error-correction coding, IEEE Trans. Inf. Theory 37 (3) (1991) 884–896.

[11] S. Mitra, Globally optimized robust systems to overcome scaled CMOS reliability challenges, in: Proceedings of the Conference on Design, Automation and Test in Europe (DATE), 2008, pp. 941–946.

[12] R.A. Shafik, B.M. Al-Hashimi, K. Chakrabarty, Soft error-aware design optimization of low power and time-constrained embedded systems, in: Proc. of Design and Test in Europe (DATE), 2010, pp. 1462–1467.
[13] J.F. Ziegler, Terrestrial cosmic rays, IBM J. Res. Dev. 40 (1) (1996) 19–39.
[14] P. Reviriego, J.A. Maestro, C. Cervantes, Reliability analysis of memories suffering multiple bit upsets, IEEE Trans. Device Mater. Reliab. 7 (4) (2007) 592–601.
[15] W.G. Fenton, T.M. McGinnity, L.P. Maguire, Fault diagnosis of electronic systems using intelligent techniques: a review, IEEE Trans. Syst. Man Cybern. C Appl. Rev. 31 (2) (2001) 269–281.
[16] Y. Zorian, E.J. Marinissen, Testing embedded-core based system chips, in: Proc. of Intl. Test Conference (ITC), 1998, pp. 130–143.
[17] J. Srinivasan, S.V. Adve, P. Bose, J.A. Rivers, The case for lifetime reliability-aware microprocessors, ACM SIGARCH Comput. Archit. News 32 (2) (2004) 276.
[18] J. Mathew, R.A. Shafik, D.K. Pradhan, Energy-Efficient Fault-Tolerant Systems, Springer, USA, 2014.
[19] S. Yang, S. Khursheed, B.M. Al-Hashimi, D. Flynn, G.V. Merrett, Improved state integrity of flip-flops for voltage scaled retention under PVT variation, IEEE Trans. Circuits Syst. Regul. Pap. 60 (11) (2013) 2953–2961.
[20] National Aeronautics and Space Administration (NASA), Scaled CMOS technology reliability users guide, Technical Report, March 2008.
[21] D.G. Platteter, Protection of LSI microprocessors using triple modular redundancy, in: Proc. of IEEE Symposium on Fault Tolerant Computing (FTCS), October 1980, pp. 1–5.
[22] A. Maheshwari, W. Burleson, R. Tessier, Trading off transient fault tolerance and power consumption in deep submicron (DSM) VLSI circuits, IEEE Trans. Very Large Scale Integr. VLSI Syst. 12 (3) (2004) 299–311.
[23] R.D. Schrimpf, D.M. Fleetwood, Radiation effects and soft errors in integrated circuits and electronic device, in: Selected Topics in Electronics and Systems, vol. 34, World Scientific USA, 2004.
[24] K. Leppälä, K. Ssr, R. Verkasalo, Protection of instrument control computers against soft and hard errors and cosmic ray effects, in: Proc. of Intl Seminar on Space Scientific Engineering, Frunze, Kirgiz SSR, USSR, 1989, pp. 1–12.
[25] S. Krishnamohan, N.R. Mahapatra, Analysis and design of Soft-error hardened latches, in: Proc. of the 15th ACM Great Lakes Symposium on VLSI (GLSVLSI), Chicago, IL, 2005, pp. 328–331.
[26] J. Sosnowski, Transient fault tolerance in digital systems, IEEE Micro. 14 (1) (1994) 24–35.
[27] K. Furutani, K. Arimoto, H. Miyamoto, T. Kobayashi, K. Yasuda, K. Mashiko, A built-in Hamming code ECC circuit for DRAMs, IEEE J. Solid-State Circuits 24 (1) (1989) 50–56.
[28] J. Kim, N. Hardavellas, K. Mai, B. Falsafi, J. Hoe, Multi-bit error tolerant caches using two-dimensional error coding, in: Proc. of the 40th Annual IEEE/ACM International Symposium on Microarchitecture (MICRO-40), IEEE Computer Society, Washington, DC, 2007, pp. 197–209.
[29] J. Chang, M. Huang, J. Shoemaker, J. Benoit, S.L. Chen, W. Chen, S. Chiu, R. Ganesan, G. Leong, V. Lukka, S. Rusu, D. Srivastava, The 65-nm 16-MB Shared On-Die L3 Cache for the Dual-Core Intel Xeon Processor 7100 Series, IEEE J. Solid-State Circuits 42 (4) (2007) 846–852.
[30] D. Bull, S. Das, K. Shivashankar, G.S. Dasika, K. Flautner, D. Blaauw, A power-efficient 32 bit ARM processor using timing-error detection and correction for transient-error tolerance and adaptation to PVT variation, IEEE J. Solid-State Circuits 46 (1) (2011) 18–31.

[31] S.-C. Lai, S.-L. Lu, K. Lai, J.-K. Peir, DITTO processor, in: Proc. of International Conference on Dependable Systems and Networks, DSN, 2002, pp. 525–534.
[32] ARM Cortex-R Processors, http://www.arm.com/products/processors/cortex-r/index.php/ (last visited on 31.01.14).
[33] Intel Xeon Processor, http://download.intel.com/design/intarch/specupdt/24967856.pdf (last visited on 31.01.14).
[34] Y. Zhang, K. Chakrabarty, Energy-aware adaptive checkpointing in embedded realtime systems, in: Proceedings of the Conference on Design, Automation and Test in Europe (DATE), IEEE Computer Society, Washington, DC, 2003, p. 10918.
[35] K. Hwang, H. Jin, R. Ho, W. Ro, Reliable cluster computing with a new checkpointing RAID-x architecture, in: Proc. of 9th Intl Heterogeneous Computing Workshop (HCW), April 2000, pp. 171–184.
[36] ARM Cortex-R4 Processor, http://www.arm.com/products/processors/cortex-r/cortex-r4.php (last visited on 31.01.14).
[37] H. Ando, Y. Yoshida, A. Inoue, I. Sugiyama, T. Asakawa, K. Morita, T. Muta, T. Motokurumada, S. Okada, H. Yamashita, Y. Satsukawa, A. Konmoto, R. Yamashita, G. Sugiyama, A 1.3 GHz fifth generation SPARC64 microprocessor, in: Technical Digest of Intl. Solid-State Circuits Conference (ISSCC), February 2003, pp. 246–491.
[38] P.T. De Sousa, F.P. Mathur, Sift-out modular redundancy, IEEE Trans. Comput. 27 (7) (1978) 624–627.
[39] C.L. Chen, M.Y. Hsiao, Error-correcting codes for semiconductor memory applications: a state-of-the-art review, IBM J. Res. Dev. 28 (2) (1984) 124–134.
[40] A. Ejlali, B.M. Al-Hashimi, M.T. Schmitz, P. Rosinger, S.G. Miremadi, Combined time and information redundancy for SEU-tolerance in energy efficient real-time systems, IEEE Trans. Very Large Scale Integr. VLSI Syst. 14 (4) (2006) 323–335.
[41] S. Mitra, N. Seifert, M. Zhang, Q. Shi, K.S. Kim, Robust system design with built-in soft-error resilience, IEEE Computers, 38 (2) (2005) 43–52.
[42] S. Yang, S. Khursheed, B.M. Al-Hashimi, D. Flynn, S. Idgunji, Reliable state retention-based embedded processors through monitoring and recovery, IEEE Trans. Comput. Aided Des. Integr. Circuits Syst. 30 (12) (2011) 1773–1785.
[43] S. Campagna, M. Violante, An hybrid architecture to detect transient faults in microprocessors: an experimental validation, in: Proc. Design, Automation & Test in Europe (DATE), 2012, pp. 1433–1438.
[44] G. Memik, M.T. Kandemir, O. Ozturk, Increasing register file immunity to transient errors, in: Proc. of Design, Automation and Test in Europe (DATE), March 7-11, 2005, pp. 586–591.
[45] D. Ernst, N.S. Kim, S. Das, S. Pant, R. Rao, T. Pham, C. Ziesler, D. Blaauw, T. Austin, K. Flautner, T. Mudge, RAZOR: a low-power pipeline based on circuit-level timing speculation, in: Proc. of 36th Annual IEEE/ACM International Symposium on Microarchitecture (MICRO-36), December 2003, pp. 7–18.
[46] P. Pop, K. Poulsen, V. Izosimov, P. Eles, Scheduling and voltage scaling for energy/reliability trade-offs in fault-tolerant time-triggered embedded systems, in: Proceedings of Intl. Conference on Hardware/Software Codesign and System Synthesis (CODES + ISSS), 2007, pp. 233–238.
[47] V. Izosimov, P. Pop, P. Eles, Z. Peng, Design optimization of time-and cost-constrained fault-tolerant distributed embedded systems, in: DATE '05: Proceedings of the Conference on Design, Automation and Test in Europe, IEEE Computer Society, Washington, DC, 2005, pp. 864–869.
[48] W. Lee, K. Patel, M. Pedram, GOP-level dynamic thermal management in mpeg-2 decoding, IEEE Trans. Very Large Scale Integr. VLSI Syst. 16 (6) (2008) 662–672.

[49] A.K. Coskun, T.S. Rosing, K.C. Gross, Temperature management in multiprocessor SoCs using online learning, in: Design Automation Conference (DAC), 2008, pp. 890–893.
[50] R. Jayaseelan, T. Mitra, Dynamic thermal management via architectural adaptation, in: Proceedings of Design Automation Conference (DAC), 2009, pp. 484–489.
[51] F. Mulas, D. Atienza, A. Acquaviva, S. Carta, L. Benini, G. De Micheli, Thermal balancing policy for multiprocessor stream computing platforms, IEEE Trans. Comput. Aided Des. Integr. Circuits Syst. 28 (12) (2009) 1870–1882.
[52] V. Hanumaiah, S. Vrudhula, Temperature-aware DVFs for hard real-time applications on multicore processors, IEEE Trans. Comput. 61 (10) (2012) 1484–1494.
[53] Y. Ge, Q. Qiu, Dynamic thermal management for multimedia applications using machine learning, in: Proc. of ACM/IEEE Design Automation Conference (DAC), 2011, pp. 95–100.
[54] T. Ebi, D. Kramer, W. Karl, J. Henkel, Economic learning for thermal-aware power budgeting in many-core architectures, in: Proceedings of International Conference on Hardware/Software Codesign and System Synthesis (CODES+ISSS), 2011, pp. 189–196.
[55] P. Mercati, A. Bartolini, F. Paterna, T.M. Rosing, L. Benini, Workload and user experience-aware dynamic reliability management in multicore processors, in: Proc. of Design Automation Conference (DAC), 2013, pp. 1–6.
[56] A. Das, R.A. Shafik, G.V. Merrett, B.M. Al-Hashimi, A. Kumar, B. Veeravalli, Reinforcement learning-based inter- and intra-application thermal optimization for lifetime improvement of multicore systems, in: Design Automation Conference (DAC), 2014.
[57] L. Huang, F. Yuan, Q. Xu, On task allocation and scheduling for lifetime extension of platform-based MPSoC designs, IEEE Trans. Parallel Distrib. Syst. 22 (12) (2011) 2088–2099.
[58] A.S. Hartman, D.E. Thomas, B.H. Meyer, A case for lifetime-aware task mapping in embedded chip multiprocessors, in: Proceedings of the Conference on Hardware/Software Codesign and System Synthesis (CODES + ISSS), ACM, Scottsdale, AZ, USA, 2010, pp. 145–154.
[59] A. Das, A. Kumar, B. Veeravalli, Communication and migration energy aware design space exploration for multicore systems with intermittent faults, in: Proceedings of the Conference on Design, Automation and Test in Europe (DATE), 2013, pp. 1631–1636.
[60] C.-L. Chou, R. Marculescu, FARM: fault-aware resource management in NoC-based multiprocessor platforms, in: Proceedings of the Conference on Design, Automation and Test in Europe (DATE), Grenoble, 2011, pp. 1–6.
[61] A. Das, A. Kumar, B. Veeravalli, Temperature aware energy-reliability tradeoffs for mapping of throughput-constrained applications on multimedia MPSoCs, in: Proceedings of the Conference on Design, Automation and Test in Europe (DATE), 2014.
[62] T. Chantem, X. Hu, R. Dick, Temperature-aware scheduling and assignment for hard real-time applications on MPSoCs, IEEE Trans. Very Large Scale Integr. VLSI Syst. 19 (10) (2011) 1884–1897.
[63] A. Coskun, T. Rosing, K. Whisnant, K. Gross, Static and dynamic temperature-aware scheduling for multiprocessor SoCs, IEEE Trans. Very Large Scale Integr. VLSI Syst. 16 (9) (2008) 1127–1140.
[64] V. Izosimov, I. Polian, P. Pop, P. Eles, Z. Peng, Analysis and optimization of fault-tolerant embedded systems with hardened processors, in: Proceedings of the Conference on Design, Automation and Test in Europe (DATE), 2009, pp. 682–687.
[65] P.K. Saraswat, P. Pop, J. Madsen, Task mapping and bandwidth reservation for mixed hard/soft fault-tolerant embedded systems, in: Proceedings of the IEEE Symposium on Real-Time and Embedded Technology and Applications (RTAS), IEEE Computer Society, Stockholm, Sweden, 2010, pp. 89–98.

[66] A. Jhumka, S. Klaus, S.A. Huss, A dependability-driven system-level design approach for embedded systems, in: Proceedings of the Conference on Design, Automation and Test in Europe (DATE), IEEE Computer Society, Munich, Germany, 2005, pp. 372–377.
[67] J. Huang, J.O. Blech, A. Raabe, C. Buckl, A. Knoll, Analysis and optimization of fault-tolerant task scheduling on multiprocessor embedded systems, in: Proceedings of the Conference on Hardware/Software Codesign and System Synthesis (CODES + ISSS), ACM, New York, 2011, pp. 247–256, Taipei, Taiwan.
[68] C. Bolchini, A. Miele, Reliability-driven system-level synthesis of embedded systems, in: IEEE International Symposium on Defect and Fault Tolerance in VLSI Systems (DFT), October 2010, pp. 35–43.
[69] M. Glaß, M. Lukasiewycz, T. Streichert, C. Haubelt, J. Teich, Reliability-aware system synthesis, in: Proceedings of the Conference on Design, Automation and Test in Europe (DATE), 2007, pp. 409–414.
[70] Y. Xie, L. Li, M. Kandemir, N. Vijaykrishnan, M.J. Irwin, Reliability-aware cosynthesis for embedded systems, J. VLSI Signal Process. Syst. 49 (1) (2007) 87–99.
[71] J. Yin, H. Song, L. Yuan, Q. Cui, A real-time fault-tolerant scheduling algorithm for software/hardware hybrid tasks, in: International Conference on Mechatronic Science, Electric Engineering and Computer (MEC), August 2011, pp. 1822–1828.
[72] A. Das, A. Kumar, B. Veeravalli, Aging-aware hardware-software task partitioning for reliable reconfigurable multiprocessor systems. in: Proceedings of the International Conference on Compilers, Architectures and Synthesis for Embedded Systems (CASES), IEEE Press, Montreal, Canada, 2013, pp. 1:1–1:10.
[73] I.A. Troxel, E. Grobelny, G. Cieslewski, J. Curreri, M. Fischer, A.D. George, Reliable management services for COTS-based space systems and applications, in: Proc. Intl. Conf. on Embedded Systems & Applications, 2006, pp. 169–175.
[74] N. Oh, S. Mitra, E.J. McCluskey, ED^4I: error detection by diverse data and duplicated instructions, IEEE Trans. Comput. 51 (2) (2002) 180–199.
[75] S. Rehman, M. Shafique, F. Kriebel, J. Henkel, RAISE: Reliability-Aware Instruction SchEduling for unreliable hardware, in: Proc. of 17th Asia and South Pacific Design Automation Conference (ASP-DAC), 2012, pp. 671–676.
[76] G. Reis, J. Chang, N. Vachharajani, R. Ranjan, D.I. August, SWIFT: software implemented fault tolerance, in: Intl. Symp. Code Generation and Optimization (CGO), March 2005, pp. 243–254.
[77] M.A. Breuer, H.H. Zhu, Error tolerance and multimedia, in: Proceedings of International Conference on Intelligent Information Hiding and Multimedia Signal Processing, December 2006, pp. 521–524.
[78] R.A. Shafik, B.M. Al-Hashimi, S. Kundu, A. Ejlali, Soft error-aware voltage scaling technique for power minimization in application-specific MPSoC, ASP J. Low Power Electron. 5 (2) (2009) 145–156.
[79] H. Esmaeilzadeh, E. Blem, R. St. Amant, K. Sankaralingam, D. Burger, Dark silicon and the end of multicore scaling, in: 2011 38th Annual International Symposium on Computer Architecture (ISCA), June 4-8, 2011, pp. 365–376.
[80] A. Sampson, W. Dietl, E. Fortuna, D. Gnanapragasam, L. Ceze, D. Grossman, EnerJ: approximate data types for safe and general low-power computation, ACM SIGPLAN Not. 46 (6) (2011) 164–174.
[81] B. Grigorian, G. Reinman, Improving coverage and reliability in approximate computing using application-specific, light-weight checks, in: Workshop on Approximate Computing Across the System Stack (WACAS), 2014.
[82] C. O'Brien, L. Bredemeyer, Final Elements and the IEC 61508 and IEC 61511 Functional Safety Standards, Exida Inc., USA, 2009.

Reliability approaches for automotive electronic systems

10

D. Medhat
Mentor Graphics, Cairo, Egypt

10.1 Introduction

"First Time, Every Time..." While this iconic advertising phrase was first used to describe the ink reliability of a ballpoint pen, it perfectly summarizes the average consumer's attitude toward automobile reliability as well. Customers don't really care how it's done, as long as everything in their car works first time, every time, even when that includes heated car seats, remote engine controls, power windows and locks, satellite radio, wireless communications, automated traction sensing controls, and the myriad of other electronics-based features present in today's vehicles. These systems encounter demanding design constructs and operating conditions that can challenge manufacturers' reliability and quality goals. Given this ever-present conflict between complexity and dependability, it's no surprise that the automotive electronics industry is constantly looking for ways to enhance the design and verification of electronic components and systems to improve their dependability, operating efficiency, and functional life span. Electronic design automation (EDA) vendors are continually developing new techniques and tool features to help designers overcome the challenges the automotive industry faces, specifically as they relate to circuit reliability verification.

10.2 Circuit reliability challenges for the automotive industry

In the automotive industry, reliability and high quality are key attributes for electronic systems and controls. However, many reliability checks, which can potentially affect a wide range of automotive integrated circuits (ICs) designs, are difficult to verify using traditional design rule checking (DRC), layout vs. schematic (LVS), and electrical rule checking (ERC) tools. Innovative reliability verification ensures the robustness of a design, at both the schematic and layout levels, by checking against various electrical and physical design rules that define IC performance standards and reduce susceptibility to premature or catastrophic electrical failures.

Automotive designers face some key operational challenges that are different from other industries. For example, automotive applications often encounter high operating voltages and high electric fields between interconnects. High-power areas in

Reliability Characterisation of Electrical and Electronic Systems. http://dx.doi.org/10.1016/B978-1-78242-221-1.00010-1

automotive frequently exceed 50 V, while logic areas can be less than 5 V. Having such high-power areas located next to these low-voltage logic areas can create electric fields that influence sensitive areas on the chip and lead to oxide breakdown.

Consequently, when designing and verifying smart power processes, designers must frequently comply with metal spacing design rules that are dependent on voltage drop, such as the following examples (expressed in pseudo-code):

- Minimum spacing between metal and poly must be "s" where voltage difference is higher than "V" volt.
- Metal1 minimum spacing can be "s1" if voltage drop across lines is up to 30 V, and it must be "s2" if voltage drop across lines is up to 80 V (where "s2" > "s1").
- It is not allowed to cross an adjacent metal level if voltage drop is higher than "V" volt.
- Shapes on a specified metal layer can't cross a specified area, based on the voltage difference.

In traditional physical and electrical verification flows, trying to implement such rules in the entire design flow, starting from layout routing implementation through DRC, is too conservative, as well as inefficient, due to a lack of voltage information on nets (in both schematic and layout). Exhaustive dynamic simulation is simply not practical, due to the turnaround time involved, and if the design is very large, it may not even be possible to simulate it in its entirety.

There are a variety of similar circuit reliability checks that the automotive industry must perform to ensure their designs will be reliable, such as voltage-aware checking, negative voltages and reverse current checking, electrostatic discharge (ESD) and latch-up checking, and electrical overstress (EOS) susceptibility analysis. Let's take a closer look at some of these challenging verification requirements in today's complex automotive systems and the limitations designers face when using traditional verification tools and techniques.

10.3 Circuit reliability checking for the automotive industry

10.3.1 Voltage-dependent checking

Metal spacing rules are normally defined in the DRC sign-off deck and are purely geometric. In voltage-dependent checking, designers must comply with spacing rules that require different spacings based on either the operating voltages or the difference in voltages between different lines running next to each other. Because voltage controls these metal spacing rules, the designer can no longer apply just one spacing rule per metal layer.

The challenge when using traditional verification tools is obtaining the voltage information on the nets needed to apply the appropriate spacing rule. To perform voltage-dependent checks, designers need to know the voltage at every internal node in the design and must manually add markers to the layout to set these voltages, a time-consuming and error-prone process. Identifying voltages with exhaustive dynamic simulation is simply not practical at the full-chip level, due to the turnaround time. If the design is a large one, it may not even be possible to simulate it in its entirety.

10.3.2 Negative voltage checking and reverse current

Another important reliability concern in the automotive area is the negative voltages that occur in automotive applications, which is also closely tied to the issue of reverse current in high-voltage ERC. To perform this type of checking, designers must apply verification on both the schematic and layout.

10.3.2.1 Schematic

Designers must specify which ports on the schematic are subjected to negative voltage. After that, they must use a static methodology to perform propagation to these negative voltages across devices, based on user-defined propagation criteria for the different device types. Propagating the negative voltages helps designers identify all the nets in that design that may become negative.

The user-defined propagation criteria may include such specifications as a voltage drop definition that will lead to the negative voltages decreasing in absolute value when they are propagated through devices. The propagation should be terminated if the internal net has positive voltage. The user must then identify all devices attached to these nets that have propagated negative voltage, consider them as the injectors, and export them to proceed with the next step on the layout side.

An effective EDA solution must support both the voltage propagation functionality and the application of the user-defined propagation criteria.

10.3.2.2 Layout

Designers must first perform a cross-reference between the schematic and the layout to capture the injectors' locations on the layout. They can then run the needed geometric operations to search for the affected sensors' locations within a certain distance from injectors and define a severity for this case for every injector and all its associated sensors. The severity assignment is based on:

- the designer's review of the distance between every injector and sensor,
- what the sensor is connected to,
- what other layout elements are between the injector and the sensor.

By taking this designer-defined severity into consideration to determine priority, the design team can decide what to fix in the layout to address the reliability concern.

All violations from both schematic side and layout side must be debugged using a results viewing environment, where the designer can highlight the results and analyze them in tandem. With the ability to use circuit reliability verification and DRC technology together, designers can create an easy, automated, and efficient flow to perform checking for both negative voltages and reverse current issues in automotive applications.

10.3.3 ESD and latch-up verification

ESD and latch-ups cause severe damage, so checking for proper ESD protection circuits is critical for reliable designs. Figure 10.1 shows some examples of ESD protection circuits.

There are many design rules that address both schematic checking and layout checking for these protection circuits [1–3]. In schematic checking, the rules are directed toward verifying the presence of the appropriate protection schemes from a topological perspective. Designers can perform checks on circuitry directly connected to pads, as well as checks on the ESD network.

For layout checking, the rules focus on verifying the point-to-point parasitic resistance between the pad and the ESD device, checking current density between pad and the ESD device (Figure 10.2), detecting pmos/nmos devices sharing the same well, detecting pmos/nmos field oxide parasitics, detecting latch-up issues, and more.

10.3.4 EOS susceptibility

EOS is one of the leading causes of IC failures, regardless of semiconductor manufacturer or design domain [4,5]. EOS events can result in a wide spectrum of outcomes, up to and including catastrophic damage, where the IC becomes permanently nonfunctional. All IC designers must deal with the EOS challenge.

Some designers call EOS checking the overvoltage conditions check. It is applied to all device types, such as metal-oxide-semiconductor (MOS) devices, diodes, bipolar devices, and capacitors. Overvoltage conditions checking can be applied to devices in a circuit to catch any occurrence of a disruptive voltage across the device terminals. For example, a reliability fault should be raised for MOS devices whenever the following condition applied on the voltage values of the drain (V_D), source (V_S), gate (V_G), and bulk (V_B) terminals is true:

$$\max(V_G) > \min(V_B, V_S, V_D) + V_{\text{breakdown}} \tag{10.1}$$

or:

$$\max(V_G) > \min(V_B, V_S, V_D) - V_{\text{breakdown}} \tag{10.2}$$

Designers can develop a strategy for circuit reliability that includes design, simulation, and verification techniques to ensure the circuit can withstand these stresses. Reliability verification can be a powerful part of this strategy—using techniques such as static voltage propagation, designers can address the circuit verification challenges while reducing simulation requirements. This helps prevent electrical circuit failure due to EOS and improves overall circuit reliability.

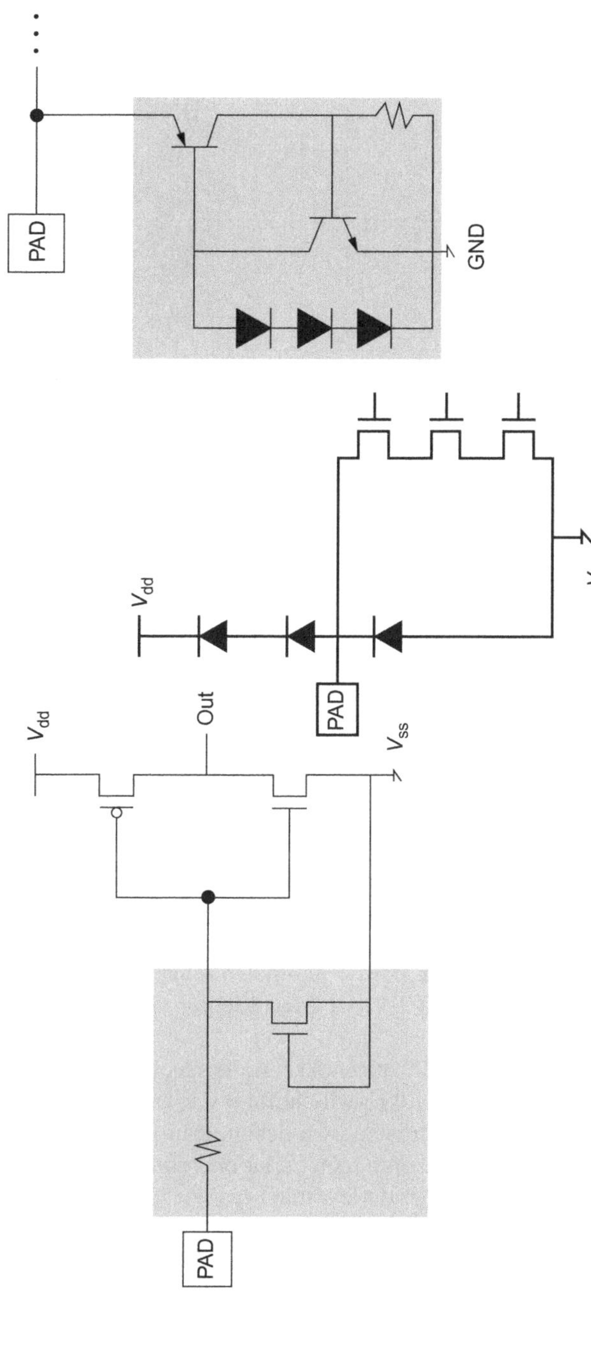

Figure 10.1 ESD protection circuits.
Source: Mentor Graphics.

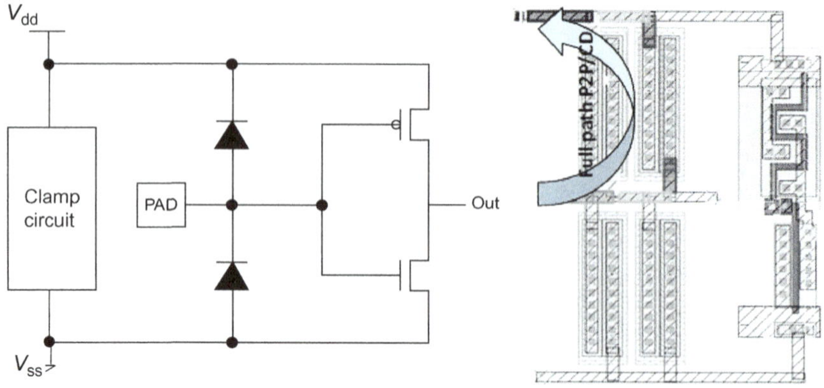

Figure 10.2 Point-to-point parasitic resistance and current density checking.
Source: Mentor Graphics.

10.4 Using advanced electronic design automation (EDA) tools

To address the circuit reliability challenges they are facing in the automotive field, designers require EDA tools that provide the features and functionalities needed to automatically perform these types of reliability checks we just discussed. EDA reliability solutions that can quickly, accurately, and automatically evaluate complex, interdependent, and customized electrical requirements can help designers achieve their goal of generating the proper net voltage information in an efficient static way. Circuit reliability verification tools that provide a voltage propagation functionality can perform voltage-dependent layout checks very efficiently while also delivering rapid turnaround, even on full-chip designs. In addition, they provide designers with unified access to all the types of design data (physical, logical, and electrical) in a single environment to enable the evaluation of topological constraints within the context of physical requirements.

In this section, we will take a look at how automated reliability verification can be performed, using the Calibre® PERC™ tool from Mentor Graphics [6]. Let's start with an overview of this tool.

The Calibre PERC reliability verification platform is specifically designed to perform a wide range of complex reliability verification tasks, using both standard rules from the foundry and custom rules created by a design team. Contrary to traditional electrical checks using a single device/pin to net relation, reliability requirements can often only be described by a topological view that combines both circuit description and physical devices. Calibre PERC uses both netlist and layout (graphic database system) information simultaneously to perform electrical checks that incorporate both layout-related parameters and circuitry-dependent checks, which enables designers to address complex verification requirements. Combining rules expressed in Standard Verification Rule Format (SVRF) and the Tool Command Language (Tcl)-based Tcl Verification Format (TVF) language across all applications provides design teams with flexibility, while ensuring compatibility with all foundries. In addition, Calibre PERC can employ topological constraints to verify that the correct structures are in place wherever circuit design rules require them. Calibre PERC can automatically identify complex circuit topology on a design netlist, either streamed from the schematic or extracted from the layout. It examines the specific constraints defined by the design team, whether they are electrical or geometric. Calibre PERC rule decks may be easily augmented to provide verification beyond standard foundry rule decks to include custom verification requirements.

Post-layout verification can incorporate complex geometric parameters into Calibre PERC checks, combining both electrical and geometric data in a single verification step. In addition, rather than running verification on an entire design, Calibre PERC's topological capabilities enable users to quickly and accurately check specific sections of designs for these types of issues.

From a debugging environment perspective, Calibre PERC output can be customized, while the Calibre RVE™ graphic user interface provides a results viewing and debug environment that can highlight results and geometries and access connectivity information.

Now, let's walk through some of the reliability challenges we discussed earlier, using Calibre PERC as our verification tool.

10.4.1 Voltage propagation

As we discussed, designers need a way to determine the voltages at all internal nodes statically. Automatic generation of voltage information on all internal nodes not only improves turnaround time and eliminates human error but also provides designers with the information needed to drive the router to apply the appropriate spacing rules during the design routing phase.

Using the voltage propagation functionality of Calibre PERC, users define input voltages (at least supply rails) and provide simple rules describing how voltages should be propagated for different device types. Using this information, the tool can quickly and automatically make voltage assignments to every node and pin in the design, based solely on netlist connectivity information (Figure 10.3). By default, all devices are considered to be ideal, with no voltage drops across them, but the user can specify a voltage drop if desired. Once the voltages are propagated, checks can be performed to determine if the circuit will function properly based upon the propagated voltage values, and user-defined rules can be applied to detect electrical design rule violations.

10.4.2 Circuit recognition

It is important for designers to be able to specify the Simulation Program with Integrated Circuit Emphasis (SPICE) circuit topology to be matched when performing circuit verification. Calibre PERC's circuit recognition functionality simplifies the task of switching between different protection schemes and enables designers to easily define complex structures. Users only need to identify their schemes, write a SPICE

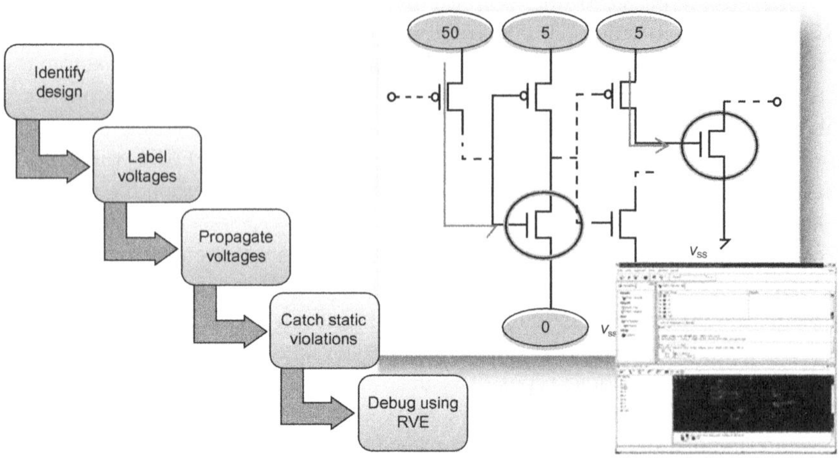

Figure 10.3 Voltage propagation functionality in Calibre PERC.
Source: Mentor Graphics.

pattern template (used as an input for Calibre PERC), and then run the tool to automatically identify violations (Figure 10.4).

10.4.3 Current density and point-to-point checking

Topology-aware current density measurements and point-to-point checking are important capabilities that, until now, have been difficult to implement. Logic-driven layout current density (LDL-CD) checking (Figure 10.5) in Calibre PERC calculates the current flow through layout interconnect layers and outputs polygon-based results that meet the specified values. (*Note*: LDL-CD checking is specifically designed for ESD current density checks—it is not intended to be used as a general current density checking tool.)

Calibre PERC LDL-CD calculates current density between target pins (or ports) in a physical layout. The pins (or ports) are obtained through Calibre PERC runtime procedures. For each pair, current density is measured from source to sink through layers that have calibrated resistance extraction statements specified in the rule file. Users specify the current value at the source and the voltage value at the sink. Resistance values between pins (or ports and pins) are calculated internally on resistance layers, and current density is then calculated for polygons on the resistance layers. Figure 10.6 illustrates an example of current density checking.

Calibre PERC logic-driven layout point-to-point resistance (LDL-P2P) checking calculates the interconnect resistance of layers specified in resistance extraction statements and outputs polygon-based results for analysis (Figure 10.7). (*Note*: LDL-P2P checking is designed specifically for ESD parasitic resistance checks—it is not intended to be used as a general resistance extraction tool.)

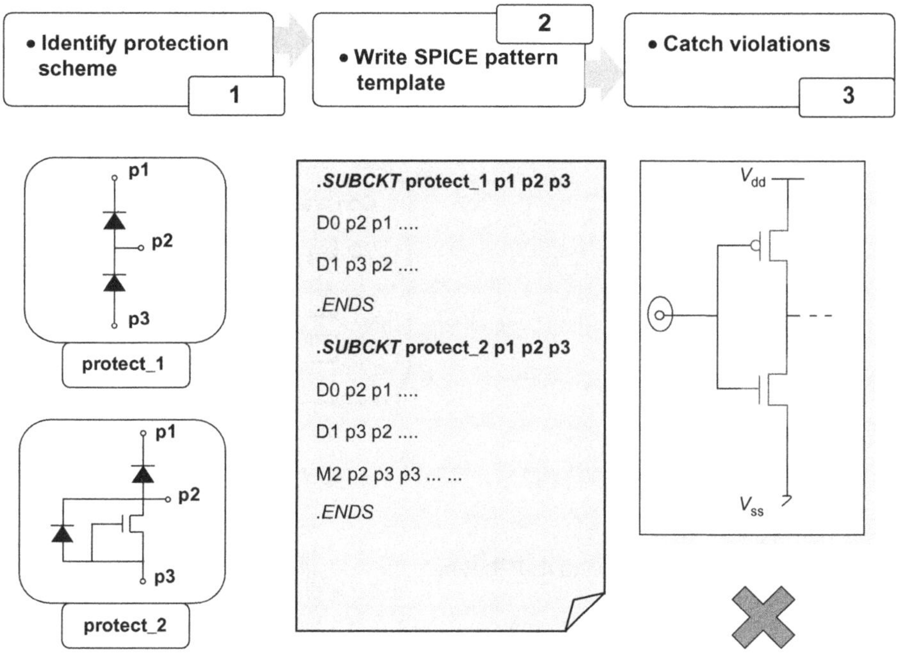

Figure 10.4 Circuit recognition functionality in Calibre PERC.
Source: Mentor Graphics.

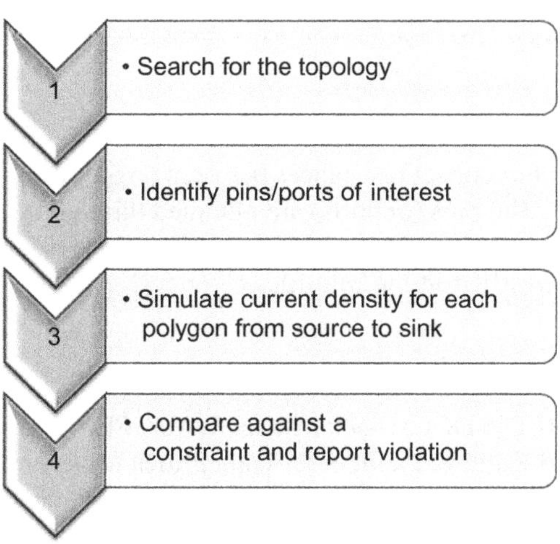

Figure 10.5 Calibre PERC LDL-CD flow.
Source: Mentor Graphics.

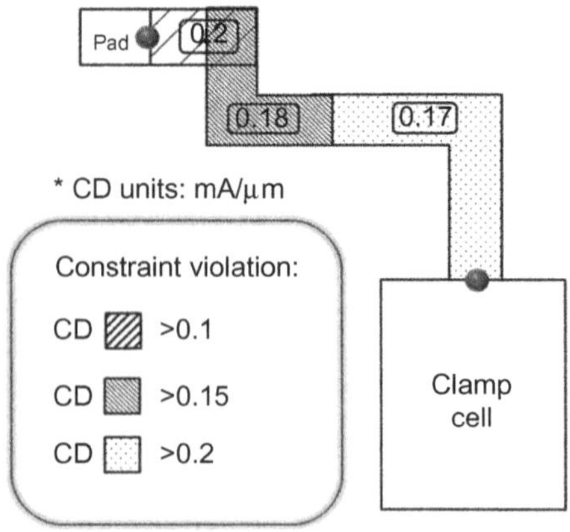

Figure 10.6 Calibre PERC LDL-CD checking calculates current density between target pins (or ports).
Source: Mentor Graphics.

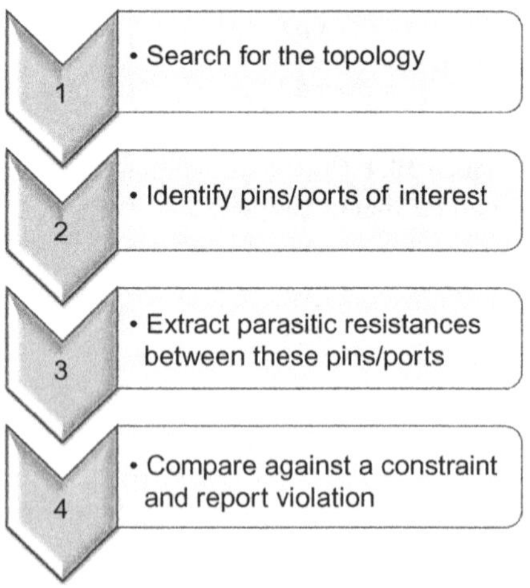

Figure 10.7 Calibre PERC LDL-P2P flow.
Source: Mentor Graphics.

Calibre PERC LDL-P2P calculates interconnect resistances between target pins (or ports) in a physical layout (Figure 10.8). The pins (or ports) are obtained through Calibre PERC runtime procedures. For each pair, resistance is measured from layers that have resistance extraction statements specified in the rule file.

10.4.4 Topology-aware geometric checking

Calibre PERC logic-driven layout DRC (LDL-DRC) checking uses Calibre PERC, Calibre nmDRC™, Calibre nmLVS™, and Calibre YieldServer capabilities to perform specialized DRC and DFM checks. LDL-DRC detects problems that cannot be identified using traditional DRC methods alone. Figure 10.9 demonstrates the LDL-DRC

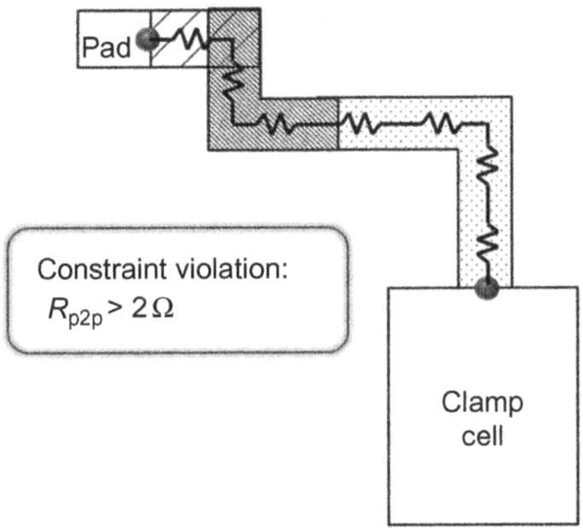

Figure 10.8 Calibre PERC LDL-P2P checking calculates interconnect resistances between target pins (or ports).
Source: Mentor Graphics.

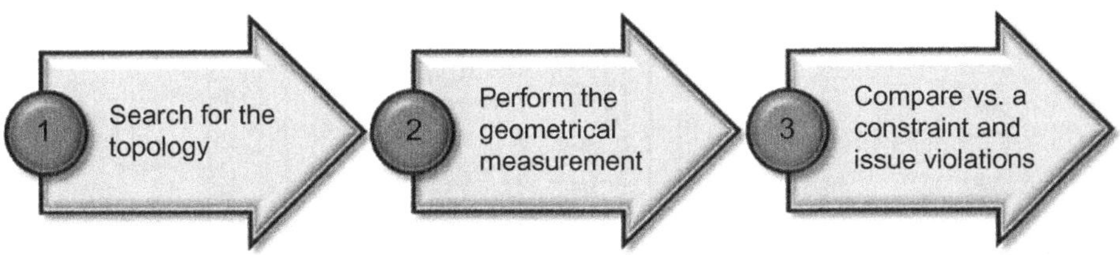

Figure 10.9 Calibre PERC LDL-DRC flow.
Source: Mentor Graphics.

flow. (*Note*: LDL-DRC is not intended to be a general DRC solution—it is designed for use with specifically targeted DRC checks that relate to circuit topology.).

A typical rule that requires LDL-DRC can be stated in pseudo-code:

- If nets cross supply domains, and if the distance between the wells is less than 10 μm, then the resistances of gate bulk pin connections to power must be within 1 Ω of each other for gates in each well, respectively.

This type of check cannot be performed using traditional DRC methods, but LDL-DRC can perform this type of check automatically (Figure 10.10).

10.4.5 Voltage-dependent DRC

As mentioned previously, in voltage-dependent DRC (VD-DRC), the spacing requirements between nets are controlled by the operating voltages. Figure 10.11 depicts an efficient automated flow for applying VD-DRC.

The flow first identifies and defines the supply voltages for the design and then uses the voltage propagation functionality to propagate supply voltages to internal design nodes (based on user-defined rules for the propagation criteria for every device type). The user can then incorporate the DRC rules in the Calibre PERC logic-driven layout (LDL) functionality [7,8] to perform the DRC checks directly or automatically generate text markers to be used later with the DRC sign-off deck [9,10].

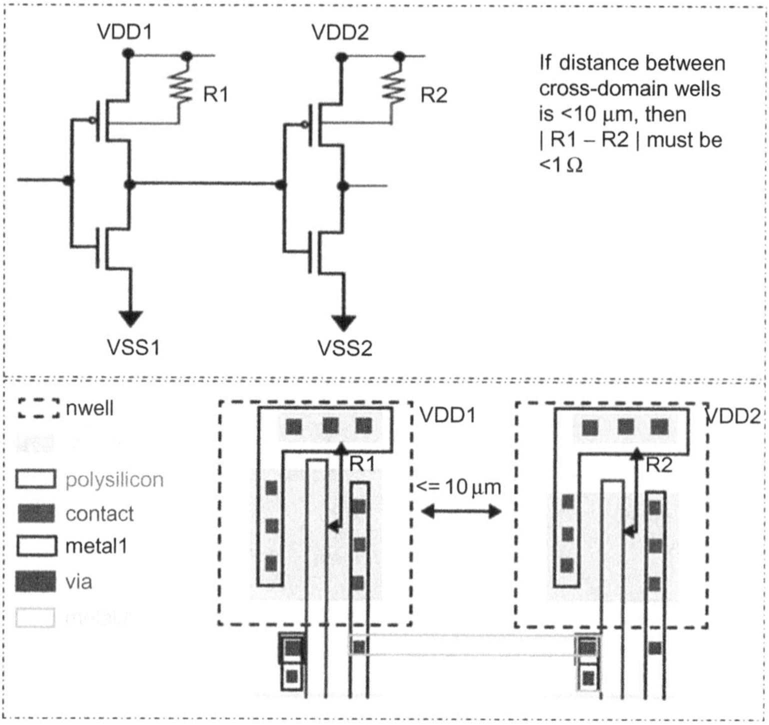

Figure 10.10 Calibre PERC LDL-DRC check.
Source: Mentor Graphics.

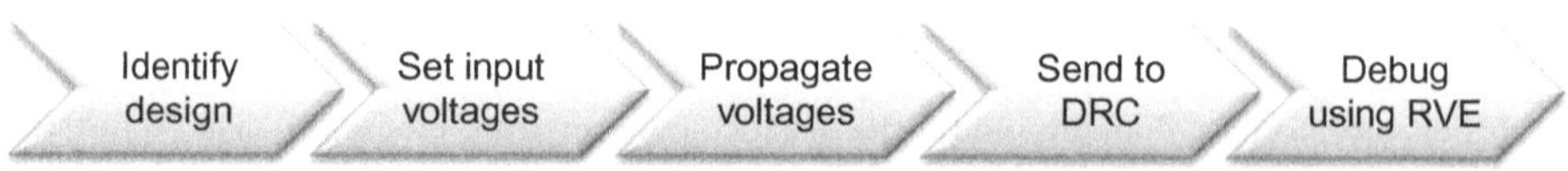

Figure 10.11 An automated flow for applying VD-DRC checking to a design.
Source: Mentor Graphics.

Errors should be debugged with a results viewing environment (such as Calibre RVE). This type of debugging environment enables the user to debug results from the geometric perspective, as well as from the topological perspective (for voltages propagated to internal nets). This is possible because the output database from this flow contains both the physical markers that identify the portions of the layout that violate the voltage-dependent spacing checks and the relevant netlist information that is required for debugging similar voltages and net names. This combination of physical and electrical information is very useful for both the designer and the layout engineer when qualifying results.

The design used as an input to this flow can be either a schematic netlist or an extracted layout netlist. By using Calibre PERC and Calibre nmDRC together to articulate this flow, designers can identify the voltages on the internal nets/devices and apply the appropriate spacing and guarding checks on the layout. The automated flow eliminates the need to manually create physical layout markers to perform voltage-aware DRC checks on the layout, which reduces both the design team workload and the chance of missing real violations or producing false violations.

In the VD-DRC flow, all voltages are computed automatically, using the voltage propagation functionality based on static rules definitions that are applied on the netlist to identify target nets/devices needed for voltage-aware DRC checking. Because the netlist information is preserved along the entire flow, the results are context-specific, making them easy to debug. This automated integration between netlist/connectivity-based voltage analysis and geometric analysis is what provides the comprehensive methodology needed to address automotive design reliability challenges. Figure 10.12 illustrates the high-level data flow for the VD-DRC process.

From a conceptual perspective, this reliability checking flow can

- implement the voltage-aware checking needed to resolve issues related to the presence of high-power areas next to logic areas, by automatically implementing the DRC variable spacing rules needed to avoid electrical fields that can influence sensitive areas on the chip;
- implement voltage-aware routing to perform the appropriate routing, taking into consideration the voltage information on all nets/devices;
- perform voltage-aware checking in the sign-off phase using the DRC sign-off deck;
- perform negative voltage checking and identify reverse current issues, enabling the user to define how to propagate the negative voltages and how to check for reverse current issues based on separation between injectors and sensors.

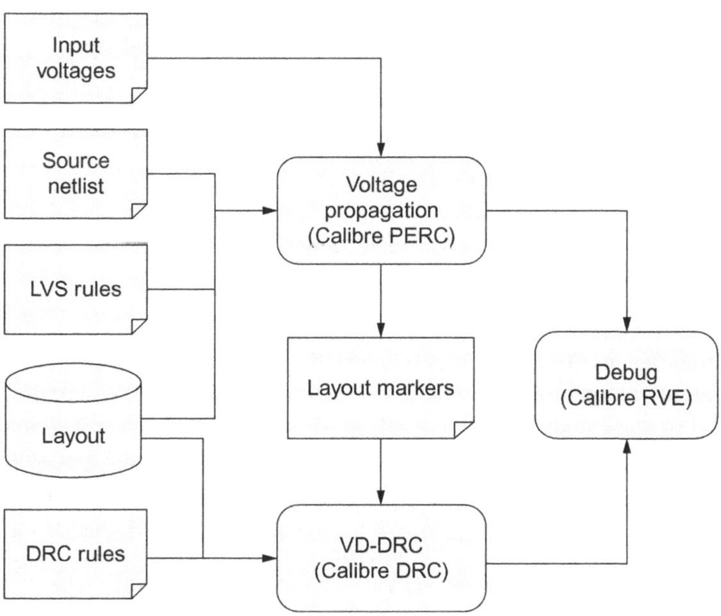

Figure 10.12 High-level VD-DRC data flow.
Source: Mentor Graphics.

Because this concept is applicable to a wide variety of reliability challenges, the rules provided for the Calibre PERC and Calibre nmDRC blocks in this automated flow will also vary. This variation is to be expected, because different reliability challenges have different input voltages, different voltage propagation criteria, and different layout DRC parameters. Consequently, the content of the rule decks will change based on the reliability issues the designer and layout engineer want to address. The bottom line, however, is that they now have an automated way both to perform static checks based on schematic netlist awareness and to link those checks to geometric rule checking.

10.5 Case studies and examples

This section presents some case studies using the Calibre PERC functionalities to address various reliability challenges for automotive industry.

10.5.1 Case study 1

Case study 1 tackles a specific voltage-dependent design rule example [9]:

> *Metal2 polygon spacing between two nets should be >=0.5 μm if the voltage difference between these two nets is greater than 40 V.*

Spotting voltage-dependent DRC violations requires several steps:

1. Define the supply voltages for the design.
2. Propagate the supply voltages through devices to the internal nodes (based on user-defined voltage propagation criteria), which results in attributes for the minimum and maximum voltage that every net can see. These attributes are reported as ranges for the internal nets.
3. Annotate these ranges to the polygons of these internal nets.
4. Apply the appropriate DRC spacing rules on the layout based on the voltage difference between the nets.

Figure 10.13 demonstrates how the results are reported from a topological perspective in the debugging environment. As shown in the results, nets "OUT2" and "OUTA" are from different voltage domains, where the difference between the maximum voltages propagated to these nodes is (50 V – 5 V = 45 V). This difference is greater than the 40 V that the rule checks for. Therefore, the spacing for metal2 polygons between these two nets must be verified against the 0.5 μm constraint.

In addition to debugging the voltages from a topological perspective, the user can also debug the results from the geometric perspective, using the results viewer. Figure 10.14 illustrates that there are two violations for our rule where the spacing was 0.25 and 0.3 μm.

Figure 10.15 illustrates the end result, a portion of the design where violations of this rule are highlighted (dotted circle).

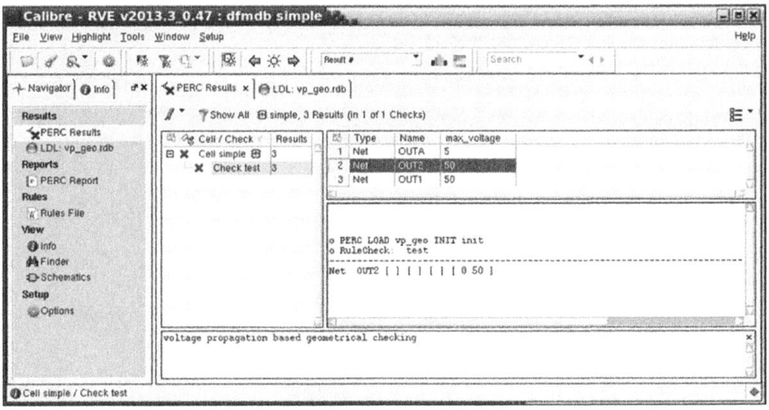

Figure 10.13 Portion of the topological results reporting.
Source: Mentor Graphics.

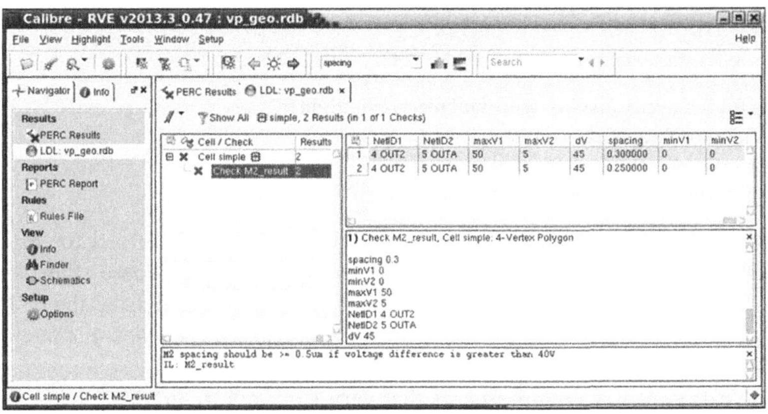

Figure 10.14 Portion of the geometric results reporting.
Source: Mentor Graphics.

10.5.2 Case study 2

Case study 2 describes a method to implement all the necessary checks to validate the topology of an ESD global network at the schematic level [1]. A complete verification flow, able to verify the presence of a correct ESD path for each pin (including checks on components, connections, and sizing), was developed using Calibre PERC. This implementation checks fulfillment against the human body model (HBM) ESD standard stress test requirements [11]. The method was developed specifically for products designed using STMicroelectronics' smart power BIPOLAR-CMOS-DMOS (BCD) process design kit (PDK), which incorporate a wide range of covered applications and a large number of different circuit configurations. The primary goal is to check

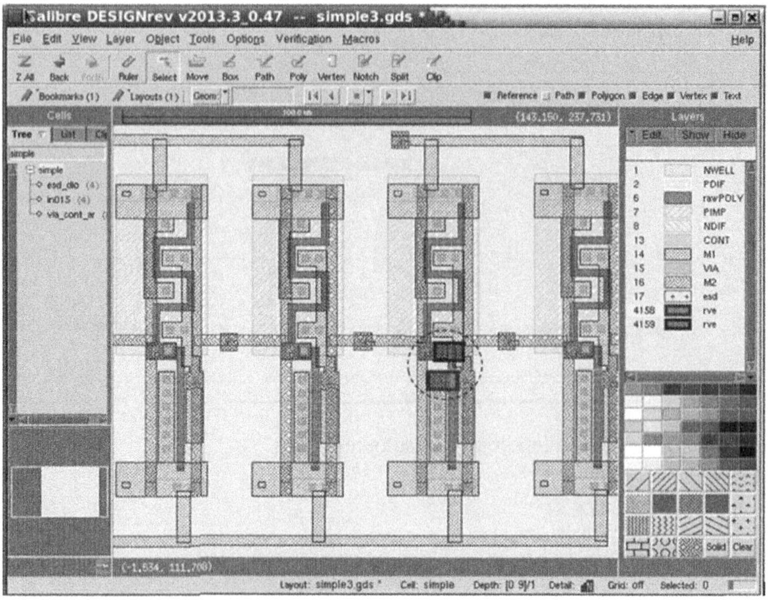

Figure 10.15 Layout display with violations highlighted.
Source: Mentor Graphics.

the robustness of the entire ESD product network without the need for SPICE-like simulation (efficiency verification is not contemplated for this case study).

This verification plan relies on an initialization method that defines the ESD path to be verified and its characteristics. The verification flow targets designs having many nonrepetitive circuital configurations, with pins characterized by several voltage ratings, and different ESD requirements that vary from pin to pin. Moreover, the check must be usable at every schematic hierarchical level. With these specific premises set, the initialization method must contain a base information set for each ESD path to be checked. The base information required to initialize the verification flow on an ESD path includes

- the path to be verified,
- the ESD configuration that must be checked,
- the operative voltage and HBM requirement of each pin.

The ideal method to initialize checks consists of dedicated "controller" components delivered together to the PDK. To identify which ESD path to verify, these controller devices must be instantiated in the schematic view of the ESD architecture to be checked, and their terminals must be connected to the specific nets to be considered. A different controller is available for each specific ESD configuration, so the choice of controller identifies the ESD configuration to check. Lastly, each controller is provided with values for the properties used to create the required set of data: maximum operating voltage, minimum operating voltage, and the ESD HBM requirement. All of this information is generated in the schematic netlist used by Calibre PERC to perform the verifications.

The global ESD network can be divided into three main hierarchical levels, as defined by the ESD EDA International Working Group [12]:

- "Cell level" (each cell is defined as an ESD protection base configuration).
- "Domain level" (composed of one or more cells belonging to the same voltage domain).
- "Top level" (formed by the composition of one or more domains—this is the global IC ESD network).

Using Calibre PERC functionalities, the design team was able to implement the required checks and validate the topology of an ESD global network at the schematic level for various products [1].

10.5.3 Case study 3

Case study 3 illustrates an efficient method for implementing an automated flow for voltage-dependent layout checks, developed by the automotive group at Infineon Technologies [10].

Electrical fields can generate undesired parasitics or cause oxide deterioration. These problems become critical in high-integration power technologies, due to the small distances between wires and the large operating voltage differences between blocks. Voltage-dependent layout checks ensure proper spacing is maintained in electrically sensitive areas.

Previously, identification of sensitive areas was performed manually. For the schematic, designers manually marked high-voltage nets with different colors or comments indicating voltage range, voltage values and voltage domains. However, these manual markers could not express net correlation (or be used to automatically transfer information to layout). For the layout, layout engineers labeled each net based on the schematic comments. For correlated nets, a special layer was used to hide false warnings.

The problems with manual identification are numerous:

- Time-consuming and tedious effort required by the layout engineer to manually label nets.
- Information must be manually copied from schematic to layout.
- Manual process is prone to errors (labels can be misplaced, leading to false warnings caused by missing correlation information and requiring multiple runs to obtain clean results).

The designers needed a process that supported automatic transfer of voltage information from schematic to layout, propagation of voltage values from top to inner hierarchy nets, and automatic detection of correlated nets. Using Calibre PERC functionalities, the desired automated flow was implemented. The outcome of this flow included

- significant reduction in the amount of input information required (100 nets labeled in schematic vs. 2000 labels in layout),
- elimination of additional work performed by the layout engineer (no manual labels and no debugging false errors due to missing net correlation information),
- minimization of communication required between designer and layout engineer,
- flexible setup (easy to add new voltage classes and distances).

Additional details about this implementation and the results are available for review [10].

10.6 Conclusion

Circuit reliability verification for automotive applications is a challenging task, which only increases as the technology advances. Advanced circuit reliability verification tools must include specific technologies to make efficient, automated circuit reliability verification practical, helping designers achieve the reliable, accurate, and comprehensive verification necessary to ensure a robust and reliable design.

Robust reliability verification is available today as a comprehensive solution. Reliability verification tools such as Calibre PERC include specific technologies to make fast, automated reliability verification practical. Reliability verification methodology comprising a single tool, a unified rule deck, and an integrated debug environment helps designers find subtle design optimization opportunities without SPICE circuit simulation while also enabling them to achieve the accurate and comprehensive verification necessary to ensure a repeatable and reliable design.

Acknowledgment

Thanks to Shelly Stalnaker for editorial support in the preparation of this chapter.

References

[1] M. Fragnoli, E. Gevinti, A. Bogani, L. Cerati, Novel initialization and implementation method for HBM ESD compliance automated check on smart power IC's, in: Automation Conference (DAC) User Track, 2012.

[2] S. Mitra, N. Palmer, R. Gauthier, M. Muhammad, R. Halbach, C. Seguin, A current density analysis tool to identify BEOL fails under ESD stress, in: ESDA Conference, 2011. Available from: http://ieeexplore.ieee.org/stamp/stamp.jsp?arnumber=06045603.

[3] M. Muhammad, R. Gauthier, J. Li, A. Ginawi, J. Montstream, S. Mitra, K. Chatty, A. Joshi, K. Henderson, N. Palmer, B. Hulse, An ESD design automation framework and tool flow for nano-scale CMOS technologies, in: Electrical Overstress/Electrostatic Discharge Symposium (EOS/ESD), 2010. Available from: http://ieeexplore.ieee.org/stamp/stamp.jsp?tp=&arnumber=5623716&isnumber=5623704.

[4] J. Lescot, V. Bligny, D. Medhat, D. Chollat-Namy, Z. Lu, S. Billy, M. Hofmann, Static low power verification at transistor level for SoC design, in: IEEE International Symposium on Low Power Electronics and Design (ISLPED), 2012, pp. 129–134. Available from: http://dl.acm.org/citation.cfm?id=2333694.

[5] C. Robertson, Calibre PERC: preventing electrical overstress failures, EE Times, (February 2012). Available from: http://www.eetimes.com/design/eda-design/4236118/Calibre-PERC–Preventing-Electrical-Overstress-Failures?Ecosystem=eda-design.

[6] Mentor Graphics, Calibre PERC, Available from: http://www.mentor.com/products/ic_nanometer_design/verification-signoff/circuit-verification/calibre-perc/.

[7] P. Gibson, F. Pikus, L. Ziyang, S. Srinivasan, A framework for logic-aware layout analysis, in: 11th International Symposium on Quality Electronic Design (ISQED), 2010, pp. 171–175, http://dx.doi.org/10.1109/ISQED.2010.5450415. Available from: http://ieeexplore.ieee.org/stamp/stamp.jsp?tp=&arnumber=5450415&isnumber=5450389.

[8] K. Kollu, T. Jackson, F. Kharas, A. Adke, Unifying design data during verification: implementing logic-driven layout analysis and debug, in: IEEE International Conference on IC Design & Technology (ICICDT), 2012, pp. 1–5, http://dx.doi.org/10.1109/ICICDT.2012.6232874. Available from: http://ieeexplore.ieee.org/stamp/stamp.jsp?tp=&arnumber=6232874&isnumber=6232832.

[9] D. Medhat, Circuit reliability challenges for the automotive industry, EE Times, (January 2013). Available from: http://www.eetimes.com/design/eda-design/4404918/Circuit-reliability-challenges-for-the-automotive-industry.

[10] R. Stoica, Automated flow for voltage-dependent layout checks, in: Mentor Graphics User2User Conference, 2012.

[11] ESDA/JEDEC, Joint JEDEC/ESDA Standard for Electrostatic Discharge Sensitivity Test—Human Body Model (HBM)—Component Level, 2012. Available from: http://www.jedec.org/standards-documents/results/js0012012.

[12] ESD Association, ESD Electronic Design Automation Checks, ESD Association, New York, 2011. Available from: http://www.esda.org/documents/ESDEDA-InCompliance_part1.pdf. http://www.esda.org/documents/ESDEDA-InCompliance_part2.pdf.

Reliability modeling and accelerated life testing for solar power generation systems

11

F. Schenkelberg
FMS Reliability, Los Gatos, CA, USA

11.1 Introduction

The ability to use sunlight to provide electricity has grown into a multibillion-dollar industry. Advances in technology and reductions in cost per kilowatt produced continue to drive the widespread adoption of solar power generation systems. Systems range from individual cell phone-charging portable panels to large-scale industrial power production. The decision to use solar power often relies on estimates of the system's cost of ownership. Because the sun provides the fuel, unlike coal- or gas-fired power plants, the cost of the fuel is eliminated. If the solar system generates enough power with minimal maintenance or component replacement costs, investing in solar power makes economic sense. It is the uncertainty of the longevity of the system that becomes the major factor in the decision to invest.

Companies in the solar power industry are regularly asked about their product's reliability. The ability to coherently discuss the risks and to clearly demonstrate an understanding of system reliability permits both buyers and funders of solar systems to fully appreciate the risks. Using the established reliability engineering tools of reliability block diagrams (RBDs) and accelerated life testing (ALT) provides the necessary information to describe a system's reliability.

The solar industry has a high installation cost and minimal operating cost. Getting the product reliability right is essential for establishing a healthy solar industry. The expected long lifetimes, the harsh operating conditions, the desire to minimize costs, and the evolving technology require a careful application of a range of reliability engineering tools. In this chapter, we will discuss the modeling and ALT tools in detail as they form a clear means to convey system reliability expectations and failure risks.

11.2 Overview

Compared to fossil fuel-based systems, solar power systems are expensive to install. The cost of the technology development, production, and installation requires an investment in the belief that the equipment will produce power without unexpected cost long enough to provide a reasonable return on investment.

Reliability Characterisation of Electrical and Electronic Systems. http://dx.doi.org/10.1016/B978-1-78242-221-1.00011-3

Estimating the durability of solar power systems relies on engineering methods, such as reliability modeling, and ALT. Research and development efforts continue to improve every aspect of solar power systems, bringing new materials, assembly processes, and failure mechanisms. Solar systems experience the full range of environments, from arctic cold to desert heat, wind, dust, and similar stresses. Furthermore, these systems are expected to operate for 30 years with little or no maintenance.

There are few solar systems that have operated for more than 25 years, and those systems have outdated technology and thus cannot provide meaningful reliability information for today's designs. The unique materials and construction do not have adequate failure mechanism models, and these models are slow to become available owing to the complexities of the designs.

A major challenge is to identify and characterize failure mechanisms sufficiently to predict the number and timing of system failures over >25 years and to perform this work within the relatively short product development process. Predicting future failures is difficult but necessary for those investing in solar systems to fully understand the investment risk.

11.2.1 Brief overview of solar power generation systems

The conversion of solar energy into electrical power occurs with the conversion of solar radiation into electricity using photovoltaic (PV) technology or with the heating of water to drive a turbine and generator. Other solar systems heat water or air rather than generate electricity. The scope of this chapter is on the solar electrical power generation systems. This includes home and industrial systems. It also includes the supporting elements for these systems, such as positioning or tracking systems, inverters, sealing and protection systems, cabling, and structural elements.

The basic idea is to convert sunlight into electricity with the least cost per unit of energy produced. Solar cells use either PV or concentrator photovoltaic (CPV) silicon or indium gallium arsenide devices, respectively, to convert sunlight into direct current (DC) power. The CPV cells are relatively small and use reflectors to focus up to 1000 times the PCV cell's area onto the CPV device.

Efficiency is the ability of the system to convert sunlight's energy into electricity. Solar cell efficiency continues to improve, starting at approximately 4% efficiency in the 1950s with the first PV cells and now achieving 44.7% efficiency with III-V multi-junction solar cells used in CPV systems. See Figure 11.1 for improvements by company and technology [1]. A 44.7% efficiency means a little less than half of the sunlight's energy is converted to electricity, with the other half being lost to reflection or heat.

Rather than directly converting sunlight to electricity, in the other method of solar energy conversion, sunlight is converted into mechanical energy to drive a turbine and generator system. Concentrating sunlight heats a fluid such as water, oil, or salt, which then transfers the solar energy into steam. Sunlight replaces other sources of heat generation in these solar power plants.

There are three basic approaches for solar energy concentration in fluid heating systems: parabolic-trough, dish-engine, and power tower or heliostat systems.

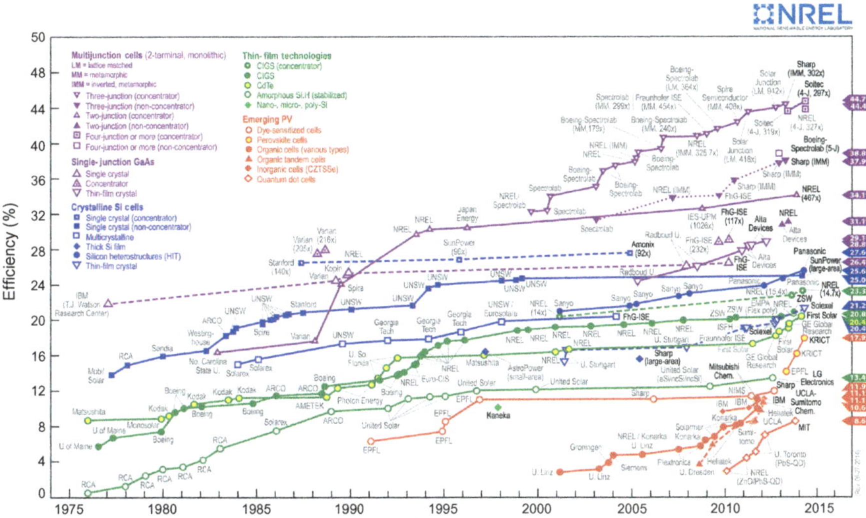

Figure 11.1 Progression of efficiencies tracked by the National Renewable Energy Laboratory. This plot is courtesy of the National Renewable Energy Laboratory, Golden, CO.

Regardless of the details, each system relies on collecting and focusing sunlight onto a target that heats the fluid.

In PV, CPV, and steam generation systems, additional improvements occur when the solar cells or solar collectors align with the sun as it moves across the sky. Tracking systems range from nonexistent or fixed mounted systems to single- or dual-axis systems. Single-axis systems may improve system efficiency by 20% or more over fixed systems. Dual-axis tracking systems improve solar conversion by 30% over fixed systems [2].

Tracking systems require some modification to the mounting structure to allow the motion, plus a motive force generally supplied by a motor and gearbox system. The added complexities of tracking systems cost more and add to the possible ways a system can fail.

Inverters convert the DC created by the PV or CPV system to alternating current used by most appliances and shared across the electrical grid. Microinverters may serve each PV panel and string inverters may serve an entire array of panels. Inverter reliability is a weak point in the reliability of solar power generation, with the lifetime of microinverters being limited to 15 years and that of string inverters to only 10 years. Other solar system elements should last 25-30 years [3].

The remainder of the solar system elements provide a means to control and transfer power or to structurally protect and support the active elements of the system. The electrical system may include bypass diodes, cabling, connectors, sensors, and monitoring systems. The structural elements may include sealants, coatings, laminates, mounting systems, and foundation elements.

11.2.2 Overview of the chapter

The solar industry continues to face challenges in bringing new technology to market. Section 11.3 will explore the effects of the pace of innovation, cost reduction pressure, time to failure uncertainty, and investment risk. Section 11.4 discusses reliability block modeling, which is a method to consider a complex system's reliability building on the reliability characterization of individual components. The modeling of failure mechanisms and systems permits estimates of system reliability, spare parts provisioning, and maintenance costs. Section 11.5 describes the basic approaches used to estimate the reliability based on testing-specific failure mechanisms using higher than expected stresses. The ALT results provide the reliability details needed for the block diagrams of system models. Sections 11.6 and 11.7 present two examples of the application of ALT. Section 11.8 explores recent advances and subjects that require attention. The intersection of solar power and reliability engineering continues to advance rapidly.

The end of the chapter has a list of recommended reading for those interested in learning more about the topics discussed. The solar industry continues to innovate and improve the ability to convert sunlight to electricity. The reliability engineering community continues to learn, model, and describe the system reliability risks and expectations. This chapter provides just an introduction to reliability engineering in the solar power industry.

11.3 Challenges

In the past 25 years, converting solar to electricity has advanced to be a viable source of energy. Starting as an expensive power source for space vehicles and those able to afford the prestige of solar power, its implementation is now becoming relatively common. Since its initial concept to today, the technology that converts solar energy to electricity has changed regularly. As with other high-technology fields, the results of innovation and improvement solve problems, improve efficiency, and reduce cost.

Reducing cost has been a major challenge and continues to drive innovation. The cost per kilowatt hour is measurable and continues to decline. The US Department of Energy total system levelized cost, the averaged cost per unit of energy over expected operating lifetime of power generation system, for PV solar systems installed in 2019 is \$0.130/kWh. However, a solar energy plant has to compete with existing power plants, such as coal- or natural gas-based plants. Investment to build a fossil fuel plant is relatively low, because the cost is primarily in the fuel, which is burned to release the energy to create electricity. For example, for a coal-fired plant, the total system levelized cost is \$0.096/kWh [4]. Thus, solar is not economically competitive at these prices yet is continuing to reduce the costs with higher efficiencies and lower installation costs.

The cost of a solar power plant lies only in the solar cells and supporting structures—the fuel cost is zero. Some maintenance to clean the panels and perform minor repairs is required, but the ongoing costs are minimal. The investment is upfront and those making the investment want to know whether the technology being used will operate as expected for 20 or more years. If it does, then the investment is worth the effort and comparable to building a coal-fired plant.

The bankability question poses a major challenge for solar technology. It is a basic reliability engineering task to determine the expected useful lifetime of the technology. How do we know a new material, construction, or device will operate for 20 years? Having a convincing body of evidence and logic enables investors to confidently invest in solar power.

11.3.1 Failures

A failure in any element of a solar power generation system causes a loss of efficiency and reduction in power generated. The challenge lies in identifying and fully understanding each failure mechanism. The consequence of a single component failure ranges from inconsequential to total loss of power generation capability. For example, a connector failure may isolate an entire string of solar panels. A string inverter failure may take an entire power plant off-line. A goal for most solar systems is to use the least expensive and most effective components for a long lifetime of maintenance or replacement-free service. Ideally, once installed, a solar power generation system will operate for 30 years with a minimum of maintenance costs and produce power efficiently.

Component and system failures generally occur for one of three reasons:

1. Faulty components, materials, or assembly (early life failures).
2. Faulty installation or failure to perform routine maintenance (e.g., failure to clean debris from PV panels and lack of maintenance failures).
3. Wear-out or degradation (wear-out failures).

Each element of a solar power generation system has many ways it can fail. For example, consider an electrical cable connector. It may fail by metal contact corrosion, faulty construction, faulty installation, internal wire breakage, lose of contact force (polymer creep or spring relaxation), fretting, etc. Each failure mechanism has a distinct and measurable probability distribution (see Figure 11.2).

Connectors provide mechanical contact for the metal pins, permitting electrical contact with low contact resistance. The choice of materials and design for an connector enclosure system provides protection from the weather, alignment, and often a fastening or locking mechanism.

Every element has many possible failure mechanisms. The design choices balance cost and reliability while providing a value to the customer. Ideally, every component will survive just as long as any other component, as in the poem *One-Hoss Shay* [5]. Unfortunately, the differences in design, manufacture, and installation inherently create variability in the expected lifetimes of the various components. Local conditions, such as daily temperature, wind loading, and airborne contaminates, also add variability to expected lifetimes. For a typical system, there may be hundreds of possible failure mechanisms competing to cause the system to fail.

Given the desired long lifetimes of solar power systems and the desire to minimize initial and maintenance costs, accelerated testing permits the exploration and understanding of the design decisions relevant for a specific system and location. Reliability

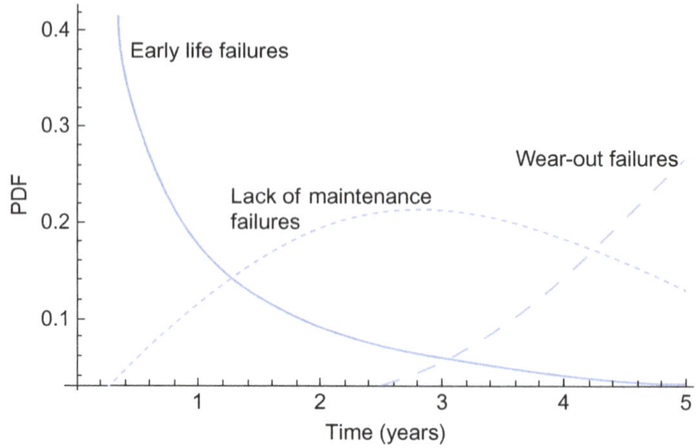

Figure 11.2 Probability distributions for different failure mechanisms.

modeling provides a means to identify system weaknesses and to estimate and describe system reliability.

11.3.2 Bankability

Deciding whether to fund a solar power generation project, whether of residential or industrial scale, relies on knowledge of the ability of the system to generate sufficient power without unexpected costs, such as premature replacement and costly repairs. The investment should provide sufficient return for those funding the project. For major projects, the funding may come from banks or investors. It is the question asked by those providing funding that explore the balance of power generation (profit) and cost over time. Given the relatively short field experience most solar technologies currently have, there is not a large body of evidence available to demonstrate whether the technologies will last sufficiently long without failure. "Bankability" is the concept used to describe whether the system in question has sufficiently low risk to warrant the funding (investment) to build it. A key element of the risk discussion is that of system reliability.

Given the many ways systems can fail and the relatively long duration of expected failure-free operation, there are many questions that require adequate responses related to the system's expected reliability. Understanding and characterizing the expected failure mechanisms permit a detailed discussion of the risks, knowns, and unknowns and how the design minimizes reliability risks.

11.3.3 Product testing

ALT provides the core of any bankability argument for solar technology. ALT helps to answer the bankability question in clear terms. In one sense, ALT provides the core arguments (description of specific failure mechanism time to failure expectations) for whether the system will provide sufficient payback for investors. Creating a compelling argument requires understanding and characterizing a wide range of possible failure mechanisms. Building the body of evidence though ALT conveys the due diligence required for bankability. Reliability modeling provides a means to organize and calculate system reliability.

Therefore, understanding the challenges facing the technology over time in the environment and in operation is essential to designing and conducting accelerated life tests that answer the bankability question.

Investor questions about the durability or longevity or cost of maintenance can be addressed by accelerated testing. The challenge is to shorten time sufficiently to make conclusions. Using ALT, we can assess the expected performance of solar cells, interconnects, motors, and most of the rest of the equipment in a solar system.

The challenge of ALT lies in understanding the failure mechanisms and how the components will respond to stress. Which set of stresses, such as temperature, humidity, and UV radiation, are important for a specific failure mechanism? We also need to know the expected stress during operation and the higher stress behavior and ability to accelerate the failure mechanism.

The pressure of competition and pace of innovation constrain the duration of ALT testing. The ideal accelerated life test entails operating the units as the customer would and monitoring for failure over the expected >20-year lifetime. For each increase in stress, and for each isolation of a subsystem or material, we add the risk that the test does not reflect the behavior in normal operation. The ability to shorten the test time increases the risk of errors, yet it also permits an estimate of the expected life performance.

If we have 2 years to conduct an ALT to examine an expected 20-year lifetime, that would require at least a $10\times$ acceleration factor, meaning that each year of testing represents 10 years of normal use. Given time constraints and competition, we may only have 1 year for ALT, so the acceleration factor now is $20\times$ and each year of testing has to represent a 20-year lifetime of exposure to the environment and use.

Solar cells are assembled into panels, which include coatings, protective and supporting layers, and electrical interconnections. The panels are assembled, transported, and installed as units. The panel is the field-replaceable unit. Besides the cell, each of the elements of the panel must also function to keep moisture out, permit sunlight to the cells, and carry the created electrical power away from the panel. Testing panels are difficult owing to their size and cost. Mounting panels outside do not provide any acceleration over use conditions. Using chambers and multiple sunlighting permits acceleration but often requires specially built chambers and test facilitates.

There is a limit to the amount of stress that can be applied when attempting to increase the acceleration factor. Increasing the temperature increases failure mechanisms related to chemical reactions, that is, oxidation, and increasing it too much may even melt one of the materials in the system. Melting only occurs at a temperature higher than that would occur under field conditions and thus is an unwanted and irrelevant failure mechanism. Increasing stress does increase the acceleration factor and has the limitation of possibly causing unwanted failure by different failure mechanisms. A careful understanding of the system, materials, and failure mechanisms permits one to design an ALT protocol that maximizes the acceleration factor and minimizes the risk of causing irrelevant failures.

New materials and processes are part of the innovation in the solar industry. The need to increase power conversion efficiency, lower costs, fabricate lighter weight components (thereby reducing both transport and installation costs), and others drive this exploration and innovation. Each new material requires study and experimentation to uncover its short- and long-term failure mechanisms. We also want to understand the relationship between higher stresses and acceleration of failures. The same is true for new assembly, transport, and installation processes. Each may change the set of stresses the solar system and component experience. Changes may create defects that lead to system failures. For example, twisting a panel during installation may cause a crack to start along a laminated edge. This crack then expands quickly during normal use and permits moisture to penetrate the solar cell, causing it to fail. Being able to determine the cause and amount of stress required to initiate a crack is part of estimating the life of a solar system.

11.4 Modeling

Modeling is a representation of a system or process. Modeling permits us to explore complex systems. Good models help guide answers to questions such as how long the system can operate, which component is likely to fail first in this environment, and how many spare or replacement parts should be stocked?

Most solar systems have a basic structure and a simple series system model is often adequate. Series systems have a design such that each element in the system must function for the system not to fail. For a simple PV system, this means the solar cells, panel, bypass diode, and inverter all must function for the system to produce power. If any one element fails, then the system fails. Of course, strings of cells within a panel may have a more complex structure including strings in parallel. In a parallel structure, an element can fail but the remaining elements continue to operate. The panel may product less power, but the system does not totally shut down.

Depending on the types of questions being asked, the reliability model should be simple and accurate enough to provide answers [6]. An example of a series system RBD is shown in Figure 11.3. The reliability of block A is 0.70 and that of block B is 0.80.

The product of the elements in series is the system reliability

$$R = \prod_{i=1}^{n} R_i \tag{11.1}$$

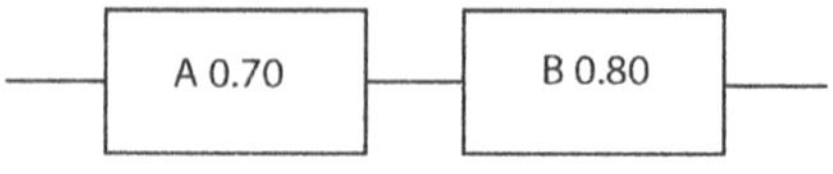

Figure 11.3 A series system block diagram.

where n is the number of elements in the series. For the simple series example illustrated in Figure 11.3, the formula becomes

$$R = R_1 \times R_2 \tag{11.2}$$

resulting in $0.70 \times 0.80 = 0.56$, or 56% reliability.

For systems with parallel elements, the RBD is shown in Figure 11.4.

The system reliability for a parallel system is

$$R = 1 - \prod_{i=1}^{n} (1 - R_i) \tag{11.3}$$

Blocks B and D both have a reliability of 0.70. The parallel structure provides a system reliability of 0.81 (or 81%). In this configuration, either block B or D is sufficient for the system to function.

For an example of three strings of five PV cells in parallel within the panel, we combine series and parallel modeling to create a simple PV panel reliability system model, as shown in Figure 11.5.

The system reliability for this combination series-parallel system is

$$R = 1 - \prod_{j=1}^{3} \left(1 - \prod_{i=i}^{5} R_i\right)_j \tag{11.4}$$

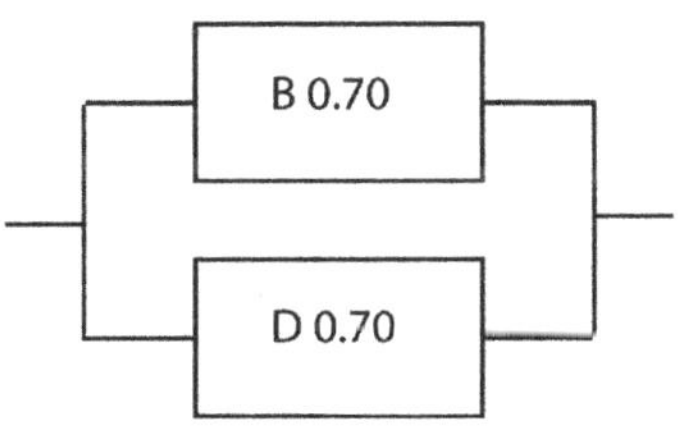

Figure 11.4 A parallel system block diagram.

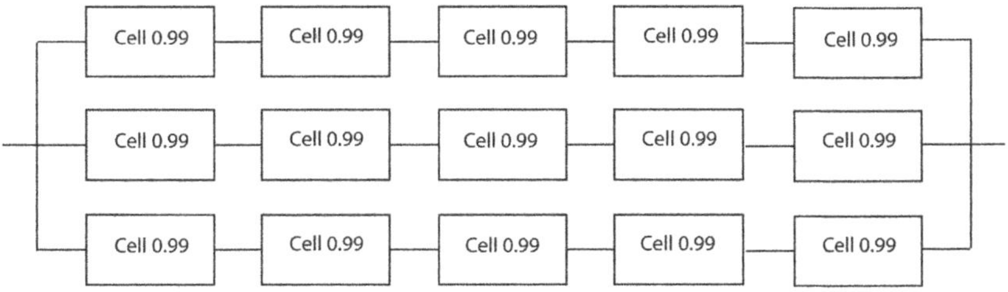

Figure 11.5 A combination series and parallel system with five cells per string and three strings in parallel.

Modeling also extends to individual failure mechanisms. Common printed circuit boards (PCBs) are made up of various electronic components, resistors, capacitors, diodes, transistors, and integrated circuits. Within solar power systems, there are monitoring, control, and inverter elements, which include a significant number of circuit boards. The installation is generally protected from direct sunlight and moisture yet does experience the full range of daily ambient temperature change. The failure of a circuit board due to solder joint failure means all or part of solar power system shuts down.

The solder joints composed of SnAgCu (SAC) solder has a failure mechanism of solder fatigue. A simple model for SAC solder fatigue is the Norris-Landzberg (N-L) equation. Understanding the failure mechanism, and how it relates to stress, permits the design using the appropriate technology and temperature controls to avoid premature solder joint failures [7].

The N-L started with the Coffin-Manson equation:

$$N\left(\Delta\varepsilon_{\mathrm{p}}\right)^{n}=C \tag{11.5}$$

where N is the number of cycles to failure, $\Delta\varepsilon_{\mathrm{p}}$ is the plastic strain range per cycle, n is an empirical material constant, and C is a proportionality factor. N-L then added two additional factors to account for the effects of cycle frequency f and the maximum temperature experienced by the solder material, $T_{\max}$. Creating a ratio of the number of cycles to failure for field and test conditions results in a formula for the acceleration factor:

$$\mathrm{AF}=\frac{N_{\text{field}}}{N_{\text{test}}}=\left(\frac{f_{\text{field}}}{f_{\text{test}}}\right)^{-m}\left(\frac{\Delta T_{\text{field}}}{\Delta T_{\text{test}}}\right)^{-n}\left(\mathrm{e}^{E_{\mathrm{a}}/k\left(\left(1/T_{\max,\ \text{field}}\right)-\left(1/T_{\max,\ \text{test}}\right)\right)}\right) \tag{11.6}$$

where N is the fatigue life under cyclic loading, i.e., the number of cycles to failure; field and test denote the use and experimental conditions, respectively; ΔT is the change in temperature over one cycle; and $T_{\max}$ is the maximum temperature in the cycle.

For solar application, every day, there is a temperature change for the electronics associated with the inverters, monitoring, and control circuit boards, from the ambient coldest temperature (typically occurring just before or at dawn) to the ambient warmest temperature (typically occurring mid to late afternoon). The warmest temperature is elevated by the temperature rise attributed to the electronics. In some applications and locations, this daily temperature swing can be as high as 25 °C.

Another set of failure mechanisms (e.g., electrolytic corrosion, metal migration, and corrosion) for electronics is caused by high temperatures along with humidity. Temperature alone causes many failure mechanisms (metal migration, charge injection, intermetallic growth and embrittlement, electromigration, etc.). In the use application, we are typically unable to isolate humidity, thus modeling and understanding its effects are important.

For solar installation with circuit boards, the lack of temperature control implies the circuit boards, connectors, and electromechanical elements will experience relatively high temperatures. Thus, the failures induced quickly by high-temperature exposure become relevant. Understanding the effect of temperature on the expected failure rate of the various elements of a solar system enables accurate reliability modeling.

To specifically model the effect of temperature and humidity on electronic microcircuits, one can use Peck's equation [8]:

$$t_f = A(\%\mathrm{RH})^n e^{(E_a/kT)} \tag{11.7}$$

where t_f is the time to failure, A is a constant, %RH is the percent relative humidity, n is −3.0, E_a is 0.9 eV, $k = 8.615 \times 10^{-5}$ eV/K is Boltzmann's constant, and T is the temperature in kelvins.

Using a ratio of time to failure for field and test conditions creates the formula for the acceleration factor:

$$\mathrm{AF} = \left(\frac{\%\mathrm{RH}_{\mathrm{field}}}{\%\mathrm{RH}_{\mathrm{test}}}\right)^{-3} \exp\left[\frac{E_a}{k}\left(\frac{1}{T_{\mathrm{field}}} - \frac{1}{T_{\mathrm{test}}}\right)\right] \tag{11.8}$$

Both models require an understanding of the operating environment stresses and how those stresses relate to failure mechanisms and may require experimentation to refine the models' estimates. Failure mechanism models are useful for initial estimates during design, for test planning during validation, and for maintenance planning during use.

Over the past 30 years, significant work has gone into developing and refining failure mechanism models for a wide range of materials and components. The advanced models may require detailed information on the construction, processes, dimensions, and environmental conditions. These detailed models are often called physics-of-failure (PoF) models as they attempt to represent the fundamental physical or chemical processes that lead to failure.

In the analysis of solar power system, understanding the failure mechanisms for each element of the system permits accurate reliability modeling. The use of PoF models starts with developing or finding suitable PoF models for the salient risks facing the solar installation. There are a number of repositories for PoF models.

The Reliability Information Analysis Center (RIAC) has created a listing of physical failure models [9]. "The objective of the WARP (a physics-of-failure repository by RIAC) effort is to collect, analyze and verify the existence and characteristics of physics-of-failure (PoF) models for electronic, electromechanical and mechanical components to provide a centralized web-based repository within the RIAC that is accessible to researchers and engineers."

The University of Maryland Center for Advanced Life Cycle Engineering has focused on creating PoF models for electronics. They have also created a software package, SARA, for the application of a range of models for electronic boards [10].

DFR Solutions has also created a software package, Sherlock, for the application of PoF modeling of electronic boards [11].

Field failure data also may benefit from modeling. Using basic empirical models to understand the rate of failures may provide information for decisions concerning product improvement, repair rate changes, and maintenance planning. Since the bulk of the cost for a solar power system is the initial installation, there is a desire to minimize maintenance and any replacements or repairs. Monitoring field failures and comparing to expectations set by the reliability models let the power plant team identify adverse reliability trends and adjust maintenance practices to minimize overall operating costs.

For repairable systems, using nonparametric models such as a mean cumulative function is often a sufficient first step. For nonrepairable elements, using a distribution analysis, such as fitting a Weibull distribution, often provides adequate information.

Warranty planning and adjustment to accruals benefit from modeling. During initial design of a product or system, the team makes estimates for repairs or element failures based on reliability models. The estimates provide information for future warranty expenses. Tracking and modeling field data provide a comparison to the initial models and a means to refine and improve warranty expense projects.

ALT is one tool for providing reliability estimates for the development team's system reliability models and is the subject of the next three sections. In the first, details of the method are presented. Two specific examples are then provided in the following sections.

11.5 Accelerated life testing (ALT)

ALT encompasses a set of techniques used to estimate the time to failure in a shorter time than if the items operated at use conditions. For most elements of a solar power plant, the expected operating lifetime is >20 years. A development project for a new solar cell may last 2-3 years, during which time the team has to estimate the expected survival rate out past 20 years.

ALT provides the tools to compress time. These ALT tools have risks both technically and statistically. Technically, the testing has to replicate the stresses and associated failure mechanisms in a way that corresponds to the actual stresses and failure mechanisms experienced during normal use. Statistically, the risks are associated with small sample sizes and extrapolating beyond the duration of the test data.

Despite the risks, the value in understanding the expected time to failure for a new inverter, solar cell, or interconnection enables the team to make informed design decisions. Careful selection and creation of an ALT protocol are essential for balancing risk and accuracy of ALT results.

There are a range of methods for conducting ALT. The most common is time compression. Creating a model and using an existing model are two other common methods. The next three sections provide definitions, an overview, and examples of the three ALT techniques.

11.5.1 Time compression

A typical solar tracking system adjusts the face of the solar panel or reflective surfaces to align with the sun as it moves across the sky. The system moves though one rotation per day. It is this relatively slow motion that enables the time-compression approach for ALT.

The time-compression ALT approach entails using the system more often than under normal conditions. For a solar tracker, we may be able to increase the speed of the tracking to experience an entire day of movement in an hour, for example. If we run the test though 24 cycles per day, the system experiences 24 days of use in 1 day. The calculation of the acceleration factor is then

$$\mathrm{AF} = \frac{\text{use cycle duration}}{\text{test cycle duration}} = \frac{24\,\mathrm{h}}{1\,\mathrm{h}} = 24 \tag{11.9}$$

Time compression in this fashion is easy to use and understand. In the above simple example, the acceleration factor is 24, or each day of testing represents 24 days of use.

One consideration is in verifying whether the failure mechanisms of interest have the same chance of occurring as when the system is in actual use. One can imagine a test that cycles the solar tracking system so fast that the heat buildup and associated failures that occur have little to do with the failures during actual use. In the design of the test, one must be careful not to introduce conditions that would alter the causes of failure, thus inducing failures that are irrelevant.

Another consideration is that the use of a system in the field experiences a wide range of environmental conditions that are difficult to reproduce in the laboratory. If dust buildup is an important element causing tracker motor failures, then including some replica of dust buildup during ALT may be necessary.

The closer the testing conditions mimic the use conditions, the better. In most cases, understanding the failure mechanisms and the primary stresses that cause the failure to occur enables the design of a meaningful accelerated life test.

11.5.2 Using three stresses to create a model

A common ALT approach is to apply higher than expected stress (such as through temperature and voltage) to the product in such a way that the failure mechanism of interest is accelerated in its progression toward failure. The trick is to relate the use and test stress in such a way that the results from the higher testing stress provide meaningful predictions at use conditions.

Without an existing acceleration model being assumed, the ALT design should produce sufficient data to determine the acceleration model. For example, using temperature as the accelerant and the Arrhenius rate equation as the form of the model (unknown activation energy), an approach could be to use three high temperatures. The plot of the time to failure relative to each temperature may provide a suitable relationship between temperature and life, allowing the extrapolation of expected life at normal usage temperatures (see Figure 11.6).

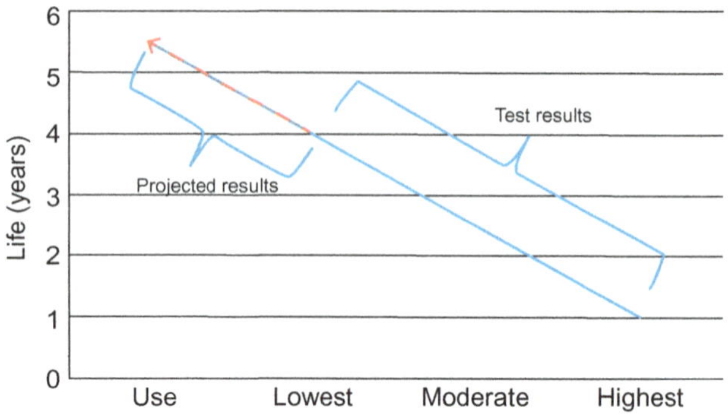

Figure 11.6 Relationship between stress and expected life.

In this approach, the model parameters are provided directly, and one does not assume that work on similar items will apply. For example, even for well-known failure mechanisms, the activation energy for the acceleration models is provided over a range. The spread of possible values reflects the many different specific design, processing, assembly, and use variations that can affect durability.

Following the temperature stress and Arrhenius model [6, p. 331], the acceleration factor may be stated as

$$\mathrm{AF} = \exp\left[\frac{E_{\mathrm{a}}}{k}\left(\frac{1}{T_{\mathrm{field}}} - \frac{1}{T_{\mathrm{test}}}\right)\right] \tag{11.10}$$

where E_a is the activation energy, k is Boltzmann's constant, and T_{field} and T_{test} are the field (use conditions) and test (experiment conditions) temperatures, respectively, in kelvins.

Two downsides of this approach are (1) the necessity of selecting appropriate high stresses that do not cause unwanted failure mechanisms and (2) the fact that higher stresses often require special equipment and setup. A major assumption is that the higher stresses cause the failure mechanism to accelerate in a fashion that is similar to when the item is under use conditions. Failure analysis is critical to validate this assumption.

A third downside is that this approach generally takes sufficient samples at multiple stress levels to fit lifetime distributions. As a general guideline, each of the three stresses should produce at least five failures, and an efficient design is to place twice as many samples in the middle stress relative to the highest stress and four times as many in the lowest stress. Placing more in the stress closest to the use conditions weights the regression with the lowest applied stress and increases the chance of quickly causing at least five failures. This 4:2:1 ratio implies at least 30 samples and precision is significantly improved with additional samples [12].

11.5.3 Using an existing model

One way to estimate system reliability is to understand and model the various component-level reliability performance. This relies on models that describe the effect of stress such as temperature, humidity, or thermal cycling on the time to failure distribution for specific failure mechanisms. In a complex solar power system, there may be dozens of failure mechanisms that will lead to failure before the decommissioning of the system.

The next two examples highlight how to use existing models found in technical literature or from vendors to calculate the reliability performance of major components. Then, use the series and parallel system reliability models to aggregate the information to estimate the system reliability performance.

11.5.3.1 Expected failure mechanisms and lifetime data

The inverter for a solar power system relies on large electrolytic capacitors for the circuit that converts the panel DC power to grid AC power. Smaller capacitors exist on most monitoring or control circuit boards as part of the power management circuit for that circuit board.

Given the large number of capacitors, understanding how to estimate the lifetime for a given location permits the design team to adjust the selection of capacitors for the installation or permits the maintenance team to anticipate failures and have suitable spare parts on hand.

A likely failure mechanism for capacitors is dielectric breakdown. This can occur as a result of manufacturing defects or power line transients and over time by dielectric degradation. If assembly process controls and adequate circuit protection are in place, the dominant life-limiting failure mechanism is dielectric degradation and eventually dielectric breakdown.

High operating voltages relative to rated voltage and to a lesser degree high operating temperatures accelerate the dielectric degradation, eventually overcoming the self-healing behavior. The ripple current contributes to the internal heating and the acceleration of dielectric degradation.

11.5.3.2 Reliability calculations

Let us work an example. Based on example component datasheets, the following reference service life (base life) information is available:

DC bus capacitors: 100,000 h at $\leq$70 °C.
AC filter capacitor: 500,000 h at $\leq$81 °C at 50% rated voltage.

The calculation of the capacitor expected lifetime distribution based on this above information is fairly straightforward. The basic idea uses the given single point (63.2 percentile point along the distribution), also called the service life or base life value; then, we make adjustments to the base life for the voltage and temperature conditions to determine the expected operating life. Finally, with knowledge about how capacitors fail over time (the slope), we can estimate a life distribution, since we have a point and a slope.

The service life, or base life provided by the vendor, is the expected duration in hours until 63% of the population has failed while operating at or near maximum operating conditions. The slope, or β value for a Weibull distribution, is best determined from testing to failure in the use environment. An estimated recommended β value from the literature is 6-8 for similar capacitors [13]. We will use a β value of 7 for our estimate.

For a film capacitor, the effects of voltage and temperature alter the provided service life value L according to the following formula [14]:

$$L = L_B \times f_1(T_M - T_C) \times f_2(V) \tag{11.11}$$

where L_B is the base life, or the service life value, as listed on the datasheet; T_M is the maximum rated temperature for the capacitor; T_C is the actual core temperature of the capacitor under operating conditions; f_1 represents the function or adjustment to the base life value for temperature; and f_2 represents the effect of voltage.

As expected, the formula for f_1 involves the Arrhenius relationship:

$$f_1 = e^{E_a/k((1/T_C)-(1/T_M))} \tag{11.12}$$

where E_a is the activation energy, which for anodic alumina is 0.94, and k is Boltzmann's constant.

The core temperatures are calculated based on operating and thermal conditions, or they can be roughly approximated by using the case temperatures. Similar formulas in the literature use the operating ambient temperature in place of the core temperature. Given the complexity in the calculation of core temperatures, using the case temperature, although not accurate, provides a first-order approximation. Furthermore, when the capacitor is operated close to the maximum operating temperature, f_1 becomes insignificant in the adjustment of the base life value.

For the DC bus capacitors, the maximum rated temperature is 85 °C and the reference service life is listed with a core temperature of ≤70 °C. The measured case temperature is 81.29 °C at a 45 °C ambient full-power condition, 480 VDC. The calculated temperature rise from the case to the core is approximately 6.5 °C, resulting in a core temperature of 87.8 °C [15]. Therefore, the resulting calculation of f_1 is 0.20. This is less than one because the expected operating hot spot temperature is higher than the reference life hot spot temperature. The actual value may be less, since the datasheet suggests that the life expectations are valid only below 70 °C.

For the AC filter capacitor, the maximum rate temperature is 85 °C and the reference service life is listed with a core temperature of ≤81 °C. The measured case temperature is 66 °C at 45 °C ambient, full-power condition, 480 VDC. The calculated temperature rise is 11 °C, resulting in a core temperature of 77 °C and an f_1 value of 1.42.

The second element of Equation (11.11) is the f_2 term related to voltage. For film capacitors [16,17],

$$f_2 = \left(\frac{E_R}{E_O}\right)^7 \tag{11.13}$$

where E_R is the maximum rated voltage and E_O is the operating voltage.

With the exponent at 7, this term dominates the overall impact of expected life when the operating voltage is less than the rated voltage. This is true for both capacitors in the system.

For the DC bus capacitor, let us use the rated voltage of 700 VDC and the expected maximum operating voltage of 480 VDC to get

$$f_2 = \left(\frac{E_R}{E_O}\right)^7 = \left(\frac{700}{480}\right)^7 = 14.0 \tag{11.14}$$

For the AC filter capacitor, the rated voltage is 530 VAC at 60 Hz and it is expected to operate at 208 VAC, so

$$f_2 = \left(\frac{E_R}{E_O}\right)^7 = \left(\frac{530}{208}\right)^7 = 697 \tag{11.15}$$

For each capacitor, using Equation (11.11) and the calculations above, we have the results shown in Table 11.1.

Note that this is an estimate of the characteristic lifetime. There is a small adjustment given that the slope is not 1, yet the correction was insignificant in this case and does not change the results.

The reliability function, $R(t)$, for the two-parameter Weibull distribution is

$$R(t) = \mathrm{e}^{-(t/\eta)^\beta} \tag{11.16}$$

where t is the time in operating hours (in this case); η is the characteristic lifetime, or the time till approximately 63.2% of units are expected to fail; and β is the slope, in this case assumed to be 7.0 [13].

For the DC bus capacitor, the expected reliability at 10 years of life, 36,500 h, is 1. This means that 100% (greater than 99.999%) of capacitors are expected to successfully operate over a 10 h/day, 10-year operating life. At 20 years, this reliability drops to 99.994%. Figure 11.7 shows the Weibull cumulative density function (CDF) based on parameters $\eta = 2.95$ (in millions of hours) and $\beta = 7$.

The AC filter capacitor has a much larger derating and resulting significant improvement in durability. AC filter capacitors are not expected to wear-out during their expected 20-year operating life, with an insignificant faction expected to fail

Table 11.1 f_1 and f_2 values and expected lifetimes of DC bus and AC filter capacitors

Capacitor	f_1	f_2	Expected lifetimes (63.2 Percentile) (h)
DC bus	0.20	14.0	2,950,000
AC filter	1.4	697	495,000,000

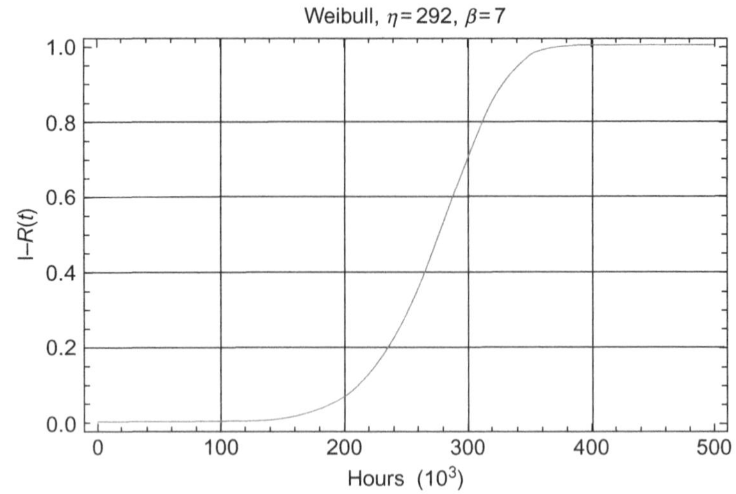

Figure 11.7 Weibull cumulative density function for DC bus capacitors.

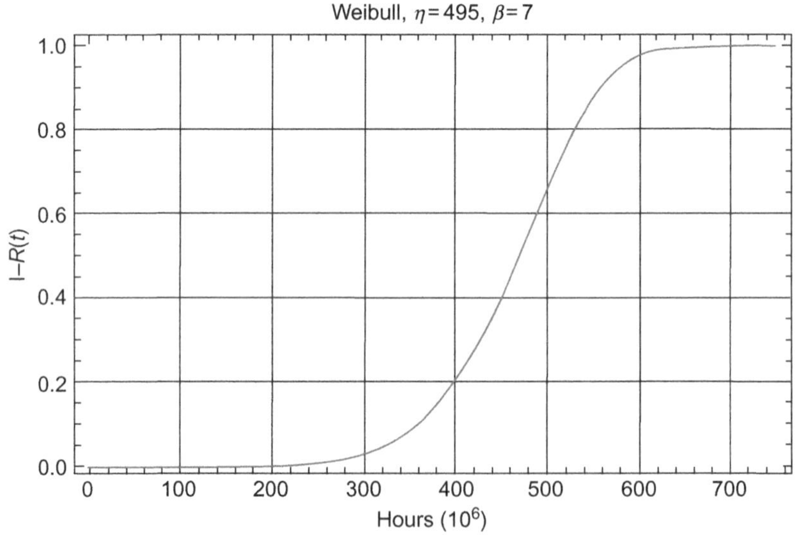

Figure 11.8 Weibull cumulative density function for AC filter capacitors.

from wear-out. Figure 11.8 shows the Weibull CDF with parameters $\eta = 495$ (in millions of hours) and $\beta = 7$.

The expected operating temperature of capacitors is at times very high and may lead to unexpected failures, yet the voltage derating significantly reduces the stress on the dielectric. Of course, other stresses (e.g., caused by lightning strikes or physical or assembly damage) may lead to capacitor failures.

DC bus capacitors are set in an electrically parallel configuration, which tends to further reduce the stresses leading to wear-out within the capacitors. The use of capacitor banks enables the reduction of ripple-current internal heating. Since we have measured temperatures as the basis for the above calculation, and the design is fixed by the number of capacitors in the bank, we will not calculate the effects of the capacitor bank on reliability. For a method to perform such a calculation, see the work by Parler (2004).

IGBT reliability calculations

Large inverters generally rely on the functionality of insulated-gate bipolar transistors (IGBTs). Because the IGBTs in a solar inverter are designed with dissimilar materials and, in operation, IGBTs generate a significant temperature rise, thermal cycling is the main stress factor that causes aging. The two most likely thermal cycling-induced failure mechanisms are bond wire fatigue and solder attach fatigue. The high operating temperatures increase the relative motion of the assembled materials because the coefficient of thermal expansion mismatch leads to material fatigue initiation and propagation.

The bridge test plan was created primarily based on temperature cycling as the main stress factor causing IGBT aging. Bond wire fatigue is more prevalent with rapid switching applications, which is not a characteristic of the expected application. Thus, solder attach failure is expected to be the driving factor. To determine the duration of test in this study, weather data for southern Ontario were studied and the acceleration factor for the ALT plan was calculated based on the Coffin-Manson equation and parameters provided from the vendor:

$$N_{f2} = 10^{11.1359}(\Delta T_C)^{-3.831} \tag{11.17}$$

Here, N is the number of cycles to 1% failure and T_C is the temperature of the IGBT copper baseplate. Using this number combined with an assumed β value of 2.3 (based on a gradual metal fatigue wear-out over time for planning purposes and on a study by Micol et al. [18]) and using a Weibull distribution, we can estimate the population time to failure distribution.

To calculate the acceleration factor using Equation (11.17) and the ratio of field to test cycles to failures, we have

$$\mathrm{AF} = \frac{N_{\mathrm{field}}}{N_{\mathrm{test}}} = \left(\frac{\Delta T_{\mathrm{field}}}{\Delta T_{\mathrm{test}}}\right)^{-3.831} \tag{11.18}$$

where N is number of cycles at field (use conditions) and test (test conditions) temperature and ΔT is the temperature cycling. Based on 20-year weather data gathered at southern Ontario and the nearby US weather stations, the calculated environmental data are summarized in Table 11.2. The weather stations considered here include Toronto, London, Gatineau, Ottawa, Peterborough, Sarnia, Sudbury, Ann Arbor, Detroit, and Flint [19].

Table 11.2 Calculation of AF based on southern Ontario region weather ΔT

Environment ΔT_C at 480 VDC	AF, $B = -3.831$	Test cycles for 10 years
Ontario ΔT_C 90th percentile, 69.6 °C	34	107
Ontario ΔT_C 50th percentile, 63.5 °C	48.6	75

Also, in running an environmental chamber from −40 °C (powered off) to +90 °C (at full power), the IGBT case temperature will cycle from −40 to 135 °C. Therefore, we will have an AF value of 34 for the 90th percentile Ontario weather data. Note that a recalculation of acceleration factor was made to take into account the effect of mini-temperature cycling of IGBTs as a result of cloud coverage; however, the overall effect proved to be minimal. To calculate the confidence level for the ALT plan, making the assumption of failure-free operation during the test period for the bridges, we use the extended-success testing exponential distribution-based sample-size formula

$$n = \frac{\ln(1-C)}{m \ln(R)} \tag{11.19}$$

where n is sample size, C is sampling confidence (type I error or alpha statistical error), m is the number of lifetimes the sample experiences in the test, and R is the reliability lower limit or probability of successfully operating over the time period. Being limited to six samples owing to cost, we can demonstrate an 80% reliability and have a 74% confidence level for our ALT plan. Thus, the ALT plan is summarized as follows: testing six IGBT subsystems within a test chamber set to cycle from −40 to 90 °C for 108 cycles simulates 10 years of temperature cycling at Ontario's 90th percentile weather conditions, which if the unit operates without failure demonstrates 80% reliability with 74% confidence.

11.5.4 Modeling reliability and availability

The ALT test results provide information useful for modeling the inverter system. We can create a simple block diagram for an inverter with seven subsystems, as shown in Figure 11.9.

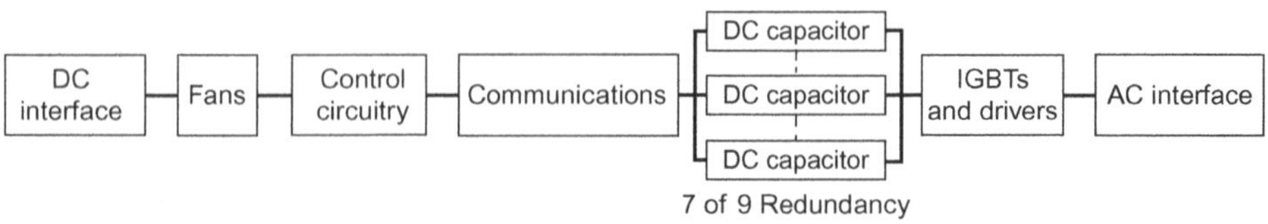

Figure 11.9 Solar inverter system reliability block diagram.

The structure of the model is essentially a series system and reflects the n-out-of-k redundancy of the DC bus capacitors. The remaining elements are all essential for the operation of the unit. There are a few elements that are not critical for the operation, yet they do require replacement or repair relatively quickly for continued safe operation of the equipment. These elements are included in the series model.

Based on the gathered reliability information, we have a mix of distributions describing the data. Therefore, we used a Weibull distribution for critical parts fans, IGBTs, and DC, and AC capacitors to make closer reliability estimations, and we used an exponential distribution for the remaining parts. The existing reliability information includes predictions based on characteristic life (η) and the slope (β) to calculate the reliability for each block based on the reliability formula:

$$R(t) = \mathrm{e}^{-(t/\eta)^{\beta}} \tag{11.20}$$

Also, to calculate availability, the mean time to repair for each block was captured from field data. Then, by entering both reliability and time to repair for each block in ReliaSoft's BlockSim™ tool and assuming inverter operation of 8 h/day, a mean availability of 99.97% was calculated over a 10-year period. The mean availability $\overline{A(t)}$ is the proportion of time during a mission or time period that the system is producing power:

$$\overline{A(t)} = \frac{1}{t}\int_0^t A(u)\,\mathrm{d}u \tag{11.21}$$

$\overline{A(t)}$ is calculated based on reliability $R(t)$ and mean time to repair, $m(u)$:

$$A(t) = R(t) + \int_0^t R(t-u)m(u)\,\mathrm{d}u \tag{11.22}$$

Figure 11.10 provides the reliability graph as a function of time, $R(t)$, based on using BlockSim™ simulations.

Note that in all calculations throughout this study, conditions of 45 °C, 480 VDC, and full power were used as the baseline; these provide conservative theoretical calculations.

11.5.4.1 Vendor data reliability calculation example

As an example of the calculations based on vendor data, let us review the fan calculations used in the RBD at this point. There are three types of fans in the inverter; here, we only consider one of the three.

The most likely failure mechanism for a properly constructed and installed fan is the failure of the bearing grease, which leads to an eventual increase in bearing friction and the loss of fan operation. Using the work of Xijin [20] on cooling fan reliability modeling, we can translate the provided vendor data to a Weibull life distribution. An AF formula suitable for cooling fans is

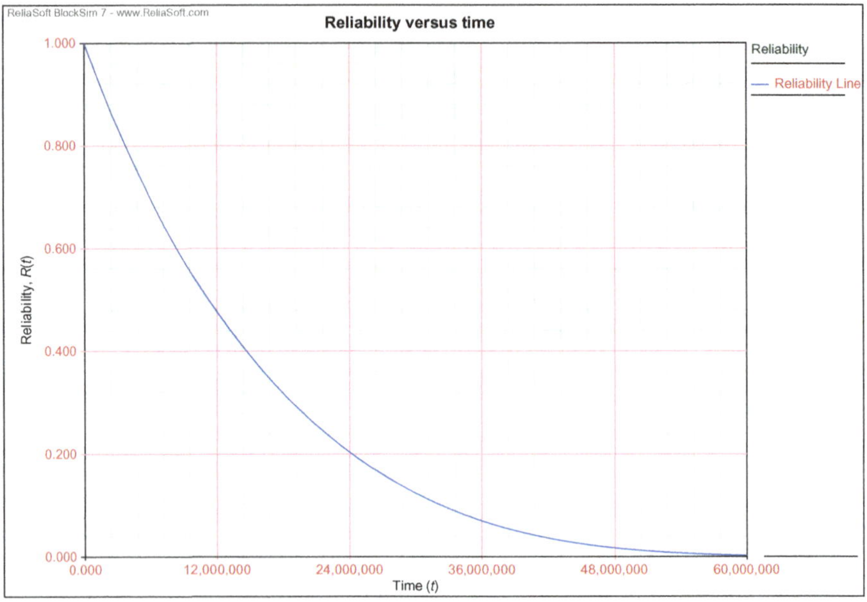

Figure 11.10 System reliability versus time.

$$AF = 1.5^{[(T_{test} - T_{field})/10]} \tag{11.23}$$

where T_{test} is from the datasheet and T_{field} is the expected use condition temperature. Let us assume the fan datasheet also provides an L_{10} value of 57,323 h, which represents the expected duration until 10% of the units are expected to fail at 40 °C. Using Equation (11.23), we determine the AF to convert the L_{10} duration to the value corresponding to 45 °C, which is 45,310 h.

With the L_{10} at the appropriate temperature, we then perform some algebra on the Weibull distribution of Equation (11.20) to determine the η value:

$$\eta = \frac{-t}{\sqrt[\beta]{\ln[R(t)]}} \tag{11.24}$$

Inputting the known $R(t)$ of 0.10 and including the L_{10} duration at 45 °C for time t, we can calculate the characteristic life, η, of the Weibull distribution for the fan operated at 45 °C.

The one missing element is β, for which Xijin recommends the use of a value of 3.0. Therefore, with β of 3.0, we can calculate the η value using the adjusted vendor L_{10} data at the expected use temperature as 95,933 h.

The fan Weibull values for the expected use temperature of 45 °C are then input for the fan within the RBD using ReliaSoft's BlockSim software package.

11.5.5 Using ALT

ALT provides a means to estimate the failure rate over time of a product without resorting to normal use conditions and the associated duration. For example, solar PV cells should operate for 25 years without failure. The product development time is less than a year for a new panel and the team wants to estimate the reliability of the cells over the 25-year duration.

ALT is a tool for the estimation of reliability for a specific failure mechanism. The results are then included in an overall system reliability model to improve reliability estimates for the system. ALT is necessary when there isn't a suitable model available. This applies in the solar power industry as innovation continues to create novel structures and systems.

A simplistic way to determine the reliability is to set up and operate solar systems for 25 years and track the number of failures over time. This isn't practical or useful.

Another method is to only test units as above for 1 year and then use the available information to make a decision. While product failures may occur (most likely due to manufacturing and installation errors), it will not be useful for delimitation of coatings or sealants, solder joint fatigue, PV cell degradation, and other longer-term failure mechanisms. While we could make an estimate, it would not be very accurate beyond the 1 year of use replicated in the testing.

Accelerated testing uses a variety of methods to compress time. Using the solar cell example, if UV radiation degrades one of the coatings, we can "accelerate" this by exposing the cells and coatings to UV radiation more often per day than in normal use. If in normal use, the panel would have 10 h of direct daylight per 24 h day, and then using UV lamps, we could expose the units 24 h per day, for a 2.4 × acceleration. Thus, instead of 25 years, we would require a little more than 10 years for the testing.

If we explore the UV damage mechanism, it may be part of a chemical reaction that clouds the coating. This chemical reaction most likely can be accelerated with temperature (the Arrhenius rate equation). With a little work, we find the expected coating temperature during use is 40 °C or less 90% of the time. By increasing the temperature (and not melting the coating), we may find another 10 × acceleration. The specifics of the acceleration are related to the activation energy of the specific chemical reaction, the testing temperature, and the use temperature.

Therefore, with a little understanding of the use conditions (10 h of daylight per day) and the specific failure mechanism in question (UV- and temperature-driven chemical reaction), we can achieve a 24 × acceleration factor. This will still take a little more than a year to fully test out to 25 years, yet the team may be able to use the data out to 24 years as a rough estimate and update when the final results become available.

This is just a short example and mainly an example of my approach to ALT. There is a lot more to ALT design, yet the most important by far is the understanding of the failure mechanism.

11.6 ALT example: how to craft a thermal cycling ALT plan for SnAgCu (SAC) solder failure mechanism

11.6.1 Objective

The objective is to provide a thermal cycling accelerated life test plan for your product to estimate a portion of the system's reliability performance. Solder joints are a part of the electronics within controllers, drivers, inverters, and other elements of a solar power system.

11.6.2 ALT plan summary

Electronic products primarily failure over time as a result of thermal cycling or thermal aging stresses. The ALT plan summary provides the thermal cycling ALT plan that will demonstrate a specified reliability.

Let's consider planning for two products, product A and product B. The resulting test plan is summarized here:

Product A ALT: To demonstrate at least 95% reliability with 90% confidence within the expected operating environment, 12 product A systems should operate within a thermal chamber cycling from 0 to 100 °C with 20 min dwells at extremes for 54 cycles.

Product B ALT: To demonstrate at least 95% reliability with 90% confidence within the expected operating environment, 45 product B systems should operate within a thermal chamber cycling from 0 to 100 °C with 20 min dwells at extremes for 18 cycles.

The differences are primarily due to the operating conditions for the two products. The discussion that follows provides a framework for the development of any ALT. Solder joints provide a convenient example as all solar power systems rely to some extent on circuit boards.

11.6.3 Background

The reliability of electronics continues to improve. Models of the dominant failure mechanisms continue to evolve. These models enable us to leverage failure mechanism knowledge to translate accelerated test results to expected normal use predictions. If the design and assembly of an electronic product are done well, the expected failure mechanisms for commercial electronic products are typically accelerated by thermal cycling or high temperature.

Thermal cycling of electronic products causes strain across any joint with dissimilar coefficients of thermal expansion. Thermal cycling may cause degradation and failure of mechanical seals, component package failures, internal component failures, PCB delamination, PCB cracking, or solder attach failures. Depending on the details of the product design, use, and technology, solder attach is often the dominate failure mechanism related to thermal cycling.

11.6.4 ALT approach

The acceleration factor estimates the equivalent damage accumulated during use thermal cycling while experiencing the ALT conditions. Considering the product's expected environment and use profile, we can determine the acceleration factor for test conditions as related to use conditions with the N-L equation. Vasudevan and Fan updated the coefficients of the N-L equation for its use on a lead-free SAC soldering system (Vasudevan and Fan, 2006).

Once we know the equivalent test duration (number of test cycles) needed to represent the expected use lifetime (design life), we can determine the sample size for the test.

The Weibayes extended-success testing sample-size formula is based on the binomial distribution when there are no failures. It also uses the shape, the β parameter from the underlying failure mechanism. For SAC solder joint fatigue, we conservatively set β to 2. The formula provides the number of samples to test in the ALT.

If no samples fail during the specified number of cycles, ALT results enable us to state with confidence C that the product will have at least reliability R over one lifetime. If there are failures during the testing, the thermal cycling should continue till at least five failures occur. Then, using conventional life data analysis techniques, we can estimate the expected use life distribution.

11.6.5 Thermal cycling ALT plan details

Product A is expected to be powered 24 h per day, 7 days a week (24/7) and active when working with product B. There is an approximate 19 °C product A internal temperature rise when it is active and a 13 °C rise when it is quiescent. Likewise, product B is battery-powered 24/7 and active when interacting with product A. Product B does not undergo a perceptible temperature rise.

11.6.6 Environmental conditions

11.6.6.1 Measured temperature rise

To estimate the use condition thermal cycling range, we include the expected temperature rise when active, 19 °C, to the high temperature of the day. For the low temperature, we assume that the unit is not active, in this case, with an approximate temperature rise of 13 °C.

11.6.6.2 Climatic data

To estimate the diurnal temperature range, we use data from a set of 20 weather stations (representing populated areas around the world) from 1 July 2005 to 1 July 2010 from the worldwide list of stations listed at the National Climatic Data Center [19]. The resulting 162,000 lines of daily data readings include the minimum and maximum temperatures. The average temperature range is 10.9 °C and the standard deviation is

5.8 °C for this data set. From these data, we estimated a 90th percentile diurnal temperature range of 19 °C and a 95th percentile diurnal temperature range of 21.1 °C.

The 90th percentile maximum temperature is 31 °C and the 10th percentile minimum temperature is 19.4 °C. Considering that additional material changes may occur with higher temperatures, we base the thermal cycling on the 90th percentile high temperature and fix the range from 31 to 12 °C for the outdoor ambient daily temperature change.

11.6.6.3 Product A average daily thermal range

We start with 31-12 °C for the outdoor ambient daily temperature change, and we add 13 °C to the lower temperature and 19 °C to the higher value to determine the range during use. We find that 50-25 °C thus provides an expected daily temperature change of 25 °C, with a maximum temperature of 50 °C.

11.6.6.4 Product B average daily thermal range

We start with 31-12 °C for the outdoor ambient daily temperature change, and since there is no appreciable temperature rise within the product, that is the expected range. Therefore, 31-12 °C thus provides an expected daily temperature change of 19 °C, with a maximum temperature of 31 °C.

11.6.7 Temperatures

The environmental temperature range and 90th percentile temperature values from the previous section provide the temperature values. For the test temperatures, a common and economical range is 0-100 °C. The lower temperatures cost more in liquid nitrogen to achieve. The high temperature may require modification if there are any performance limiters or safeguards or any susceptible component materials.

11.6.8 Dwell times

Under operating conditions, the temperature varies daily, and the temperature extremes may occur for approximately 4 h (240 min) each day. Under testing conditions, we want sufficient dwell time to permit damage to occur. A common value for SAC solder dwell time is 20 min or more.

11.6.9 Acceleration factor determination

11.6.9.1 Product A acceleration factor

Using the N-L equation (11.6) with the following values:

$T_{\text{max,o}} = 50$ °C (or 323.15 K),
$T_{\text{min,o}} = 25$ °C,
$\Delta T_{\text{o}} = 25$ °C,
$T_{\text{max,t}} = 100$ °C (or 373.15 K),

$T_{min,t} = 0\ °C$,
$\Delta T_t = 100\ °C$,
$D_o = 240$ min,
and
$D_t = 20$ min.
we obtain
AF = 69.5.

For a 5-year expected useful life, and a daily temperature change, that would be $365 \times 5 = 1825$ cycles. It will take approximately 27 test cycles to replicate a 5-year operating life.

11.6.9.2 Product B acceleration factor

Using the N-L equation (11.6) with the following values:

$T_{max,o} = 31\ °C$ (or 304.15 K),
$T_{min,o} = 12\ °C$,
$\Delta T_o = 19\ °C$,
$T_{max,t} = 100\ °C$ (or 373.15 K),
$T_{min,t} = 0\ °C$,
$\Delta T_t = 100\ °C$,
$D_o = 240$ min,
and
$D_t = 20$ min.
we obtain
AF = 219.5.

For a 5-year expected useful life, and a daily temperature change, that would be $365 \times 5 = 1825$ cycles. It will take approximately nine test cycles to replicate a 5-year operating life. (Note that with very high acceleration factors, it is advisable to double the number of test cycles to protect against any nonlinearity or errors in the model; therefore one would use 18 cycles to represent one lifetime.)

11.6.10 Sample-size determination

The basic test is a pass/fail test after a simulated 5 years of use under cyclic stress. Based on the binomial distribution and assuming no failures allow the determination of the sample size *n* for a given shape parameter (β), test duration, confidence *C*, and reliability *R* (proportion passing) [21],

$$n = \frac{\ln(1-C)}{m^{\beta} \ln(R)} \tag{11.25}$$

Here, *C* is the type I statistical confidence, which is kept at or above 0.6 and less than one; *R* is the proportion still operating at the designated duration (a value between zero and one with higher being more reliable); m is the number of lifetimes the test simulates (kept less than 5 unless there is a very clear understanding of the failure mechanism and no other failure mechanism occurs over the extended test time); and for SAC solder, we set $\beta=2$.

11.6.10.1 Product A sample size

Product A is expensive and there are limited prototypes available for testing. Given that the acceleration factor is reasonably large, it is possible to run the test for the equivalent of two lifetimes: $m=2$. Since the lifetime m is raised to the β value, it can significantly reduce the number of required samples. Using the following values:

$C=0.9$,
$m=2$ (meaning that the test should run for $27\times2=54$ cycles),
$\beta=2$ (a conservative value based on experience and the literature for SAC thermal cycling life),
and
$R=0.95$.
we obtain
$n=12$.

Thus, product A ALT can demonstrate a 90% confidence of at least a 95% reliability after 5 years of thermal cycling use conditions with 12 samples that all pass the final performance criteria.

11.6.10.2 Product B sample size

Product B is inexpensive and there are prototypes available for testing. Using

$C=0.9$,
$m=1$ (meaning that the test should run for 18 cycles),
$\beta=2$ (a conservative value based on experience and the literature for SAC thermal cycling life),
and
$R=0.95$.
we obtain
$n=45$.

Using more samples for one lifetime is preferable for both statistical and assumption benefits.

The product B ALT can demonstrate a 90% confidence of at least a 95% reliability after 5 years of thermal cycling use conditions with 45 samples that all pass the final performance criteria.

11.6.11 Assumptions

The dominant failure mode induced by thermal cycling is SAC solder joint fatigue. The daily temperature changes are dominated by the ambient air diurnal temperature changes.

11.7 How to craft a temperature, humidity, and bias ALT plan for CMOS metallization corrosion

11.7.1 Objective

The objective is to provide a temperature, humidity, and bias (THB) ALT plan for your product to estimate a portion of the system's reliability performance. Complementary metal-oxide-semiconductor (CMOS) is a common construction for microprocessors, memory and other digital logic, it is also a base structure for silicon-based solar cells. Second only to thermal cycling as the leading cause of electronics failure is the exposure to relatively high temperatures and humidity while operating. Since most solar power system electronics receive little protection from ambient temperature and humidity, understanding the expected lifetime for components susceptible to THB-induced failures will improve the system reliability model.

The following discussion provides another example of a framework for the creation of a meaningful ALT study. Again, we will use the fictitious product A and product B to illustrate that one test is not suitable for all situations.

11.7.2 ALT plan summary

Electronic products primarily fail over time as a result of thermal cycling or thermal aging stresses. The ALT plan summary provides the THB ALT plan that will demonstrate a specified reliability.

To demonstrate at least 95% reliability over 5 years with 90% confidence within the expected operating environment, 23 product As should operate within a thermal- and humidity-controlled chamber set to expose the circuit board within the product to 90 °C and 90%RH for 1750 h.

To demonstrate at least 95% reliability over 5 years with 90% confidence within the expected operating environment, 45 product Bs should operate within a thermal- and humidity-controlled chamber set to expose the circuit board within the product to 90 °C and 90%RH for 232 h.

11.7.3 Background

The combination of THB provides the stresses to accelerate a wide range of failure mechanisms. Some are primarily accelerated by temperature alone; with the addition of humidity and bias, a wider range of failure mechanisms become accelerated. For example, some metal migration or corrosion mechanisms require a voltage gradient to occur.

11.7.4 ALT approach

Peck and Hallberg and Peck studied the effects of THB over many experiments. They then fit an Eyring model with two stress items to the combined data. The resulting Peck's equation has an inverse power law model for humidity and an Arrhenius model for temperature. Note that the units under test are powered and active in order to create voltage gradients across as many elements of the product as possible. Peck's equation permits the calculation of the THB test's acceleration factor. The acceleration factor estimates the equivalent damage accumulated during use conditions while experiencing the ALT conditions.

Considering the product's expected environment and use profile, we can determine the acceleration factor for test conditions as related to use conditions with Peck's equation. Once we know the equivalent test duration needed to represent the expected use lifetime (design life), we can determine the sample size for the test.

The Weibayes extended-success testing sample-size formula is based on the binomial distribution when there are no failures. It also uses the shape (β) parameter from the underlying failure mechanism. For metallization corrosion, we conservatively set β to 1. The formula provides the number of samples to test in the ALT protocol.

If no samples fail during the specified number of cycles, the ALT results allow us to state with confidence C that the product will have at least reliability R over one lifetime. If there are failures during the testing, the thermal cycling should continue till at least five failures occur. Then, using conventional life data analysis techniques, we can estimate the expected use life distribution.

11.7.5 THB ALT plan details

Product A is expected to be powered 24/7 and is active when working with product B. There is an approximate 19 °C product A internal temperature rise when it is active and a 13 °C rise when it is quiescent. Likewise, product B is battery-powered 24/7 and is active when interacting with product A. Product B does not exhibit a perceptible temperature rise.

11.7.6 Environmental conditions

11.7.6.1 Measured temperature rise

To estimate the use condition thermal cycling range, we add the expected temperature rise when active, 19 °C, to the high temperature of the day. Since higher temperature promotes the failure mechanisms, we will assume the unit is always at the higher temperature. This is conservative and provides some margin for experimental error, model nonlinearities, etc.

11.7.6.2 Climatic data

To estimate temperature and humidity values, we used data from a set of 20 weather stations (representing populated areas around the world) from 1 July 2005 to 1 July 2010 from the worldwide list of stations listed at the National Climatic Data Center [19]. The resulting 162,000 lines of daily data readings include the minimum and maximum temperatures and dew point. The estimated 90th percentile maximum temperature is 31 °C. The estimated 90th percentile relative humidity is 80%RH.

The 90th percentile maximum temperature is 31 °C and the 10th percentile minimum temperature is 19.4 °C. Considering that additional material changes may occur with higher temperatures, we base the maximum temperature on the 90th percentile high temperature and add the internal temperature rise expected.

11.7.6.3 Product A average daily thermal range

The estimated 90th percentile maximum temperature of 31 °C plus the product A operating internal temperature rise of 19 °C brings the value to 50 °C. The estimated 90th percentile relative humidity is 80%RH. Because the unit is powered 24/7, the local RH within the unit will be less; therefore, using 80%RH will be conservative.

11.7.6.4 Product B average daily thermal range

The estimated 90th percentile maximum temperature is 31 °C. The estimated 90th percentile relative humidity is 80%RH. Because there is no appreciable internal temperature rise within product B, there is no adjustment to these values.

11.7.7 Acceleration factor determination

The 90th percentile temperature and humidity values from the previous section provide the use weather values. For the test temperatures and humidity, common and economical values are 90 °C and 90%RH. The test temperature may require modification if there are any performance limiters or safeguards or any susceptible component materials. The test temperature is near the circuit board operating temperature while in the chamber. The chamber set point may be lower to accommodate the internal temperature rise.

11.7.7.1 Product A acceleration factor

Using Peck's equation (11.8) with the following values:

$T_o = 50$ °C (or 323.15 K),
$RH_o = 80\%RH$,
$T_t = 90$ °C (or 363.15 K),
and
$RH_t = 90\%RH$.
we obtain
$AF = 50$.

For a 5-year expected useful life, and under full-time high-humidity and high-temperature conditions, that would be $365 \times 5 \times 24 = 43{,}800$ h. Therefore, it will take approximately 875 h at 90 °C and 90%RH to replicate a 5-year operating life.

11.7.7.2 Product B acceleration factor

Using Peck's equation with the following values:

$T_o = 31$ °C (or 304.15 K),
$RH_o = 80\%RH$,
$T_t = 90$ °C (or 363.15 K),
and
$RH_t = 90\%RH$.
we obtain
$AF = 377$.

For a 5-year expected useful life, and under full-time high-humidity and high-temperature conditions, that would be $365 \times 5 \times 24 = 43{,}800$ h. Therefore, it will take approximately 116 h at 90 °C and 90%RH to replicate a 5-year operating life. With a large AF value, it is advisable to double the duration of the test to provide margin for any nonlinearity in the acceleration model, thus increasing the test time to 232 h.

11.7.8 Sample-size determination

The basic test is a pass/fail test after a simulated 5 years of use under cyclic stress. The formula in Equation (11.25), based on the binomial distribution and with no failures assumed, enables the determination of the sample size for a given shape parameter (β), test duration, confidence, and reliability (proportion passing).

11.7.8.1 Product A sample size

Product A is expensive and there are limited prototypes available for testing. Given that the acceleration factor is reasonably large, it is possible to run the test for the equivalent of two lifetimes, $m = 2$. Using

$C = 0.9$,
$m = 2$ (meaning that the test should run for $875 \times 2 = 1750$ h),
$\beta = 1$ (a conservative value based on experience and the literature for metallization corrosion life),
and
$R = 0.95$.
we obtain
$n = 23$.

The product A ALT can demonstrate a 90% confidence of at least a 95% reliability after 5 years of thermal cycling use conditions with 23 samples that all pass the final performance criteria.

11.7.8.2 Product B sample size

Product B is inexpensive and there are prototypes available for testing. Using

$C = 0.9$,
$m = 1$ (meaning that the test should run for 232 h),

$\beta = 1$ (a conservative value based on experience and the literature for metallization corrosion life),

and

$R = 0.95$.

we obtain

$n = 45$.

Using more samples for one lifetime is preferable for both statistical and assumption benefits.

The product B ALT can demonstrate a 90% confidence of at least a 95% reliability after 5 years of thermal cycling use conditions with 45 samples that all pass the final performance criteria.

11.7.9 Assumptions

The dominant failure mode induced by THB is CMOS metallization corrosion.

11.8 Developments and opportunities

Commercially available solar systems have only a few installations that have operated for 20+ years. While the information about system failures has been valuable, it is often with technology that has been replaced and improved. The pace of development continues to focus on improving efficiency and lowering cost per watt. This continues to reset the reliability experience to nil based on installed systems for major elements of systems.

The development of PoF models for failure mechanisms continues. These models provide accurate time to failure behaviors for specific failure mechanisms. These models supply the needed information for RBDs of complete systems.

The opportunity for major improvements relies on validating the long-term reliability performance of models and ALT results. The risk of missing a failure mechanism remains relatively high until we have the experience of time.

Discovering failure mechanisms for new technologies is another opportunity. Currently, this information is considered propriety, yet the sharing of failure mechanism data may lead to significant breakthroughs in modeling, life estimation, and design improvements.

International standards provide some structure and do not provide meaningful support for communicating system and component reliability. The pace of technology development will keep the standards organization behind, yet improvements in how we communicate reliability models, objectives and assessments would be helpful.

The pace of installation of solar power generation systems continues to set records. According to the Energy Information Administration, in 2012, there was 3.5 million MWh of electricity generated by solar PV panels. In 2013, that more than doubled to

8.3 million MWh, which come to think of it, a decade ago the United States generated just 6000 MWh from solar PV [22].

The pace of installation and the rate of technical development both require improvements in system and component reliability knowledge.

11.9 Conclusions

Basic models and accurate time to failure estimates enable decision makers from banking institutions to system operators to make informed decisions. The role of reliability engineering across the life cycle of solar systems continues to grow in importance for the success of the industry.

Using the modeling structures outlined in this chapter along with carefully executed accelerated life tests is an excellent start for the clear communication of system reliability performance.

As the installed systems age, we learn how the systems fail over time. We learn which models and estimates were correct or not. We continue to make improvements.

As the installed base of systems grows and as the pace of technology eventually slows, we will have the experience of validated models to provide accurate life estimates. The ability to forecast system life and costs further reduces the risks associated with the installation of solar power generation systems.

11.10 Sources of further information

The essential textbooks for the reliability engineering elements of this chapter are:

- P.D.T. O'Connor, A. Kleyner, Practical Reliability Engineering, John Wiley and Sons, Chichester, 2012.
- W. Nelson, Accelerated testing: statistical models, test plans, and data analysis, in: S.S. Wilks Samuel (Ed.), Wiley Series in Probability and Mathematical Statistics, John Wiley & Sons, New York, 1990.

Online references for reliability and statistical topics include:

- NIST/SEMATECH, e-Handbook of Statistical Methods, http://www.itl.nist.gov/div898/handbook/, April 28, 2014.
- ReliaSoft Reliawiki, http://www.reliawiki.org/index.php/Main_Page, April 28, 2014.
- The Reliability Calendar, http://reliabilitycalendar.org, April 28, 2014.

References

[1] Fraunhofer ISE, World record solar cell with 44.7% efficiency, press release (September 23), Available from: http://www.ise.fraunhofer.de/en/press-and-media/press-releases/presseinformationen-2013/world-record-solar-cell-with-44.7-efficiency, 2013 (accessed 14.10.13).

[2] M. Cichon, Solar tracking systems gain ground, Renewable Energy World.com, Available from: http://www.renewableenergyworld.com/rea/news/article/2013/06/on-track-to-succeed-with-solar-tracking-systems, 2013 (accessed 11.11.13).
[3] Sandia National Laboratories, Inverter Reliability Program, Sandia, New Mexico, 2012. Available from: http://energy.sandia.gov/?page_id=12998. http://news.pv-insider.com/photovoltaics/improving-pv-plant-inverter-reliability (accessed 15.10.13).
[4] U.S. Energy Information Administration, Table 1. Estimated levelized cost of electricity (LCOE) for new generation resources, 2019, Annual Energy Outlook 2014 (2014). Available from: http://www.eia.gov/forecasts/aeo/electricity_generation.cfm (accessed 28.04.14).
[5] O.W. Holmes, The One Hoss Shay, Houghton, Mifflin and Company, Boston and New York, 1858. For a complete copy of the poem see http://www.fmsreliability.com/education/one-hoss-shay/.
[6] P.D.T. O'Connor, A. Kleyner, Practical Reliability Engineering, fifth ed., John Wiley and Sons, Chichester, 2012.
[7] V. Vasudevan, X. Fan, An acceleration model for lead-free (SAC) solder joint reliability under thermal cycling, in: 58th Electronic Components and Technology Conference, IEEE, Orlando, 2008, pp. 139–145.
[8] O. Hallberg, D.S. Peck, Recent humidity accelerations, a base for testing standards, Qual. Reliab. Eng. Int. 7 (3) (1991) 169–180.
[9] Reliability Information Analysis Center, WARP, Department of Defense Information Analysis Center, Available from: http://www.theriac.org/WARP/index.php, 2013 (accessed 12.12.13).
[10] CALCE, Simulation Assisted Reliability Assessment (SARA®), University of Maryland, College Park, MD, 2013. Available from:http://www.calce.umd.edu/software/ (accessed 12.12.13).
[11] DFR Solutions, Discover Sherlock!, DFR Solutions, College Park, MD, 2013. Available from: http://www.dfrsolutions.com/software (accessed 12.12.13).
[12] W.M. Meeker, G.H. Hahn, How to Plan an Accelerated Life Test, ASQC Quality Press, Milwaukee, WI, 1985.
[13] C. Hillman, CEO DFR Solutions, Personal communication, Phone discussion about beta based on DFR Solutions testing of similar capacitors and finding relatively high Weibull beta values, 7–8, 2010.
[14] S.G. Parler Jr., Deriving life multipliers for electrolytic capacitors, IEEE Power Electron. Soc. Newslett. 16 (1) (2004) 11–12.
[15] M. Stoike, Development Engineer, Electronicon, Personal communication, Email message to author dated December 14, 2010.
[16] CapSite, Reliability issues, Available from: http://my.execpc.com/~endlr/reliability.html, 2009 (accessed 17.02.11).
[17] R.A. Cooper, J.B. Ennis, D.L. Smith, ZR Marx capacitor life test and production statistics, in: Pulsed Power Conference, IEEE, Monterey, CA, 2005, pp. 1219–1222.
[18] A.C. Micol, C. Martin, O. Dalverny, M. Mermet-Guyennet, M. Karama, Reliability of lead-free solder in power module with stochastic uncertainty, Microelectron. Reliab. 49 (6) (2009) 631–641.
[19] NCDC, National Climatic Data Center, U.S. Department of Commerce, Available from: http://www7.ncdc.noaa.gov/CDO/cdoselect.cmd?datasetabby=GSOD&countryabby=&georegionabby=, 2012 (accessed 06.07.12).
[20] T. Xijin, Cooling fan reliability: failure criteria, accelerated life testing, modeling and qualification, in: Annual Reliability and Maintainability Symposium, RAMS '06, IEEE, Newport Beach, CA, 2006, pp. 380–384.

[21] G.S. Wasserman, Reliability Verification, Testing, and Analysis in Engineering Design, Marcel Dekker, New York, 2003.

[22] EA, Renewable sources: total (all sectors), 2004-February 2014, Table 1.1.A, Net Generation from Electric Power Monthly, U.S. Energy Administration, Available from: http://www.eia.gov/electricity/monthly/epm_table_grapher.cfm?t=epmt_1_01_a, 2014 (accessed 13.03.14).

Index

Note: Page numbers followed by *f* indicate figures and *t* indicate tables.

A

Accelerated life testing (ALT)
  availability modeling, 234–236, 234*f*, 236*f*
  block modeling, 9
  expected failure mechanisms, 229
  extrapolation, 85
  HALT, 19
  lifetime data, 229
  optoelectronic devices, 84–85
  PoF, 29–30, 36
  reliability calculations, 229–234
  stress and expected life, 227–228
  thermal cycling, 238–242
  time compression, 227
  vendor data reliability calculation, 235–236
Acceleration factor determination
  CMOS metallization corrosion, 245–246
  SAC solder failure mechanism, 240–241
AFM. *See* Atomic force microscopy (AFM)
ALT. *See* Accelerated life testing (ALT)
ARM's Cortex-R4 processors, 178, 178*f*
Atomic force microscopy (AFM), 60–61
Automotive electronics industry, 8
Automotive integrated circuits (ICs) designs, 195

B

Back-end failure mechanisms
  dielectric breakdown, 156, 157*f*
  electromigration, 154–155, 155*f*
  SIV, 155–156, 156*f*
  stress migration, 155–156, 156*f*
Ball grid array (BGA), 36–37
BH. *See* Buried heterostructure (BH)
Bias temperature instabilities (BTI)
  impact of, 123–124, 124*f*
  NBTI, 76, 121–123, 122*f*
  PBTI, 123
Bipolar junction transistor (BJT), 75–76
BJT. *See* Bipolar junction transistor (BJT)
BTI. *See* Bias temperature instabilities (BTI)
Buried heterostructure (BH)
  activation energy, 88, 89*f*
  BH interface effects, 96
  degradation patterns, 90–91, 91*f*
  ESD threshold, 107–108, 109, 109*f*
  input electrical power, 87
  melt-back process, 86–87
  MOCVD, 86–87
  n-type InP cladding, 86–87
  optical output power, 87
  potential and dominant factors, 91
  power change data, 88–89, 90*f*
  p-type InGaAs layer, 86–87
  random failure, 86–87
  schematic diagram, 86–87, 86*f*
  soft and hard ESD degradation, 106–107, 106*f*, 107*f*
  substrate quality effects, 97–98, 97*f*, 98*f*
  telecommunication tunable lasers, 86–87
  thermal characteristics, 88, 88*t*
  threshold current *vs.* life-test aging, 85–86, 86*f*
  wall plug efficiency, 87–88
  wavelength-drift specification, 89, 90*t*
  wavelength shift data, 88–89, 89*f*, 90*f*
  WDM network, 87

C

calcePWA software, 34–35
Calibre PERC tool
  advantages, 200–201
  circuit recognition, 201–202, 203*f*
  current density, 202–204, 203*f*, 204*f*
  LDL-CD, 202, 203*f*, 204*f*
  LDL-DRC, 204–205, 205*f*, 206*f*
  LDL-P2P, 202, 204, 204*f*, 205*f*
  point-to-point checking, 202–204, 204*f*, 205*f*
  voltage-dependent DRC, 205–208, 206*f*, 207*f*

Calibre PERC tool *(Continued)*
voltage propagation, 201, 202*f*
Chemical analytic techniques
infrared spectroscopy, 46–48
ion chromatography, 46
mass spectroscopy, 49–50
Raman spectroscopy, 48–49
CMOS metallization corrosion
acceleration factor determination, 245–246
ALT plan summary, 243
environmental conditions, 244–245
failure mechanisms, 243
THB ALT plan details, 244
THB effects, 243–244
Coffin-Manson equation, 30–31, 224, 233
Component and system failures, 219
Concurrent engineering (CE), 64

D

Design for reliability (DfR), 7, 64–65, 226
Design rule checking (DRC) tools, 195
Destructive techniques
chemical analysis (*see* Chemical analytic techniques)
cutting, 44
depotting, 45
focused ion beam sample preparation, 51
grinding and polishing, 44, 45
jet etching, 45
potting/encapsulation, 44
SEM imaging, 50–51
TEM, 52
DfR. *See* Design for reliability (DfR)
Discrete electronic components, 6, 7
burn-in and screening tests, 65–66
DfR, 64–65
ESD-induced failure, 78
ESD robust systems, 78–79
failure modes (*see* Failure mechanisms (FM))
process reliability, 65
DRC tools. *See* Design rule checking (DRC) tools

E

EDA. *See* Electronic design automation (EDA)
Electrical and electronic equipment (EEE), 3, 6
AFM, 60–61
conductivity/resistivity measurement, 58–59
degradation, 58
destructive (*see* Destructive techniques)
nondestructive (*see* Nondestructive techniques)
passive component measurement, 59
profilometer surface roughness, 60
Electrical overstress (EOS) susceptibility, 101, 198–199
Electrical rule checking (ERC) tools, 195
Electromigration (EM), 171, 172–173
Al interconnects, 126–127, 128*f*
back-end failure mechanisms, 154–155, 155*f*
damascene Cu/SiO_2 and Cu/low-*k* structures, 126–127, 127*f*, 128–129, 128*f*
failure time, 127–128
grain structure, 127–128
MTTF, 31–32
optoelectronic devices, 83
Schottky diodes, 73
test structures, 127
thin-film IC interconnects, 126–127
Electronic design automation (EDA)
current density checking, 198, 200*f*
ESD protection circuits, 198, 199*f*
latch-ups, 198
negative voltage checking, 197–198
operational challenges, 195–196
reverse current checking, 197–198
voltage-dependent checking, 196–197
Electrostatic discharge (ESD)
BH and RWG lasers, 109, 109*f*
configuration, 209–211
damage characteristics, 102–103, 102*f*, 103*f*
DFB lasers, 101
discrete electronic components, 78–79
DWDM application, 101
EOS, 101
failure threshold (*see* Polarity effect, ESD)
hard degradation, 106–107, 106*f*, 107*f*
and latch-up verification, 198
sensitivity classifications, 101–102, 102*t*

size effect, 107–108, 108*f*
soft degradation, 106–107, 106*f*
TR-NWT-000870, 101–102
EM. *See* Electromigration (EM)
Embedded systems, 8
application checkpointing, 177, 177*f*
audio decoder, 188, 188*f*
critical charge, 173, 173*f*
cross-layer system adaptation, 187–188, 187*f*
design trade-offs, 179–180, 180*t*
EDAC codes, 169, 176–177, 176*f*
failure rates, 172–173, 173*f*
hardware design automation, 189–190
hardware faults, 170–173, 170*f*
latches/flip-flops fault detection, 181, 181*f*
programming models, 189
reexecution, 177, 177*f*
state monitoring and recovery, 181–182, 182*f*
technology scaling, 172–173, 173*f*
video decoder, 188, 188*f*
voltage scaling, 173, 173*f*
webpage rendering, 188, 188*f*
ERC tools. *See* Electrical rule checking (ERC) tools
Error detection and correction (EDAC), 169
Error hardening methods, 175–176
ESD. *See* Electrostatic discharge (ESD)
Extrapolation model, 85, 98–100, 160–161

F

Failure mechanisms (FM)
aluminum electrolytic capacitors, 67–68
Avalanche diodes, 72
back-end (*see* Back-end failure mechanisms)
BIT, 75–76
cooling fans, 235–236
discrete electronic components, 67
front-end (*see* Front-end failure mechanisms)
gunn diodes, 74
HBT, 77
HEMT, 77–78
IMPATT diodes, 73–74
integrated circuits, 148–149, 149*f*
memory technology, 158–159, 159*f*
modeling (*see* Reliability modeling)
MOS, 76–77
package-level, 157–158, 157*f*
PIN diodes, 74
PoF models, 247
probability distribution, 219, 220*f*
reliability modeling, 224
SAC solder (*see* SAC solder failure mechanism)
Schottky diodes, 72–73
stress effects, 228
switching diode, 72
tantalum capacitor, 69
temperature effects, 221–222
THB, 243
thermal cycling-induced effects, 233
tunnel diodes, 74
varactor, 71–72
wear-out (*see* Wear-out failure mechanisms)
Z diodes, 72
Failure modes, mechanisms, and effects analysis (FMMEA), 29, 32
definition, 32–33
documentation process, 34
fundamental unit, 33
high-risk and low-risk ratings, 34
identification, 33–34
loading conditions, 33
methodology, 32–33, 33*f*
FM. *See* Failure mechanisms (FM)
Fowler-Nordheim conduction, 32, 120
Front-end failure mechanisms
EELS, 151–152, 152*f*
hard breakdown, 150–151, 151*f*
HCI, 154, 154*f*
NBTI, 152–154, 153*f*
PBD, 150–151
PBTI, 152–154, 153*f*
percolation path, 149–150, 150*f*
physical defect signature, 151–152, 152*f*
RTN fluctuations, 149–150, 150*f*
soft breakdown, 150–151, 151*f*
TDDB, 149–150, 150*f*
Fujitsu SPARC64 processors, 178–179, 179*f*

H

Haitz's law, 83
Heterojunction bipolar transistors (HBT), 77
High electron mobility transistors (HEMT), 77–78
Highly accelerated life testing (HALT), 19
Hot carrier injection (HCI), 154

I

ILD. *See* Interlevel dielectric (ILD)
Impact ionization avalanche transit-time (IMPATT) diodes, 73–74
Infrared spectroscopy, 46–49
Insulated-gate bipolar transistors (IGBT), 233–234
Interlevel dielectric (ILD)
  Al interconnects, 129–130
  debate models, 130–131
  failure time distributions, 130, 131*f*
  Frenkel-Poole mechanism, 130–131
  low-*k* dielectric and dielectric capping layer, 129–130, 130*f*
  RC performance requirements, 131
Ion chromatography, 46

J

Jet etching, 45

L

Layout *vs.* schematic (LVS) tools, 195

M

Mass spectrometers (MSs)
  detector, 49
  ionizer, 49
  mass analyzer, 49
  pulsed glow discharge, 50
  TOF-SIMS, 50
Mean time between failures (MTBF), 18
Metal organic chemical vapor deposition (MOCVD), 86–87

N

Nanodevices, 7. *See also* Nanoelectronic devices
  architectural evolution, 148, 148*f*
  logic unit, 144*f*, 145–146
  material evolution, 146–147, 147*f*
  memory unit, 143–146, 145*f*
  p-type and n-type transistors, 143
Nanoelectronic devices
  accelerated and field-stress conditions, 161–162, 161*f*
  extrapolation, 160–161
  failure mechanisms (*see* Failure mechanisms)
  intrinsic and extrinsic issues, 160
  PHM, 162
  physical analysis, 162
  SM-EM, 160
  statistical models, 160
  TDDB and NBTI, 160
  time-dependent variability, 162
Negative bias temperature instability (NBTI), 76, 152–154, 171, 172–173
N-modular redundancy (NMR) method, 174–175
Nondestructive techniques
  acoustic microscopy, 57–58
  digital and integrated camera systems, 54
  infrared thermography, 55–56
  metallurgical microscopes, 53
  Nomarski interference techniques, 54
  stereo microscopes, 53
  substage and vertical illumination sources, 53–54
  transmission microscopes, 53
  transmitted and reflected light, 54
  visual inspection, 52
  X-ray imaging, 54–55
  XRF analysis, 56–57
Norris-Landzberg (N-L) equation, 224

O

Optoelectronic devices, 7
  ALT, 84–85
  vs. electronic devices, 83, 84*t*
  extrapolation, 85
  failure mechanisms (*see* Electrostatic discharge (ESD))
  Haitz's law, 83
  history, 83
  laser diodes and LED, 83
  light *vs.* forward drive current, 85
  Moor's law, 83
  photonic *vs.* electronic devices, 84–85, 84*t*

semiconductor laser (*see* Buried heterostructure (BH); Ridge waveguide (RWG))
sublinear model extrapolation, 98–100, 99*f*, 99*t*
temperature and current accelerations, 100, 101*f*
test structures, 85

P

Package-level failure mechanisms, 157–158, 157*f*
Parallel system RBD, 223, 223*f*
PBTI. *See* Positive bias temperature instability (PBTI)
Peck's equation, 225, 243–244, 245, 246
Permanent logic faults, 171
PHM. *See* Prognostics and health management (PHM)
Physics-of-failure (PoF) models, 5, 225, 226
accelerated tests, 28, 35, 36
BGA constructions, 36–37
canary device, 36
Coffin-Manson equation, 30–31
development, 247–248
DfR, 65
ESTs, 34
failure site, 29
fatigue failure, 29–30
Fowler-Nordheim conduction, 32
MTTF, 31–32
overstress failure mechanism, 29–30, 30*t*
PCB, 29
PHM, 36
reliability, 27–28
reliability modeling, 225
RUP prediction, 37
time-dependent dielectric breakdown, 32
TTF, 29
unreliable products, 28
vibration-induced stresses, 30
virtual qualification, 34–35
wear-out mechanism, 29–30, 30*t*, 31*t*
p-Metal contact effects
alloy contact, 91–92, 92*f*, 93–95, 93*f*, 94*f*
contact layer thickness, 96
InGaAs thickness, 96
plasma damage, 95, 95*f*
Schottky-type contacts, 91–92
PoF. *See* Physics-of-failure (PoF) models
Point-to-point parasitic resistance, 198, 200*f*
Polarity effect, ESD
Au/Ti/Pt/Au wafers, 104–105, 104*t*
Au/Zn wafers, 104–105, 104*t*
BH lasers, 105, 105*f*
forward, reverse, and forward/reverse polarities, 103–104, 104*f*, 105–106
Joule heating, 105, 106*t*
Positive bias temperature instability (PBTI), 123, 152–154, 153*f*
Prognostics and health management (PHM), 36, 162

R

Raman spectroscopy, 48–49
RAZOR flip-flop, 182–183, 182*f*
RBD. *See* Reliability block diagrams (RBDs)
Reactive ion etching (RIE), 95
Reliability block diagrams (RBDs), 215, 222, 222*f*, 223, 223*f*, 235, 236, 247
Reliability building. *See* Discrete electronic components
Reliability calculations
Arrhenius relationship, 230
IGBT, 233–234
maximum operating voltage, 230, 231
maximum rated temperature, 230
reliability function, 231
service life and base life information, 229–230
Weibull cumulative density function, 231–232, 232*f*
Reliability engineering
Barnard identification, 4
conventional viewpoint, 22–23, 23*f*
correct viewpoint, 23–24, 23*f*
design and production verification, 11–12, 12*f*
failure *vs.* success testing, 19
FMEA process, 21
ignoring cause and effect relationship, 17–18
inadequate integration, 13–14, 14*f*
incorrect personnel, 20–21

Reliability engineering *(Continued)*
incorrect time, 19–20
MTBF, 18
nonvalue-adding activities, 22
probability, 15–16
product reliability, 12
quantification, 16–17
valuable activities, 21–22
Reliability Information Analysis Center (RIAC), 225
Reliability modeling
Coffin-Manson equation, 224
failure mechanisms, 224
field failure data, 226
Norris-Landzberg (N-L) equation, 224
parallel system RBD, 223, 223*f*
Peck's equation, 225
PoF models, 225
Reliability Information Analysis Center (RIAC), 225
repairable systems, 226
series-parallel system RBD, 223–224, 223*f*
series system RBD, 222–223, 222*f*
Ridge waveguide (RWG)
ESD, 108*f*, 109, 109*f*
schematic diagram, 85–86, 86*f*
thermal characteristics, 88, 88*t*
wall plug efficiency, 88
RIE. *See* Reactive ion etching (RIE)

S

SAC solder failure mechanism
acceleration factor determination, 240–241
ALT plan summary, 238
dwell times, 240
environmental conditions, 239–240
expected use life distribution estimation, 239
product details, 239
temperatures, 240
thermal cycling, 238
Sample-size determination
CMOS metallization corrosion, 246–247
SAC solder failure mechanism, 241–242
Semiconductor laser
BH laser (*see* Buried heterostructure (BH))
RWG laser (*see* Ridge waveguide (RWG))
SER. *See* Soft errors (SER)
Series-parallel system RBD, 223–224, 223*f*
Series system RBD, 222–223, 222*f*
Sherlock software package, 226
Silicon integrated circuits, 7
Al wires and Au package interconnects, 125–126
bathtub curve, 117, 117*f*
Cu damascene trench, 125–126
defect-related failures, 117–118
deterministic approach, 116
geometry reduction, 125–126
historical progression, 125–126, 125*f*
long-term failure rate, 115, 116*f*
Moore's law, 115
probabilistic approach, 116
screening techniques, 117–118
wear-out (*see* Wear-out failure mechanisms)
SIV. *See* Stress-induced voiding (SIV)
Soft errors (SER)
DRAM package, 132
ECC, 132
scaling trends, 132–133, 132*f*
single event upsets, 131–132
SRAM and flip-flops, 132–133
virtual qualification, 133, 134*f*
Solar power generation system reliabilty
bankability, 220
costing, 215, 218–219
durability, 216
efficiencies, 216, 217*f*
failures, 219–220, 220*f*
fluid heating systems, 216–217
inverter reliability, 217
modeling (*see* Reliability modeling)
product testing, 220–222
tracking systems, 217
Standby-sparing method, 174–175, 175*f*, 179–180
Stress and expected life model, 227–228, 228*f*
Stress-induced voiding (SIV), 155–156, 156*f*
Stress migration (SM), 155–156, 156*f*
Stylus profilometry, 60
Sublinear model extrapolation

experimental curve, 98
life-test curve fittings, 99–100, 99*f*, 99*t*

T

TDDB. *See* Time-dependent dielectric breakdown (TDDB)
Technology readiness levels (TRL), 2–3, 2*f*
Temperature, humidity, and bias (THB) ALT plan. *See* CMOS metallization corrosion
Thermal cycling ALT plan. *See* SAC solder failure mechanism
Time-compression ALT approach, 227
Time-dependent dielectric breakdown (TDDB)
electrochemical model, 120
electromigration, 171
failure time *vs.* gate voltage, 120, 121*f*
hardware faults, 172–173
percolation theory, 120
soft and hard breakdown, 119–121, 119*f*
Timing faults, 171–172, 172*f*, 182–183
TMR. *See* Triple modular redundancy (TMR) method
Transient logic faults, 170–171
Transistor degradation
BTI, impact of, 123–124, 124*f*
historical progression, 118–119, 118*f*
hot carrier aging, 125
NBTI, 121–123, 122*f*
PBTI, 123
plasma damage, 118–119
TDDB, 119–121
threshold voltage, 118–119
Triple modular redundancy (TMR) method
applications, 174, 174*f*
limitations, 174, 174*f*
schematic diagram, 173–174, 174*f*
working principle, 173–174, 174*f*

V

Verification techniques, 9

W

Wear-out failure mechanisms
electromigration failures (*see* Electromigration (EM))
ILD, 129–131, 130*f*, 131*f*
PoF, 29–30, 30*t*, 31*t*
SER, 131–133, 132*f*, 134*f*
stress voiding, 129
transistors (*see* Transistor degradation)
Weibull cumulative density function, 232*f*

Lightning Source UK Ltd.
Milton Keynes UK
UKOW07n0620140116

266331UK00002BA/4/P

9 781782 422211